T0206243

Electrochemical Devices
for Energy Storage Applications

Electrochemical Devices for Energy Storage Applications

Edited by
Mesfin A. Kebede
Fabian I. Ezema

CRC Press
Taylor & Francis Group
Boca Raton London New York

CRC Press is an imprint of the
Taylor & Francis Group, an **informa** business

CRC Press
Taylor & Francis Group
6000 Broken Sound Parkway NW, Suite 300
Boca Raton, FL 33487-2742

First issued in paperback 2021

© 2020 by Taylor & Francis Group, LLC
CRC Press is an imprint of Taylor & Francis Group, an Informa business

No claim to original U.S. Government works

ISBN-13: 978-0-367-42567-8 (hbk)
ISBN-13: 978-1-03-217610-9 (pbk)
DOI: 10.1201/9780367855116

Visit the Taylor & Francis Web site at
http://www.taylorandfrancis.com

and the CRC Press Web site at
http://www.crcpress.com

Contents

Preface.. vii
Editors ix
Contributors xi

1. **Layered, Spinel, Olivine, and Silicate as Cathode Materials for Lithium-Ion Battery** 1
   Mesfin A. Kebede, Nithyadharseni Palaniyandy, and Lehlohonolo F. Koao

2. **Metal Oxide-Based Anode Materials for Lithium-Ion Battery** 19
   Mesfin A. Kebede and Nithyadharseni Palaniyandy

3. **Sodium-Ion Battery Anode Materials and Its Future Prospects and Challenges** 41
   Nithyadharseni Palaniyandy and Mesfin A. Kebede

4. **Cathode Materials for Sodium-Ion-Based Energy Storage Batteries** 59
   Assumpta C. Nwanya, Mesfin A. Kebede, Fabian I. Ezema, and M. Maaza

5. **Magnesium Battery** 81
   E. Sheha

6. **Graphene-Based Electrode Materials for Supercapacitor Applications** 101
   Moshawe J. Madito, Katlego Makgopa, Christopher B. Mtshali, and Abdulhakeem Bello

7. **Transition Metal Oxide-Based Nanomaterials for High Energy and Power Density
   Supercapacitor** 131
   *Raphael M. Obodo, Assumpta C. Nwanya, Tabassum Hassina, Mesfin A. Kebede,
   Ishaq Ahmad, M. Maaza, and Fabian I. Ezema*

8. **The Role of Modelling and Simulation in the Achievement of Next-Generation
   Electrochemical Capacitors** 151
   Innocent S. Ike, Iakovos J. Sigalas, Sunny E. Iyuke, and Egwu E. Kalu

9. **Cerium Oxide: Synthesis, Structural, Morphology, and Applications
   in Electrochemical Energy Devices** 181
   Ugochi K. Chime, M. Maaza, and Fabian I. Ezema

10. **Multifunctional Energy Storage: Piezoelectric Self-charging Cell** 197
    Blessing N. Ezealigo, M. Maaza, and Fabian I. Ezema

11. **The Contributions of Electrolytes in Achieving the Performance Index
    of Next-Generation Electrochemical Capacitors (ECs)** 215
    Innocent S. Ike, Iakovos J. Sigalas, Sunny E. Iyuke, and Egwu E. Kalu

Index... 249

Preface

There is the need to tackle the issues arising from the use of traditional fossil energy sources, which produce global pollution in order to produce a cleaner environment and maintain sustainable development. This can only be addressed using clean energy technologies. Electrochemical energy storage devices are considered to play a vital role in addressing the situation. Nowadays, it is realised that the combination of solar, wind, and battery have to become cost competitive to the traditional energy sources. The persistent research and development efforts over the last few decades have started to pay off in achieving a competitive greener energy alternative that is leading to lower environmental pollution.

The book addressed two important aspects of electrochemical energy storage systems. Chapters 1 through 5 look at various aspects of batteries ranging from the lithium-ion battery, sodium ion battery, and magnesium battery, while Chapters 6 through 11 discuss the issues that border on supercapacitors and related areas of electrochemical studies. The articles are written in such a way as to assist beginners in the field of electrochemical science, while opening up a greater understanding of the field for experts.

More specifically, Chapter 1 describes the major types of lithium-ion battery cathode materials: layered $LiMO_2$, spinel $LiMn_2O_4$, olivine $LiFePO_4$, and silicates Li_2MSiO_4 structures and reviews their advantages as well as associated intrinsic challenges in terms of capacity retention and rate capabilities. Chapter 2 discusses the three groups of anodes for lithium ion batteries, namely, intercalation/de-intercalation, alloying/de-alloying, and conversion type according to their electrochemical reaction mechanism. Chapter 3 focuses on the different types of anode materials (carbon-based, transition metal oxide-based, alloy-based, organic- and phosphorus-based) for sodium ion batteries and their electrochemistry mechanism and associated challenges.

Chapter 4 reviews the working principle of sodium ion batteries (SIBs), the stability windows, and capacities of some of the cathode materials.

Chapter 5 discusses the magnesium battery, and it covers (1) electrolyte design and synthesis; (2) development of intercalation-type cathodes; and (3) understanding of the mechanisms of Mg intercalation or chemical interaction of the electrolyte with the electrodes.

Chapter 6 summarises comprehensively the recent research efforts on graphene-based electrodes for enhancing the performance of the supercapacitor device.

Chapter 7 presents the recent research achievements on the use of transition metal compounds in supercapacitors and the associated challenges with the performance of the supercapacitor device.

Chapter 8 describes the role of modelling and simulation in the area of electrode material synthesis, optimization and fabrication, electrolytes synthesis and optimizations, and separators synthesis and fabrications.

Chapter 9 covers the application of cerium oxide in electrochemical energy devices such as supercapacitors, fuel cells, and water splitting.

Chapter 10 presents the piezoelectric self-charging cell as a multi-functional energy storage system.

Chapter 11 outlines the effect of electrolytes in achieving the performance index of next-generation electrochemical capacitors (ECs).

Finally, as the book covers a wide range of energy storage devices, we hope it helps the readers to access the important insights.

Mesfin A. Kebede
Pretoria, South Africa

Fabian I. Ezema
Nsukka, Nigeria

Editors

Mesfin A. Kebede completed his PhD in Materials Science and Engineering from Inha University, South Korea, in 2009. He is currently a principal researcher at the Energy Centre, Council of Scientific and Industrial Research (CSIR), in South Africa. He has been studying and researching the properties and applications of nanostructured materials as electrodes for lithium-ion battery, gas sensors, luminescent and photoluminescence and photocatalysis for more than a decade. In his prior career, he worked as an assistant professor at Hawassa University, Ethiopia, where he supervised and mentored MSc and PhD students' research projects. He has authored two book chapters and more than 50 papers in reputed journals with an h-index of 14 with over 590 citations and has served as reviewer for several high-impact journals and as an editorial board member.

Fabian I. Ezema is a professor at the University of Nigeria, Nsukka. He earned a PhD in Physics and Astronomy from the University of Nigeria, Nsukka. His research focused on several areas of Materials Science, from synthesis and characterizations of particles and thin-film materials through chemical routes with emphasis on energy applications. For the last 15 years, he has been working on energy conversion and storage (cathodes, anodes, supercapacitors, solar cells, among others), including novel methods of synthesis, characterization and evaluation of the electrochemical and optical properties. He has published about 180 papers in various international journals and given over 50 talks at various conferences. His h-index is 21 with over 1500 citations and he has served as reviewer for several high-impact journals and as an editorial board member.

Contributors

Ishaq Ahmad
National Centre for Physics
Islamabad, Pakistan

Abdulhakeem Bello
Department of Materials Science and Engineering
African University of Science and Technology
Abuja, Nigeria

Ugochi K. Chime
Department of Physics and Astronomy
University of Nigeria
Nsukka, Nigeria

Fabian I. Ezema
Department of Physics and Astronomy
University of Nigeria
Nsukka, Nigeria

and

Department of Physics
Faculty of Natural and Applied Sciences
Coal City University
Enugu, Nigeria

and

UNESCO-UNISA Africa Chair in
    Nanosciences-Nanotechnology
College of Graduate Studies
University of South Africa
Pretoria, South Africa

and

Material Research Department
iThemba LABS-National Research Foundation
Somerset West, South Africa

and

University of the Western Cape
Life Science Building, Biotechnology Department
Bellville, South Africa

Blessing N. Ezealigo
Department of Mechanical, Chemical and
    Materials Engineering
University of Cagliari
Cagliari, Italy

Tabassum Hassina
School of Physics
Peking University
Beijing, China

Innocent S. Ike
Department of Chemical Engineering
Federal University of Technology
Owerri, Nigeria
and
African Centre of Excellence in Future Energies
    and Electrochemical Systems (ACE-FUELS)
Federal University of Technology
Owerri, Nigeria

and

School of Chemical and Metallurgical
    Engineering
University of the Witwatersrand
Johannesburg, South Africa
and
DST/NRF Centre of Excellence in Strong
    Materials (COE-SM)
University of the Witwatersrand
Johannesburg, South Africa

Sunny E. Iyuke
School of Chemical and Metallurgical
    Engineering
University of the Witwatersrand
Johannesburg, South Africa

and

Petroleum Training Institute (PTI)
Warri, Nigeria

Egwu E. Kalu
Department of Chemical Engineering
Federal University of Technology
Owerri, Nigeria
and
African Centre of Excellence in Future Energies
    and Elcctrochemical Systems (ACE-FUELS)
Federal University of Technology
Owerri, Nigeria

and

FAMU-FSU Engineering
Tallahassee, Florida

Mesfin A. Kebede
Energy Centre, Smart Places
Council for Scientific & Industrial Research
    (CSIR)
Pretoria, South Africa

Lehlohonolo F. Koao
Department of Physics
University of the Free State
Phuthaditjhaba, South Africa

M. Maaza
Material Research Department
iThemba LABS-National Research Foundation
Somerset West, South Africa
and
UNESCO-UNISA Africa Chair in Nanosciences/
    Nanotechnology
College of Graduate Studies
University of South Africa (UNISA)
Pretoria, South Africa

and

University of the Western Cape
Life Science Building, Biotechnology Department
Bellville, South Africa

Moshawe J. Madito
iThemba LABS
National Research Foundation
Cape Town, South Africa

Katlego Makgopa
Department of Chemistry
Faculty of Science
Tshwane University of Technology
Pretoria, South Africa

Christopher B. Mtshali
iThemba LABS
National Research Foundation
Cape Town, South Africa

Assumpta C. Nwanya
Department of Physics and Astronomy
University of Nigeria
Nsukka, Nigeria

and

UNESCO-UNISA Africa Chair in
    Nanosciences-Nanotechnology
College of Graduate Studies
University of South Africa
Pretoria, South Africa

Raphael M. Obodo
Department of Physics and Astronomy
University of Nigeria
Nsukka, Nigeria

Nithyadharseni Palaniyandy
Energy Centre, Smart Places
Council for Scientific & Industrial Research
Pretoria, South Africa

E. Sheha
Department of Physics
Benha University
Benha, Egypt

Iakovos J. Sigalas
School of Chemical and Metallurgical Engineering
University of the Witwatersrand
Johannesburg, South Africa

and

DST/NRF Centre of Excellence in Strong
    Materials (COE-SM)
University of the Witwatersrand
Johannesburg, South Africa

1

Layered, Spinel, Olivine, and Silicate as Cathode Materials for Lithium-Ion Battery

Mesfin A. Kebede, Nithyadharseni Palaniyandy, and Lehlohonolo F. Koao

CONTENTS

1.1 Introduction...1
    1.1.1 The Working Principles of LIB..2
1.2 Layered $LiCoO_2$, $LiMnO_2$, $LiNiO_2$, $LiCo_{0.33}Mn_{0.33}Ni_{0.33}O_2$, $xLi_2MnO_3(1-x)LiMO_2$
    (M = Transition Metals), Li_2MnO_2..3
    1.2.1 Layered $LiCoO_2$, $LiMnO_2$, $LiNiO_2$, $LiCo_{0.33}Mn_{0.33}Ni_{0.33}O_2$ 4
        1.2.1.1 $LiCoO_2$..4
        1.2.1.2 $LiMnO_2$..5
        1.2.1.3 $LiNiO_2$...6
        1.2.1.4 $LiCo_{0.33}Mn_{0.33}Ni_{0.33}O_2$ 6
    1.2.2 Lithium- and Manganese-Rich Layered Structure Materials..................................6
    1.2.3 Nickel-Rich, $LiNi_xCo_yMn_zO_2$ ($x > 0.6$)..7
1.3 Spinel $LiMn_2O_4$ and $LiMn_{1.5}Ni_{0.5}O_4$ (LMNO)...7
    1.3.1 $LiMn_2O_4$..7
    1.3.2 $LiMn_{1.5}Ni_{0.5}O_4$..7
1.4 Olivine Cathodes...8
    1.4.1 $LiFePO_4$..8
    1.4.2 $LiMnPO_4$..9
    1.4.3 $LiNiPO_4$...10
1.5 Silicate Cathode: Li_2MSiO_4, M = Fe and Mn..11
    1.5.1 Li_2FeSiO_4...11
    1.5.2 Li_2MnSiO_4...11
1.6 Conclusions...13
References...13

1.1 Introduction

Rechargeable lithium-ion battery (LIB) not only has played a substantial role of powering portable electronic gadgets throughout the past decades, but also contributed in revolutionizing electronics technology advancement since its introduction by Sony in 1990. LIB is so attractive owing to the fact that it could provide a high-energy density with small size, has longer cycle life, has no memory effect, etc. as compared to its predecessor rechargeable battery chemistries (lead acid, Ni–Cd, Ni MH, etc.), as shown in Figure 1.1 (Tarascon and Armand 2011, 171–79).

Along with continuing market expansion and increasing demand, apart from the famous dominance on electronics, LIB is being implemented in various areas of technology such as to power electric bicycles and electric vehicles (EVs). Moreover, the use and significance of LIB keeps on gaining tremendous momentum than ever before, as its use becomes multifunctional and very diverse along with the rise

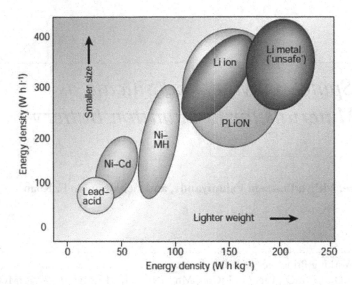

FIGURE 1.1 Energy density of different batteries. (Reprinted by permission from Macmillan publishers Ltd. *Issues and Challenges Facing Rechargeable Lithium Batteries*, Tarascon, J.-M. and Armand, M., 2011, Copyright 2001.)

of the 4th Industrial Revolution (4IR) characterised by hyper-connectivity which encompasses artificial intelligence, Internet of Things, big data, etc. The other breakthrough is wearable devices such as smartwatches, smart eyeglasses, fitness-tracking bands, etc. All these aforementioned technology advancements require and consume more power with less volume acquiring. For this, LIB is still the best energy source option. The other recent research hotspot is the flexible lithium ion batteries, by improving designs and functions.

1.1.1 The Working Principles of LIB

A LIB is an electrochemical cell (transducer) which functions as the energy storage device by converting electric energy into electrochemical energy and vice versa. The schematic diagram in Figure 1.2 represents the very basic fundamental working principles of LIB. LIB basically consists of three main components that are cathode (positive electrode), the anode (negative electrode), and electrolyte. Normally, between the cathode and anode, there will be a separator which mechanically separates the positive and negative electrodes to prevent short circuiting, which could lead to explosion (Zhang 2007, 351–64; Hao et al. 2013, 11–16; Liu et al. 2018, 265–75). This separator is designed in such a way with the appropriate size of porosity to allow maximum ionic conductivity of the lithium ion-containing electrolyte (Weber et al. 2014, 66–81).

Typically, rechargeable batteries are characterised and evaluated by their energy density (Wh kg^{-1}), specific energy (Wh L^{-1}), specific power (W kg^{-1}), cycle life, efficiency, and cost. This in turn is determined by the performance of the components that make up the battery cell: negative and positive electrodes, as well as separator and electrolyte. Importantly, the capacity and voltage of the electrodes determine the energy density and specific energy of the cell, as seen in equations (1.1) and (1.2).

$$C_{cell} = \frac{C_a C_c}{C_a + C_c} \tag{1.1}$$

$$\text{Energy density} = C_{cell} * V_{cell}. \tag{1.2}$$

Interestingly, though the cathode electrodes usually provide a low discharge capacity as compared to the anode, they actually dictate the overall parameters of the cell, such as the voltage, the capacity,

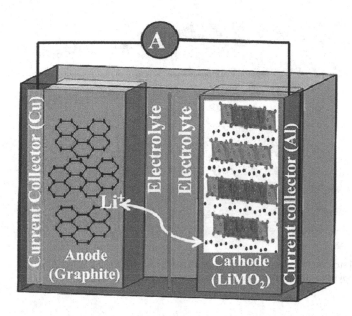

FIGURE 1.2 Schematic of lithium-ion battery. (Reprinted with permission from Daniel, C. et al., *AIP Conference Proceedings*, 1597, 26–43, 2014. Copyright 2014 by the American Institute of Physics.)

and energy density. Thus, the developments of high-energy and high-power density cathode materials become extremely crucial and have received much attention in recent times.

There are various types of cathode materials being researched for LIB applications. However, the scope of this chapter only focuses on the most famous and commercialized categories of cathodes of lithium ion batteries, namely, layered $LiMO_2$ (M=Co, Mn, Ni), spinel $LiMn_2O_4$, olivine $LiFePO_4$, and silicate Li_2MSiO_4 structures which are also commercially utilized in the global market. In this chapter, we discuss them in detail as topics of the chapter.

1.2 Layered $LiCoO_2$, $LiMnO_2$, $LiNiO_2$, $LiCo_{0.33}Mn_{0.33}Ni_{0.33}O_2$, $xLi_2MnO_3(1 - x)LiMO_2$ (M = Transition Metals), Li_2MnO_2

The most notable cathode for LIB, layered $LiCoO_2$ (LCO) is isostructural to the α-$NaFeO_2$-type structure that has rhombohedral structure and has trigonal symmetry with the R–3m space group (No. 166), which has lithium and cobalt ions located in octahedral 3a and 3b sites of cubic close-packed (ccp) of the oxygen stacks, respectively. There are three slabs of edge-sharing CoO_6 octahedra separated by interstitial layers of Li in the unit cell of the layered $LiCoO_2$ as shown in Figure 1.3. As lithium ions intercalate/deintercalate during charging and discharging, the crystal structure dynamics change. When $LiCoO_2$ is in a fully lithiated state, it remains in layered structures with hexagonal unit cells. During removal of Li from the layered crystal lattice, as charging implemented the nonstoichiometric $Li_{1-x}CoO_2$ compounds to form, and, correspondingly, the oxidation of Co^{3+} changes to Co^{4+} as the charge compensation process takes place.

Currently, the large family of lithium intercalation cathode materials ($LiCoO_2$, $LiMnO_2$, $LiNiO_2$, $LiCo_{0.33}Mn_{0.33}Ni_{0.33}O_2$, etc.), in relation to a layered rhombohedral structure (R–3m symmetry), are being introduced and are believed to be the next-generation cathode materials, as they have potential to provide relatively high and reasonable capacity and high voltage. When considered for practical usage in electric vehicles and portable power devices, each end-member comes with its own advantages and shortcomings. For instance, the manganese-based $LiMnO_2$ showed promise due to its lower cost, lower toxicity, higher safety, and comparable high theoretical capacity ($\sim$285 mA h g^{-1}). Nevertheless, the challenge

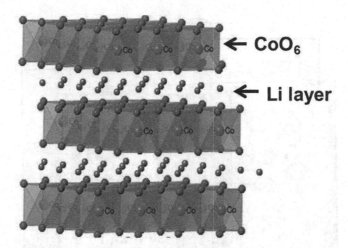

FIGURE 1.3 Crystal structure of LiCoO$_2$. (Reprinted with permission from Daniel, C. et al., *AIP Conference Proceedings*, 1597, 26–43, 2014. Copyright 2014 by the American Institute of Physics.)

of this layered cathode material is found to be metastable thermodynamically, converting to the cubic spinel structure LiMn$_2$O$_4$ (Fd3m symmetry) during electrochemical deintercalation/intercalation. The LiMnO$_2$ does not easily form a layered R–3m structure (Armstrong and Bruce 1996, 499), instead, it forms the orthorhombic phase structure that causes low rate capability and capacity (Amatucci et al. 1997, 11–25).

The other individual end-member is LiNiO$_2$ (LNO). Dyer et al. initially synthesised LNO by bubbling oxygen through lithium hydroxide at about 800°C in 1954 (Dyer, Borie Jr, and Smith 1954, 1499–503). In the late 1980s, Dahn et al. extensively studied LNO to use it as an alternative to LCO. LNO is isostructural to LCO (R(–)3m), and thus structurally speaking, it should easily allow for lithium intercalation. The benefit of LNO is that nickel is more abundant than cobalt and would make for a cheaper alternative. Some of the drawbacks of LNO are the synthesis of LNO is very difficult due to the instability of the trivalent nickel at elevated temperature and strongly dependant on appropriate conditions (depends on precursors, oxygen atmosphere, annealing temperature). The other drawback is LiNiO$_2$ suffers Ni^{2+} mixing in the Li$^+$ layer that provides different capacities for different synthesis parameters (Ellis, Lee, and Nazar 2010, 691–714).

1.2.1 Layered LiCoO$_2$, LiMnO$_2$, LiNiO$_2$, LiCo$_{0.33}$Mn$_{0.33}$Ni$_{0.33}$O$_2$

1.2.1.1 LiCoO$_2$

LiCoO$_2$ was the first and historic cathode material implemented by Sony in 1990, which is still dominating the commercial lithium ion batteries market globally. The mechanism of the intercalation of lithium-ions in the structure of a layered LiCoO$_2$ structure was first explained/discovered by renowned lithium-ion battery scientist John Goodenough. In the crystal structure of LiCoO$_2$, the Li atoms are sitting in between layers of the hosting CoO$_2$ slabs (as shown in Figure 1.4), and they leave the cathode structure during charging and return during discharging. The very interesting thing about LiCoO$_2$ cathode material is that it has a relatively high theoretical specific capacity of 274 mA h g^{-1}, and high theoretical volumetric capacity of 1363 mA h cm^{-3}, low self-discharge, high discharge voltage, and stable cycling performance. Its most commonly acknowledged drawbacks are cost, its toxic tendency, and safety issue. However, the practical experimentally achievable capacity delivered by layered LCO is only about 140 mA h g^{-1}, which is almost half of its theoretical specific capacity of 274 mA h g^{-1}. The main reason for getting such limited capacity of 140 mA h g^{-1}, that normally leads to capacity fading, is because of the oxygen loss that occurs below lithium content $x = 0.5$ for electrochemically charged Li$_{1-x}$CoO$_2$ samples.

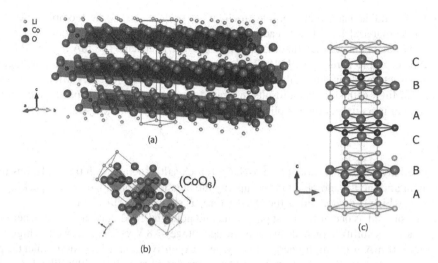

FIGURE 1.4 (a) Layered R(-)3m LiCoO$_2$, (b) the structure of octahedral CoO$_6$, (c) stacking arrangement of the layers. (Reprinted with permission from *J. Appl. Crystallogr.*, 44, Momma, K. and Izumi, F., "VESTA 3 for Three-Dimensional Visualization of Crystal, Volumetric and Morphology Data," 1272–1276; Reprinted with permission from *J. Electrochem. Soc.*, 164 (1), Schipper, F. et al., "Recent Advances and Remaining Challenges for Lithium Ion Battery Cathodes I. Nickel-Rich, LiNi$_x$Co$_y$Mn$_z$O$_2$," A6220– A6228, Copyright 2017, ECS.)

1.2.1.2 LiMnO$_2$

The other promising layered cathode material is LiMnO$_2$, which has a high theoretical capacity of 285 mA h g^{-1}, about twice that of the spinel LiMn$_2$O$_4$ capacity 148 mA h g^{-1}, abundant in the earth's crust, cheaper and environmentally benign (Ellis, Lee, and Nazar 2010, 691–714). However, layered LiMnO$_2$ cathode drastically loses its capacity due to the high possibility of phase transformation to the spinel-like phase, (6) which is associated with Mn dissolution, and Jahn-Teller distortion of Mn^{3+} (Cho et al. 2001, 18–20; Amatucci et al. 1997, 11–25).

The layered LiMnO$_2$ normally has two crystal structures, orthorhombic and monoclinic (Li, Su, and Wang 2018, 2182–89) as represented in Figure 1.5a and b), of which the orthorhombic LiMnO$_2$ is thermodynamically more stable than monoclinic LiMnO$_2$ (Zhou et al. 2016, 248–54). Consequently,

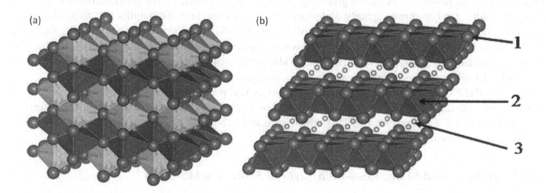

FIGURE 1.5 (a) Orthorhombic LiMnO$_2$ (Pmmn), (b) monoclinic LiMnO$_2$ (C2 m–1) (1: oxygen, 2: manganese, 3: lithium). (Reprinted with permission from J. Appl. Crystallogr., 44 (6), Momma, K. and Izumi, F., "VESTA 3 for Three-Dimensional Visualization of Crystal, Volumetric and Morphology Data," 1272–1276; Reprinted with permission from J. Electrochem. Soc., 164 (1), Schipper, F. et al., "Recent Advances and Remaining Challenges for Lithium Ion Battery Cathodes I. Nickel-Rich, LiNi$_x$Co$_y$Mn$_z$O$_2$," A6220– A6228, copyright 2017, ECS.)

researchers have studied intensively the electrochemical properties of orthorhombic $LiMnO_2$ as the potential cathode material for LIBs because of its potential to offer a high theoretical capacity of 285 mA h g^{-1}, which is about twice that of the spinel $LiMn_2O_4$ capacity 148 mA h g^{-1} within the same Mn^{4+}/Mn^{3+} redox couple (Bruce 1997, 1817–24; Thackeray 1997, 1–71). The issue of huge volume expansion and structural strain during the intercalation/deintercalation process is still a challenge for the orthorhombic $LiMnO_2$ cathode, which results in a rapid decrease of specific capacity and poor cycling stability, which hampers its practical application in the high-energy storage system.

1.2.1.3 LiNiO₂

The nickel-based layered $LiNiO_2$ adopts the α-$NaFeO_2$ rock salt structure with the oxide ions forming a ccp arrangement and the Li^+ and Ni^{3+} ions occupying the octahedral interstices in this packing alternating (111) planes, which is designated as the O_3 structure, as shown in Figure 1.1a.

$LiNiO_2$ is considered as one of the most promising and potential candidates as a next generation cathode material due to its ability to provide high operating voltage ~3.8 V vs Li/Li^+ as well as high reversible capacity ~200–250 mA h g^{-1}, high-energy density, less expensive, and a low environmental pollution impact as compared to $LiCoO_2$ cathode material (Armstrong and Bruce 1996, 499; Ellis, Lee, and Nazar 2010, 691–714). The primary challenge associated with getting stoichiometric $LiNiO_2$ is difficult to synthesize because of a high-temperature treatment of $LiNiO_2$ leads to the decomposition from $LiNiO_2$ to nonstoichiometric composition $Li_{1-x}Ni_{1+x}O_2$ ($x > 0$), which has a partially disordered cation distribution at the lithium sites. The stoichiometry of $LiNiO_2$ affects the electrochemical properties. Therefore, it is necessary to prepare $LiNiO_2$ under optimum conditions in order to have high performances. However, its intrinsic cycling instability prevented commercial applicability. Generally, to achieve high-energy density LIB to address the electric vehicle's (EV) requirement to fulfil a drive range of ~500 Km (300 miles) between charges, the Ni fraction has to be increased progressively.

1.2.1.4 LiCo₀.₃₃Mn₀.₃₃Ni₀.₃₃O₂

As the research on individual and their secondary layered materials progressed, it headed to the development of ternary materials $LiNi_{1-x-y}Co_xMn_yO_2$ (Liu, Yu, and Lee 1999, 416–19) and eventually the symmetrical $LiNi_{0.33}Co_{0.33}Mn_{0.33}O_2$ (NCM333) was studied by Ohzuku and Makimura (2001, 744–45). The ternary cathode material NCM333 could give a capacity of ~150 mA h g^{-1} in the voltage range of 2.5–4.3 V (Ohzuku and Makimura 2001, 744–45) and could deliver up to 200 mA h g^{-1} at a higher voltage cut off of 4.6 V vs Li^+/Li, but with the sacrifice of capacity stability (Yabuuchi and Ohzuku 2003, 171–74). Each end-member transition metal plays their specific role, to improve the specific capacity of the electrode, increasing the nickel amount is important as nickel gives higher favourable capacity, cobalt provides improved Li-ion kinetics resulting in better rate capacity, and manganese is known to improve stability and safety. The transition metal cations found in the tertiary positive materials $LiNi_{1-x-y}Co_xMn_yO_2$ are Mn^{4+}, Co^{3+}, and Ni^{2+}, but in most of the cases, $Ni^{2+/4+}$ and $Co^{2+/3+}$ redox couples are electrochemically active during lithium-ion intercalation/deintercalation reaction, while the oxidation state of manganese +IV, which is inactive electrochemically will not change during cycling. Thus, manganese can act as a structural stabiliser.

The material NCM333 is characterised by a low power rate performance due to the decreasing content of Co in the structure. The conventional reason for a low power rate performance is mainly due to similar ionic radii of Li (0.74°A) and Ni^{2+} (0.69°A), which leads to Li^+/Ni^{2+} cation mixing, Ni^{2+} can easily occupy the Li^+ position in $LiMn_{0.33}Co_{0.33}Ni_{0.33}O_2$ material.

1.2.2 Lithium- and Manganese-Rich Layered Structure Materials

The lithium and manganese rich, (Li-rich) × Li_2MnO_3·(1 – x), $LiNi_aCo_bMn_cO_2$ cathode materials are given high attention due to their ability to deliver high capacity, ca. 250 mA h g^{-1}, and require low production cost since the cost of Mn is less expensive than Co or even Ni. Accordingly, they are considered as likely candidates for the next generation of cathodes following Ni-rich (Thackeray et al. 2005, 2257–67; Erickson, Ghanty, and Aurbach 2014, 3313–24; Yan, Liu, and Li 2014, 63268–84; Berg et al. 2015, A2468–75; Erickson et al. 2015, A2424–38; Wagner et al. 2016, 11359–71). Unfortunately, the main difficulties to overcome for commercialization of Li-rich and manganese-rich materials are their extreme capacity and voltage fading. Especially the fading of voltage is a seriously unwanted detrimental effect for portable electronics, as they require stable voltage sources (Erickson, Ghanty, and Aurbach 2014, 3313–24; Yan, Liu, and Li 2014, 63268–84; Rozier and Tarascon 2015, A2490–99).

1.2.3 Nickel-Rich, $LiNi_xCo_yMn_zO_2$ ($x > 0.6$)

The intention and strategy of developing the nickel-rich cathode materials are to gain benefits from the individual advantages of nickel, cobalt, and manganese of the layered structure by trying to optimise it with varying concentrations ($LiNi_{1-x-y}Co_xMn_yO_2$) to achieve high-energy density cathode materials. A high Ni content is typically important to achieve high reversible capacity. On the other hand, Co improves the layered ordering as well as rate capability, while Mn in the Mn^{4+} electrochemical inactive state could enhance the structural and thermal stability at the deeply delithiated state of the material.

Though the application of higher nickel contents is still facing numerous challenges that need to be addressed, at present, some of the EVs have already started to use NCM523 cathode LIB as their power sources. To meet the demand of a targeted ~500 km driving range per charge, one of the most technologically advanced materials is layered nickel-rich $LiNi_xCo_yMn_zO_2$ (NCM, $x > 0.6$).

1.3 Spinel $LiMn_2O_4$ and $LiMn_{1.5}Ni_{0.5}O_4$ (LMNO)

1.3.1 $LiMn_2O_4$

The spinel $LiMn_2O_4$ is the second most famous cathode material commercialised next to the popular $LiCoO_2$ cathode, which was first commercialised as a cathode for LIB used by Sony in 1990. The LMO cathode is recognised by providing an operating voltage of ~4.1 V and a practical capacity of 120 mA h g^{-1} (theoretical capacity, 148 mA h g^{-1}). This cathode material works by enabling the Li$^+$ ion to have a diffusion path in three dimension (3D). The conventional unit cell of the spinel $LiMn_2O_4$ (LMO) has 8 Li, 16 Mn, and 32 O atoms and 56 atoms in total. The 8 Li atoms sit at the 8a tetrahedral sites, the Mn atoms place at the 16d sites, and form a cage with O atoms sitting at the 32e sites to host Li$^+$ ions during the charging/discharging (Kebede et al. 2014, 44–49) at ambient temperature, and it can also be written as $(Li^+)_{8a}[Mn^{3+}Mn^{4+}]_{16d}O_4^{2-}$.

The spinel cathode could be synthesised by various synthesis techniques, such as solution combustion (Kebede et al. 2015, 51–57), thermo-polymerization (Kebede et al. 2017, A3259–65), Pechini (Jafta et al. 2013, 7592–98), aqueous reduction (Kunjuzwa et al. 2016, 111882–88), molten salt (Kebede and Ozoemena 2017, 025030), etc. To get a well crystalised LMO cathode, it normally requires heat treatment of above 700°C.

The electrochemical performance challenge of LMO is severe capacity fading during repetitive charge/discharge cycling especially at an elevated temperature above 60°C. There are various mechanisms that have been proposed and have revealed in regards to capacity fading the Jahn-Teller effect, manganese dissolution, and disproportion reaction.

Various capacity fading mechanisms have been proposed, such as electrochemical reaction with an electrolyte at high voltage (Pistoia et al. 1996, 2683–89), instability of the two-phase structure in the charged state, leading to a more stable single-phase structure, phase transformation from a cubic spinel to a tetragonal rock-salt structure due to nonequilibrium lithiation (Thackeray et al. 1998, 7–9), loss of crystallinity during cycling (Shin and Manthiram 2002, A55–58), and manganese dissolution. Some of the various strategies to address the capacity fading challenges are: (i) cation doping, (ii) nano-sizing of the cathodes (Kebede and Ozoemena 2017, 025030), and (iii) microwave treatment (Kebede et al. 2017, A3259–65).

1.3.2 $LiMn_{1.5}Ni_{0.5}O_4$

The high voltage LMNO has two phases, the ordered and disordered, which cannot be identified by X-ray diffractometer; however, Furrier Transfer Infrared can confirm it. In the disordered, the Ni and Mn ions are randomly positioned as shown in Figure 1.6a, while in the ordered, the Ni ions have their

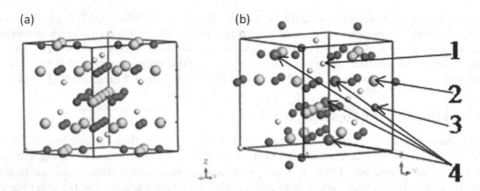

FIGURE 1.6 The crystal structures for (a) disordered Fd-3m (#227) (b) ordered P4343 (#212), where (1) -Li, (2) - Mn, (3) - O, and (4) - Ni generated using Materials Studio 2019 software.

specific cites to sit, as shown in Figure 1.6b. In the disordered Fd3m (#227) LMNO, the Mn ion and Ni ion will occupy the 16d octahedral sites randomly distributed, the Li ion will occupy 8a, and O will occupy the 32e sites. In the ordered P4332 (#212) LMNO, Mn is assigned to 12b and Ni to 4a octahedral sites. Practically, in as prepared sample, both types of LMNO exist, but the dominating one dictates its electronic property.

1.4 Olivine Cathodes

1.4.1 LiFePO$_4$

The crystal structure of olivine-type lithium iron phosphate (LiFePO$_4$, LFP) has been widely studied by many authors as a cathode material for lithium ion batteries. The LFP crystals occupy the orthorhombic system (Pnma space group, No. 62), in which the oxygen is forming in a hexagonal close-packed framework, Li and Fe are located in half of the octahedral sites, and P ions are located in one-eighth of the tetrahedral sites. Figure 1.7a illustrates the LFP crystal structure, in which the FeO$_6$ octahedra are distorted and through which the channels lithium ions can be removed. The corner-shared FeO$_6$ octahedra are linked together in the bc-plane, while LiO$_6$ octahedra form edge-sharing chains along the b-axis. Upon removal of Li, the material converts to the ferric form of FePO$_4$. The tetrahedral PO$_4$ units bridge with neighbouring layers of FeO$_6$ octahedra by sharing a common edge with one FeO$_6$ octahedron and two edges with LiO$_6$ octahedron. These strong bridges give LFP's remarkable high theoretical capacity (~170 mA h g^{-1}, with the nominal output of LFP is 3.2 V), low cost, ultra-long lifetime, high operating voltage (~ 3.4 V vs Li/Li$^+$), high safety, and no environmental pollution. LiFePO$_4$ is an intrinsically safer cathode material than LiCoO$_2$ and manganese spinel (see below). The LFP is intrinsically safer than LCO and LMO material due to its Fe–P–O bond, which is stronger than the Co–O bond, therefore when abused (short-circuited, overheated, etc.), the oxygen atoms are much harder to remove. However, this type of material suffered from a low electronic conductivity (~ 10^{-9} S cm^{-1}) and low ionic diffusion (~10^{-14} cm^2 s^{-1}) rate due to its olivine structure, which has a 1-dimensional channel for Li$^+$ transport ions (Tang et al. 2019, 677–82). This type of crystal structure leads to poor electrochemical performance and low power density. To circumvent this issue, many efforts have been made, such as coating, foreign metal doping, and morphology controlling. The carbon-coated LFP has remarkably enhanced the electrical conduction between the particles, which ensures high rate capability and avoiding the agglomeration of the particles (Z. Xu et al. 2016, A2600–2610; Zou et al. 2016, 1601549). For example, Fischer et al. (2018, 1646–53) developed polymer templated LiFePO$_4$/C nanonetworks as the anode for LIBs. This material demonstrates excellent cyclability with the highest capacity retention of 92% over 1000 cycles at 1°C, which is shown in Figure 1.7b.

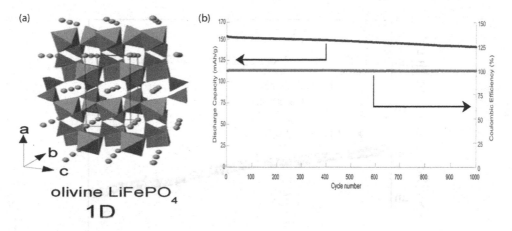

FIGURE 1.7 Crystal structure of LiFePO$_4$ (olivine 1D-type) (a) and capacity retention and Coulombic efficiency of the polymer-templated LFP/C nanonetwork over 1000 cycles at 1°C (b). (Reprinted with permission from Fischer, M.G. et al., *ACS Appl. Mater. Interfaces*, 10, 1646–1653, 2018. Copyright 2018 American Chemical Society.)

Xiong et al. fabricated a nitrogen-doped carbon-coated LiFePO$_4$ nanosphere synthesised by a hydrothermal and chemical polymerization method. When utilised as a cathode for LIBs, the N–C@LFP nanospheres deliver excellent electrochemical performance, better reversibility, and smaller polarization as compared to the pure LFP nanospheres. The N–C@LFP (Xiong et al. 2018, 377–82) electrode exhibits a superior specific capacity of 158.4 mA h g^{-1} over 200 cycles at the rate of 1°C and good rate capability (107.5 mA h g^{-1} at the rate of 30°C). The enhanced electrochemical performance is ascribed to the conductive carbon and nitrogen layer. Saroha et al. (2017, 25–36) developed a ZnO-doped LiFePO$_4$/C composite synthesised using the sol-gel-assisted ball-milling route. The cyclic performance was improved due to the metal oxide doping.

1.4.2 LiMnPO$_4$

The same theoretical capacity of LiFePO$_4$ and different electrochemical performances of another type of olivine, LiMnPO$_4$, was developed due to its high operating voltage ~ 4.1 V, low cost, higher energy density of ~700 W h kg^{-1}, and high safety (Chen et al. 2016, 4069–75). However, LiMnPO$_4$ has an electrochemical insulator [low electronic conductivity (<10–10 S cm^{-1}) and ionic diffusivity (<10^{-16} cm^2 s^{-1})], and a large volume change of (~10.7%) occurred during the extraction of Li transition from LiMnPO$_4$ (302.9 Å) to MnPO$_4$ (270.4 Å), which limits the reaction kinetics rates (0.45 moles Li per formula unit) and low energy density. That could be due to the Jahn-Teller effect associated with the Mn^{3+} (Fu et al. 2017, 272–82). To tackle this issue, many researchers have devoted their research to synthesise a carbon-coated LiMnPO$_4$ composite (there ~0.94 Li per formula unit was extracted) and metal-doped LiMnPO$_4$ materials, in order to enhance the kinetic properties of lithium ions. For example, Wen et al. (2016, 85–93) developed mesoporous LiMnPO$_4$/C nanoparticles as a cathode for LIB applications, in the presence of 1 M LiPF$_6$ in ethylene carbonate (EC) and dimethyl carbonate (DMC) (1:1, V V^{-1}). The initial discharge capacity of 156 mA h g^{-1} was observed, and after that, the discharge capacity was gradually decreased to ~135 mA h g^{-1} after 100 cycles, with the highest capacity retention of 83%. In 2019, Junzhe et al. fabricated a hierarchical carbon-coated LiMnPO$_4$ microsphere by the mixing of raw materials such as MnCO$_3$, PVP, and LiH2PO$_4$ dissolved in distilled water and freeze-dried, followed by calcination treatment (J. Li et al. 2019, A118–24). The cycling results of carbon-coated LiMnPO$_4$ (S–LMP/C) exhibit a stable and high reversible capacity of 102 mA h g^{-1} at the 200th cycle, corresponding to 98% of the capacity retention at 5C, which is apparently higher than the pristine material (shown in Figure 1.8). Yttrium, iron, and titanium metals were doped into

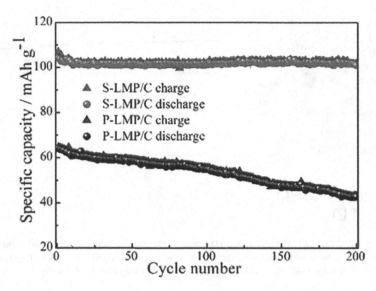

FIGURE 1.8 Cycling performances of carbon-coated $LiMnPO_4$ (S-LMP/C) and pristine $LiMnPO_4$ (P-LMP/C). (Reprinted with permission from J. Li et al., 2019, A118–A124. Copyright 2019, ECS).

a Mn site, in order to improve the electrochemical kinetics of $LiMnPO_4/C$ material. It is observed that the reversible capacity was enhanced up to a limited amount of yttrium (0.01%) and 0.2% Fe in the Mn site was observed for $LiMnPO_4/C$ materials (H. Xu et al. 2016, 27164–69; Huang et al. 2016, 11348–54; J. Zhang et al. 2017, 3189–94).

1.4.3 LiNiPO$_4$

$LiNiPO_4$ is considered as a high voltage cathode material (~ 5.1 V vs Li) for lithium ion batteries. However, the commercial application of this material was hampered due to its limited oxidative stability in the presently available conventional carbonate-based electrolyte [higher than the other olivine ($LiFePO_4$ and $LiMnPO_4$)], in which the electrolyte is degraded and resulted in poor cyclic performances (Devaraju et al. 2015, 11041). The strong P–O bonding in $LiNiPO_4$ stabilizes the Ni^{2+} at low temperature, but to access the Ni^{2+}/Ni^{3+}, reduction couple occurred at 5.1 V vs Li. Moreover, $LiNiPO_4$ material has low electronic conductivity and limited lithium diffusion, and resulted in poor electrochemical performances. Also, indeed, it is difficult to synthesise pure phase $LiNiPO_4$. That is, the direct synthesis of pure phase $LiNiPO_4$ at a low temperature (450°C) shows about 10% of anti-site defects, which will damage the lithium transport properties in the 1D channel olivine. Moreover, increasing the synthesis temperature does not help; in situ neutron experiments have shown that the concentration of anti-sites increases with temperature, and reaches 15% at 900°C (Mauger et al. 2017, 63–69). Therefore, in order to overcome this issue, several approaches have been attempted, such as reducing the particle size from micro to nanosize, doping, and surface treatment. For example, Thong et al. (2018, 625–31) prepared the $LiNiPO_4$ (LNP) with high surface area, crystallinity, and size-controlled by the sol-gel method, followed by calcining the gel at 400°C, 600°C, and 800°C. The 600°C showed to be the optimal calcination temperature to produce pure $LiNiPO_4$ with an average crystallite size of 43 nm and surface area of 27.47 $m^2 g^{-1}$. Örnek (2017, 524–34) fabricated a high voltage $LiNiPO_4$@C cathode material that was synthesized by microwave and solvothermal methodology using different solvents such as ethylene glycol, isopropanol, and isobutanol for LIB applications. The $LiNiPO_4$@C cathode material produced in an isopropanol environment exhibited the best electrochemical properties, achieved a discharge capacity of 157 mA h g^{-1} at 0.1°C, and displayed almost 81% capacity retention after the 80th cycle.

1.5 Silicate Cathode: Li_2MSiO_4, M = Fe and Mn

The olivine (polyanion-type) and silicate [Li-orthosilicates Li_2MSiO_4 (M = Fe and Mn)] cathode materials have attracted remarkable attention owing to their robust crystal structure, feasible redox transitions, high thermal stability due to the strong Si–O covalent bonding, naturally abundant, environmentally friendly, and better safety. Another noticeable feature is that two lithium ions can be intercalated per formula unit during the charge/discharge process offering a much higher theoretical specific capacity (333 mA h g^{-1}) than that of commercial conventional cathodes ($LiCoO_2$, $LiMn_2O_4$, $LiFePO_4$). The electrochemistry mechanism was obtained as follows:

$$LiM_3 + SiO_4 \rightarrow M_4 + SiO_4 + e^- + Li^+ \left(M = Fe \text{ and } Mn\right) \quad (1.3)$$

1.5.1 Li_2FeSiO_4

Li_2FeSiO_4 is considered as one of the most promising cathode materials due to its low cost, resource abundance, stable structure, and environmental benignity. Li_2FeSiO_4 has two redox peaks, one at 2.8 V (Fe^{2+}/Fe^{3+}) and the other at 4.0–4.8 V (Fe^{3+}/Fe^{4+} vs Li^+/Li) (Ni et al. 2017, 1771–81). Also, Li_2FeSiO_4 has three different crystal structures, such as orthorhombic phase and monoclinic phase (Wei et al. 2018, 7458–67), which are shown in Figure 1.9. The former one could be obtained during the low-temperature (below 500°C) synthesis with space group Pmn21. During the high-temperature synthesis, two phases were involved such as monoclinic with the space group P21/n and orthorhombic phase with the space group Pmnb. Most of the recent works have been focused on high-temperature synthesis due to its direct formation of pure phase Li_2FeSiO_4 and also to achieve hybridization of carbon-coated material, such as solid-state reaction, etc. However, during cycling, the P21/n phases transform to a more stable inverse-Pmn21 phase. This cycled inverse-Pmn21 phase is isostructural with Pmn21, except that all the Fe has exchanged sites with half of Li (Zeng et al. 2019, 1592–1602). However, these types of crystal structures inherently have low ionic ($\sim 1 \times 10^{-14}$ cm^2 s^{-1}) and electronic ($\sim 6 \times 10^{-14}$ S cm^{-1}) conductivities as well as sluggish kinetics during cycling. To mitigate these issues, various efforts have been dedicated to facilitating electron and/or ion transport through coating with conductive materials, reducing the particle size to the nanoscale, and metals doping into the Fe site (Xu et al. 2015, 340–47; L. Zhang et al. 2015, 11782–86; Qiu et al. 2016, 1870–77). For example, Ding et al. (2016, 297–304) developed a three-dimensional ordered macroporous Li_2FeSiO_4/C composite by adopting a hard-soft template method. The composite exhibits a high reversible capacity of 237 mA h g^{-1} and good rate capability (~ 110 mA h g^{-1} at 10 C) for over 400 cycles, and it was found that around ~ 1.50 Li^+ ions inserted and de-inserted during cycling. This enhanced electrochemical performance could be due to the interconnected carbon framework that improves the electronic conductivity and 3D structure that could offer the short Li ion diffusion pathways and restrain volumetric changes. L. Li et al. (2019, 24–32) fabricated Ni or Pb substituted on the Fe and Si sites of the Li_2FeSiO_4/C material. This study indicates that the Ni or Pb doping at the Fe site offers more stable electrochemical properties and materials doped at the Si site have high initial charge and discharge capacity, while the large charge transfer resistance of materials indicates poor reversibility and unstable electrochemical performance.

1.5.2 Li_2MnSiO_4

The Li_2MnSiO_4 has drawn much more attention than Li_2FeSiO_4 due to its two redox couple reactions that are Mn^{2+}/Mn^{3+} and Mn^{3+}/Mn^{4+}, which deliver a higher voltage of 4.5 V vs Li/Li$^+$ than Li_2FeSiO_4, and also Mn^{4+} in the Li_2MnSiO_4 structure is more stable than Fe^{4+} in Li_2FeSiO_4 (Cheng et al. 2017, 10772–97). Moreover, Li_2MnSiO_4 has four main polymorphs which are characterised by its different distributions and orientations of LiO_4, MnO_4, and SiO_4. These polymorphs are orthorhombic, in space groups Pmn21 and Pmnb, and monoclinic, in space groups P21/n and Pn. The four polymorphs of Li_2MnSiO_4 are all related to the β and γ families of so-called tetrahedral structures based on Li_3PO_4

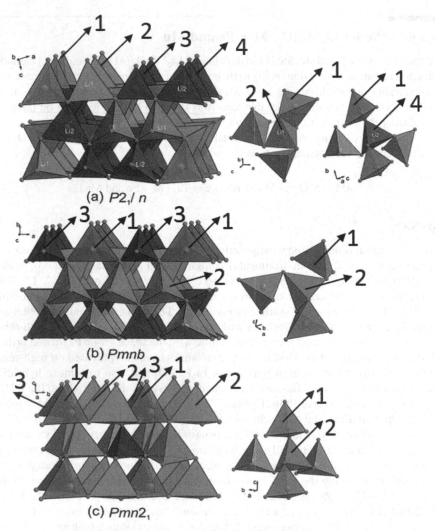

FIGURE 1.9 Structures of Li_2FeSiO_4 polymorphs: (a) space group P21/n, (b) space group Pmnb, and (c) space group Pmn21. In the image, 1 is related to FeO_4 tetrahedra, 2 and 4 are related to LiO_4 tetrahedra, and 3 is related to SiO_4 tetrahedra. The inset shows the linkage channel of LiO_4 (2 and 4) with the surrounding FeO_4 (1). (Reprinted with permission from Wei, H. et al., *ACS Sustain. Chem. Eng.*, 6, 7458–7467, 2018. Copyright 2018 American Chemical Society.)

(Feng et al. 2018, 3223–31), which are shown in Figure 1.10 (Dong, Hull, and West 2018, 715–23). However, Li_2MnSiO_4 has a large irreversible capacity and poor cycle stability due to its extremely low electrical conductivity [10^{-14} S cm^{-1} at room temperature and Li$^+$ diffusion rate (10^{-17}–10^{-18} cm^2 s^{-1})] (Mancini et al. 2017, 1561–70; Ding et al. 2018, 6309–16).

The intrinsic structural instability of Li_2MnSiO_4 material leads to severe capacity fading upon cycling, which causes by Jahn-Teller distortion of Mn^{3+} and dissolution of Mn^{2+} in the Li$^+$ insertion/extraction progress. To tackle these issues, numerous approaches have been developed, such as hybridization with conductive carbonaceous materials (Peng et al. 2018, 1–8) and doping with foreign ions (Wagner et al. 2016, 11359–71; Zhu et al. 2019, 956–65). For example, Zhu et al. (2019, 956–65) fabricated the nitrogen-doped carbon layer-coated Li_2MnSiO_4 (Li_2MnSiO_4/NC) by the sol-gel process and it was examined as a cathode material for LIBs. This material exhibits excellent electrochemical property, which was mainly attributed to the nitrogen doping and carbon layer coating changes, the electronic structure, and improved the electrical conductivity of Li_2MnSiO_4.

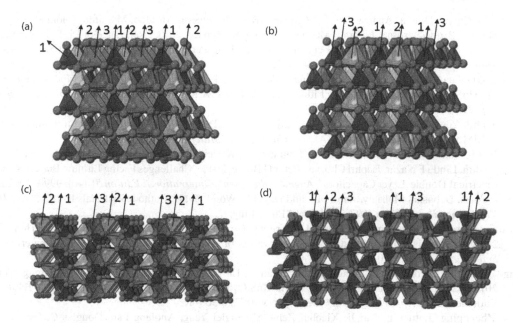

FIGURE 1.10 Four different polymorphs of Li_2MnSiO_4: (a) Pmn21, (b) Pn, (c) P21/n, and (d) Pmnb. (1, 2, and 3) Grey, green, and blue colour represents SiO_4, MnO_4, and LiO_4 tetrahedra units, respectively, and grey spheres represent oxygen. (Reprinted with permission from Feng, Y., et al., *Inorg. Chem.*, 57, 3223–3231, 2018. Copyright 2018 American Chemical Society.)

1.6 Conclusions

In summary, the pioneering and commercialised cathode materials such as $LiCoO_2$, $LiMn_2O_4$, and $LiFePO_4$ are well explored in applications of mobile technologies, stationary utilities, and electrifying of vehicles. The nickel-rich, manganese-rich layered cathode materials have shown a high prospect in providing high-energy density by addressing their challenges. The different categories of phosphates and silicate cathodes are also created huge hope in the electrochemical energy storage industry.

As compared to that of commercial conventional cathodes ($LiCoO_2$, $LiMn_2O_4$, $LiFePO_4$), the olivine (polyanion-type) and silicate [Li-orthosilicates Li_2MSiO_4 (M = Fe and Mn)] cathode materials have attracted remarkable attention due to their robust crystal structure, feasible redox transitions, and high thermal stability due to the strong Si–O covalent bonding, naturally abundant, environmentally friendly, and better safety. They are also preferable owing to offering a much higher theoretical specific capacity (333 mA h g^{-1}) during the charge/discharge process. Lithium-ion battery cathodes are therefore so crucial being the main component of the well-matured and preferred rechargeable battery technologies.

REFERENCES

Amatucci, G. G., C. N. Schmutz, A. Blyr, C. Sigala, A. S. Gozdz, D. Larcher, and J. M. Tarascon. 1997. "Materials' Effects on the Elevated and Room Temperature Performance of $CLiMn_2O_4$ Li-Ion Batteries." *Journal of Power Sources* 69 (1/2): 11–25.

Armstrong, Robert A., and Peter G. Bruce. 1996. "Synthesis of Layered $LiMnO_2$ as an Electrode for Rechargeable Lithium Batteries." *Nature* 381 (6582): 499.

Berg, Erik J., Claire Villevieille, Daniel Streich, Sigita Trabesinger, and Petr Novák. 2015. "Rechargeable Batteries: Grasping for the Limits of Chemistry." *Journal of the Electrochemical Society* 162 (14): A2468–75.

Bruce, P. G. 1997. "Solid-State Chemistry of Lithium Power Sources." *Chemical Communications-Chemical Society*, 1997 (19): 1817–24.

Chen, Lin, Enrico Dilena, Andrea Paolella, Giovanni Bertoni, Alberto Ansaldo, Massimo Colombo, Sergio Marras, Bruno Scrosati, Liberato Manna, and Simone Monaco. 2016. "Relevance of LiPF$_6$ as Etching Agent of LiMnPO$_4$ Colloidal Nanocrystals for High Rate Performing Li-Ion Battery Cathodes." *ACS Applied Materials & Interfaces* 8 (6): 4069–75.

Cheng, Qiaohuan, Wen He, Xudong Zhang, Mei Li, and Lianzhou Wang. 2017. "Modification of Li$_2$ MnSiO$_4$ Cathode Materials for Lithium-Ion Batteries: A Review." *Journal of Materials Chemistry A* 5 (22): 10772–97.

Cho, Jaephil, Yong Jeong Kim, Tae-Joon Kim, and Byungwoo Park. 2001. "Enhanced Structural Stability of O-LiMnO$_2$ by Sol–Gel Coating of Al$_2$O$_3$." *Chemistry of Materials* 13 (1): 18–20.

Choi, Nam-Soon, Zonghai Chen, Stefan A. Freunberger, Xiulei Ji, Yang-Kook Sun, Khalil Amine, Gleb Yushin, Linda F Nazar, Jaephil Cho, and Peter G Bruce. 2012. "Challenges Facing Lithium Batteries and Electrical Double-Layer Capacitors." *Angewandte Chemie International Edition* 51 (40): 9994–10024.

Daniel, Claus, Debasish Mohanty, Jianlin Li, and David L. Wood. 2014. "Cathode Materials Review." In *AIP Conference Proceedings 1597*, 26–43. AIP Publishing.

Devaraju, Murukanahally Kempaiah, Quang Duc Truong, Hiroshi Hyodo, Yoshikazu Sasaki, and Itaru Honma. 2015. "Synthesis, Characterization and Observation of Antisite Defects in LiNiPO$_4$ Nanomaterials." *Scientific Reports* 5: 11041.

Ding, Zhengping, Yiming Feng, Datong Zhang, Ran Ji, Libao Chen, Douglas G. Ivey, and Weifeng Wei. 2018. "Crystallographic Habit Tuning of Li$_2$MnSiO$_4$ Nanoplates for High-Capacity Lithium Battery Cathodes." *ACS Applied Materials & Interfaces* 10 (7): 6309–16.

Ding, Zhengping, Jiatu Liu, Ran Ji, Xiaohui Zeng, Shuanglei Yang, Anqiang Pan, Douglas G. Ivey, and Weifeng Wei. 2016. "Three-Dimensionally Ordered Macroporous Li$_2$FeSiO$_4$/C Composite as a High Performance Cathode for Advanced Lithium Ion Batteries." *Journal of Power Sources* 329: 297–304.

Dong, Bo, Stephen Hull, and Anthony R. West. 2018. "Phase Formation, Crystallography, and Ionic Conductivity of Lithium Manganese Orthosilicates." *Inorganic Chemistry* 58 (1): 715–23.

Dyer, Lawrence D., Bernard S. Borie Jr., and G. Pedro Smith. 1954. "Alkali Metal-Nickel Oxides of the Type MNiO$_2$." *Journal of the American Chemical Society* 76 (6): 1499–503.

Ellis, Brian L., Kyu Tae Lee, and Linda F Nazar. 2010. "Positive Electrode Materials for Li-Ion and Li-Batteries." *Chemistry of Materials* 22 (3): 691–714.

Erickson, Evan M., Chandan Ghanty, and Doron Aurbach. 2014. "New Horizons for Conventional Lithium Ion Battery Technology." *The Journal of Physical Chemistry Letters* 5 (19): 3313–24.

Erickson, Evan M., Elena Markevich, Gregory Salitra, Daniel Sharon, Daniel Hirshberg, Ezequiel de la Llave, Ivgeni Shterenberg, Ariel Rosenman, Aryeh Frimer, and Doron Aurbach. 2015. "Development of Advanced Rechargeable Batteries: A Continuous Challenge in the Choice of Suitable Electrolyte Solutions." *Journal of the Electrochemical Society* 162 (14): A2424–38.

Feng, Yiming, Ran Ji, Zhengping Ding, Datong Zhang, Chaoping Liang, Libao Chen, Douglas G Ivey, and Weifeng Wei. 2018. "Understanding the Improved Kinetics and Cyclability of a Li$_2$MnSiO$_4$ Cathode with Calcium Substitution." *Inorganic Chemistry* 57 (6): 3223–31.

Fischer, Michael G., Xiao Hua, Bodo D. Wilts, Elizabeth Castillo-Martínez, and Ullrich Steiner. 2018. "Polymer-Templated LiFePO$_4$/C Nanonetworks as High-Performance Cathode Materials for Lithium-Ion Batteries." *ACS Applied Materials & Interfaces* 10 (2): 1646–53.

Fu, Xiaoning, Kun Chang, Bao Li, Hongwei Tang, Enbo Shangguan, and Zhaorong Chang. 2017. "Low-Temperature Synthesis of LiMnPO$_4$/RGO Cathode Material with Excellent Voltage Platform and Cycle Performance." *Electrochimica Acta* 225: 272–82.

Hao, Jinglei, Gangtie Lei, Zhaohui Li, Lijun Wu, Qizhen Xiao, and Li Wang. 2013. "A Novel Polyethylene Terephthalate Nonwoven Separator Based on Electrospinning Technique for Lithium Ion Battery." *Journal of Membrane Science* 428: 11–16.

Huang, Qiao-Ying, Zhi Wu, Jing Su, Yun-Fei Long, Xiao-Yan Lv, and Yan-Xuan Wen. 2016. "Synthesis and Electrochemical Performance of Ti–Fe Co-Doped LiMnPO$_4$/C as Cathode Material for Lithium-Ion Batteries." *Ceramics International* 42 (9): 11348–54.

Jafta, Charl J., Mkhulu K. Mathe, Ncholu Manyala, Wiets D. Roos, and Kenneth I. Ozoemena. 2013. "Microwave-Assisted Synthesis of High-Voltage Nanostructured LiMn$_{1.5}$Ni$_{0.5}$O$_4$ Spinel: Tuning the Mn$_3$ Content and Electrochemical Performance." *ACS Applied Materials & Interfaces* 5 (15): 7592–98.

Kebede, Mesfin A., and Kenneth I. Ozoemena. 2017. "Molten Salt-Directed Synthesis Method for $LiMn_2O_4$ Nanorods as a Cathode Material for a Lithium-Ion Battery with Superior Cyclability." *Materials Research Express* 4 (2): 025030.

Kebede, Mesfin A., Maje J. Phasha, Niki Kunjuzwa, Lukas J. Le Roux, Donald Mkhonto, Kenneth I. Ozoemena, and Mkhulu K. Mathe. 2014. "Structural and Electrochemical Properties of Aluminium Doped $LiMn_2O_4$ Cathode Materials for Li Battery: Experimental and Ab Initio Calculations." *Sustainable Energy Technologies and Assessments* 5: 44–49.

Kebede, Mesfin A., Maje J. Phasha, Niki Kunjuzwa, Mkhulu K. Mathe, and Kenneth I. Ozoemena. 2015. "Solution-Combustion Synthesized Aluminium-Doped Spinel ($LiAl_xMn_{2-x}O_4$) as a High-Performance Lithium-Ion Battery Cathode Material." *Applied Physics A* 121 (1): 51–57.

Kebede, Mesfin A., Spyros N. Yannopoulos, Labrini Sygellou, and Kenneth I. Ozoemena. 2017. "High-Voltage $LiNi_{0.5}Mn_{1.5}O_4.\delta$ Spinel Material Synthesized by Microwave-Assisted Thermo-Polymerization: Some Insights into the Microwave-Enhancing Physico-Chemistry." *Journal of the Electrochemical Society* 164 (13): A3259–65.

Kunjuzwa, Niki, Mesfin A. Kebede, Kenneth I. Ozoemena, and Mkhulu K. Mathe. 2016. "Stable Nickel-Substituted Spinel Cathode Material ($LiMn_{1.9}Ni_{0.1}O_4$) for Lithium-Ion Batteries Obtained by Using a Low Temperature Aqueous Reduction Technique." *Rsc Advances* 6 (113): 111882–88.

Li, Junzhe, Shao-hua Luo, Qing Wang, Shengxue Yan, Jian Feng, Xueyong Ding, Ping He, and Lingbo Zong. 2019. "Facile Fabrication of Hierarchical $LiMnPO_4$ Microspheres for High-Performance Lithium-Ion Batteries Cathode." *Journal of the Electrochemical Society* 166 (2): A118–24.

Li, Ling, Enshan Han, Chen Mi, Lingzhi Zhu, Lijun Dou, and YaKe Shi. 2019. "The Effect of Ni or Pb Substitution on the Electrochemical Performance of Li_2FeSiO_4/C Cathode Materials." *Solid State Ionics* 330: 24–32.

Li, Xiaohui, Zhi Su, and Yingbo Wang. 2018. "Electrochemical Properties of Monoclinic and Orthorhombic $LiMnO_2$ Synthesized by a One-Step Hydrothermal Method." *Journal of Alloys and Compounds* 735: 2182–89.

Liu, Jian, Yanbo Liu, Wenxiu Yang, Qian Ren, Fangying Li, and Zheng Huang. 2018. "Lithium Ion Battery Separator with High Performance and High Safety Enabled by Tri-Layered $SiO_2@$ PI/m-PE/$SiO_2@$ PI Nanofiber Composite Membrane." *Journal of Power Sources* 396: 265–75.

Liu, Zhaolin, Aishui Yu, and Jim Y. Lee. 1999. "Synthesis and Characterization of $LiNi_{1-x-y}Co_xMn_yO_2$ as the Cathode Materials of Secondary Lithium Batteries." *Journal of Power Sources* 81: 416–19.

Mancini, Marilena, Meike Fleischhammer, Stephanie Fleischmann, Thomas Diemant, Rolf J. Behm, Peter Axmann, and Margret Wohlfahrt-Mehrens. 2017. "Investigation on the Thermal Stability of Li_2MnSiO_4-Based Cathodes for Li-Ion Batteries: Effect of Electrolyte and State of Charge." *Energy Technology* 5 (9): 1561–70.

Mauger, A., C. M. Julien, M. Armand, J. B. Goodenough, and K. Zaghib. 2017. "Li(Ni, Co)PO_4 as Cathode Materials for Lithium Batteries: Will the Dream Come True?" *Current Opinion in Electrochemistry* 6 (1): 63–69.

Momma, Koichi, and Fujio Izumi. 2011. "VESTA 3 for Three-Dimensional Visualization of Crystal, Volumetric and Morphology Data." *Journal of Applied Crystallography* 44 (6): 1272–76.

Ni, Jiangfeng, Yu Jiang, Xuanxuan Bi, Liang Li, and Jun Lu. 2017. "Lithium Iron Orthosilicate Cathode: Progress and Perspectives." *ACS Energy Letters* 2 (8): 1771–81.

Ohzuku, Tsutomu, and Yoshinari Makimura. 2001. "Layered Lithium Insertion Material of $LiNi_{1/2}Mn_{1/2}O_2$: A Possible Alternative to $LiCoO_2$ for Advanced Lithium-Ion Batteries." *Chemistry Letters* 30 (8): 744–45.

Örnek, Ahmet. 2017. "Influences of Different Reaction Mediums on the Properties of High-Voltage $LiNiPO_4@$C Cathode Material in Terms of Dielectric Heating Efficiency." *Electrochimica Acta* 258: 524–34.

Peng, Tao, Wei Guo, Qi Zhang, Yingge Zhang, Miao Chen, Yinghui Wang, Hailong Yan, Yang Lu, and Yongsong Luo. 2018. "Uniform Coaxial CNT@$Li_2MnSiO_4@$C as Advanced Cathode Material for Lithium-Ion Battery." *Electrochimica Acta* 291: 1–8.

Pistoia, G., A. Antonini, R. Rosati, and D. Zane. 1996. "Storage Characteristics of Cathodes for Li-Ion Batteries." *Electrochimica Acta* 41 (17): 2683–89.

Qiu, Hailong, Huijuan Yue, Tong Zhang, Tingting Li, Chunzhong Wang, Gang Chen, Yingjin Wei, and Dong Zhang. 2016. "Enhanced Electrochemical Performance of Li_2FeSiO_4/C Cathode Materials by Surface Modification with $AlPO_4$ Nanosheets." *Electrochimica Acta* 222: 1870–77.

Rozier, Patrick, and Jean Marie Tarascon. 2015. "Li-Rich Layered Oxide Cathodes for Next-Generation Li-Ion Batteries: Chances and Challenges." *Journal of the Electrochemical Society* 162 (14): A2490–99.

Saroha, Rakesh, Amrish K. Panwar, Yogesh Sharma, Pawan K. Tyagi, and Sudipto Ghosh. 2017. "Development of Surface Functionalized ZnO-Doped $LiFePO_4$/C Composites as Alternative Cathode Material for Lithium Ion Batteries." *Applied Surface Science* 394: 25–36.

Schipper, Florian, Evan M. Erickson, Christoph Erk, Ji-Yong Shin, Frederick Francois Chesneau, and Doron Aurbach. 2017. "Recent Advances and Remaining Challenges for Lithium Ion Battery Cathodes I. Nickel-Rich, $LiNi_xCo_yMn_zO_2$." *Journal of the Electrochemical Society* 164 (1): A6220–28.

Shin, Youngjoon, and Arumugam Manthiram. 2002. "Microstrain and Capacity Fade in Spinel Manganese Oxides." *Electrochemical and Solid-State Letters* 5 (3): A55–58.

Tang, Jun, Xiongwei Zhong, Haiqiao Li, Yan Li, Feng Pan, and Baomin Xu. 2019. "In-Situ and Selectively Laser Reduced Graphene Oxide Sheets as Excellent Conductive Additive for High Rate Capability $LiFePO_4$ Lithium Ion Batteries." *Journal of Power Sources* 412: 677–82.

Tarascon, J.-M., and Michel Armand. 2011. *Issues and Challenges Facing Rechargeable Lithium Batteries. Materials for Sustainable Energy.* A Collection of Peer-Reviewed Research and Review Articles from Nature Publishing Group. London, UK: World Scientific.

Thackeray, Michael M. 1997. "Manganese Oxides for Lithium Batteries." *Progress in Solid State Chemistry* 25 (1–2): 1–71.

Thackeray, Michael M., Christopher S. Johnson, John T. Vaughey, N. Li, and Stephen A. Hackney. 2005. "Advances in Manganese-Oxide 'Composite' Electrodes for Lithium-Ion Batteries." *Journal of Materials Chemistry* 15 (23): 2257–67.

Thackeray, Michael M., Yang Shao-Horn, Arthur J. Kahaian, Keith D. Kepler, Eric Skinner, John T. Vaughey, and Stephen A. Hackney. 1998. "Structural Fatigue in Spinel Electrodes in High Voltage (4 V) Li/ $Li_xMn_2O_4$ Cells." *Electrochemical and Solid-State Letters* 1 (1): 7–9.

Thong, Ying Jie, Jeng Hua Beh, Jau Choy Lai, and Teck Hock Lim. 2018. "Synthesis and Characterization of Alginate-Based Sol–Gel Synthesis of Lithium Nickel Phosphate with Surface Area Control." *Industrial & Engineering Chemistry Research* 58 (2): 625–31.

Wagner, Nils P., Per Erik Vullum, Magnus Kristofer Nord, Ann Mari Svensson, and Fride Vullum-Bruer. 2016. "Vanadium Substitution in Li_2MnSiO_4/C as Positive Electrode for Li Ion Batteries." *The Journal of Physical Chemistry C* 120 (21): 11359–71.

Weber, Christoph J., Sigrid Geiger, Sandra Falusi, and Michael Roth. 2014. "Material Review of Li Ion Battery Separators." In *AIP Conference Proceedings 1597*, 66–81. AIP Publishing.

Wei, Huijing, Xia Lu, Hsien-Chieh Chiu, Bin Wei, Raynald Gauvin, Zachary Arthur, Vincent Emond, De-Tong Jiang, Karim Zaghib, and George P. Demopoulos. 2018. "Ethylenediamine-Enabled Sustainable Synthesis of Mesoporous Nanostructured $Li_2FeIISiO_4$ Particles from Fe (III) Aqueous Solution for Li-Ion Battery Application." *ACS Sustainable Chemistry & Engineering* 6 (6): 7458–67.

Wen, Fang, Hongbo Shu, Yuanyuan Zhang, Jiajia Wan, Weihua Huang, Xiukang Yang, Ruizhi Yu, Li Liu, and Xianyou Wang. 2016. "Mesoporous $LiMnPO_4$/C Nanoparticles as High Performance Cathode Material for Lithium Ion Batteries." *Electrochimica Acta* 214: 85–93.

Xiong, Q. Q., J. J. Lou, X. J. Teng, X. X. Lu, S. Y. Liu, H. Z. Chi, and Z. G. Ji. 2018. "Controllable Synthesis of NC@ $LiFePO_4$ Nanospheres as Advanced Cathode of Lithium Ion Batteries." *Journal of Alloys and Compounds* 743: 377–82.

Xu, Han, Jun Zong, Fei Ding, Zhi-wei Lu, Wei Li, and Xing-jiang Liu. 2016. "Effects of Fe 2 Ion Doping on $LiMnPO_4$ Nanomaterial for Lithium Ion Batteries." *RSC Advances* 6 (32): 27164–69.

Xu, Yimeng, Wei Shen, Cong Wang, Aili Zhang, Qunjie Xu, Haimei Liu, Yonggang Wang, and Yongyao Xia. 2015. "Hydrothermal Synthesis and Electrochemical Performance of Nanoparticle Li_2FeSiO_4/C Cathode Materials for Lithium Ion Batteries." *Electrochimica Acta* 167: 340–47.

Xu, Zhengrui, Libin Gao, Yijing Liu, and Le Li. 2016. "Recent Developments in the Doped $LiFePO_4$ Cathode Materials for Power Lithium Ion Batteries." *Journal of the Electrochemical Society* 163 (13): A2600–10.

Yabuuchi, Naoaki, and Tsutomu Ohzuku. 2003. "Novel Lithium Insertion Material of $LiCo_{1/3}Ni_{1/3}Mn_{1/3}O_2$ for Advanced Lithium-Ion Batteries." *Journal of Power Sources* 119: 171–74.

Yan, Jianhua, Xingbo Liu, and Bingyun Li. 2014. "Recent Progress in Li-Rich Layered Oxides as Cathode Materials for Li-Ion Batteries." *RSC Advances* 4 (108): 63268–84.

Zeng, Yan, Hsien-Chieh Chiu, Majid Rasool, Nicolas Brodusch, Raynald Gauvin, De-Tong Jiang, Dominic H. Ryan, Karim Zaghib, and George P. Demopoulos. 2019. "Hydrothermal Crystallization of Pmn21 Li_2FeSiO_4 Hollow Mesocrystals for Li-Ion Cathode Application." *Chemical Engineering Journal* 359: 1592–602.

Zhang, Jun, Shaohua Luo, Qing Wang, Zhiyuan Wang, Yahui Zhang, Aimin Hao, Yanguo Liu, Qian Xu, and Yuchun Zhai. 2017. "Yttrium Substituting in Mn Site to Improve Electrochemical Kinetics Activity of Sol-Gel Synthesized $LiMnPO_4$/C as Cathode for Lithium Ion Battery." *Journal of Solid State Electrochemistry* 21 (11): 3189–94.

Zhang, Ling, Jiangfeng Ni, Wencong Wang, Jun Guo, and Liang Li. 2015. "3D Porous Hierarchical Li_2FeSiO_4/C for Rechargeable Lithium Batteries." *Journal of Materials Chemistry A* 3 (22): 11782–86.

Zhang, Sheng Shui. 2007. "A Review on the Separators of Liquid Electrolyte Li-Ion Batteries." *Journal of Power Sources* 164 (1): 351–64.

Zhou, Hui, Yingshun Li, Jiaolong Zhang, Wenpei Kang, and Y. W. Denis. 2016. "Low-Temperature Direct Synthesis of Layered m-$LiMnO_2$ for Lithium-Ion Battery Applications." *Journal of Alloys and Compounds* 659: 248–54.

Zhu, Hai, Weina Deng, Liang Chen, and Shiying Zhang. 2019. "Nitrogen Doped Carbon Layer of Li_2MnSiO_4 with Enhanced Electrochemical Performance for Lithium Ion Batteries." *Electrochimica Acta* 295: 956–65.

Zou, Yihui, Shuai Chen, Xianfeng Yang, Na Ma, Yanzhi Xia, Dongjiang Yang, and Shaojun Guo. 2016. "Suppressing Fe–Li Antisite Defects in $LiFePO_4$/Carbon Hybrid Microtube to Enhance the Lithium Ion Storage." *Advanced Energy Materials* 6 (24): 1601549.

2

Metal Oxide-Based Anode Materials for Lithium-Ion Battery

Mesfin A. Kebede and Nithyadharseni Palaniyandy

CONTENTS

2.1 Introduction 19
2.2 Metal Oxide Anodes: TiO_2, $Li_4Ti_5O_{12}$, SnO_2, and SnO 20
    2.2.1 TiO_2 Anode for LIB 20
    2.2.2 $Li_4Ti_5O_{12}$ 21
    2.2.3 SnO_2 and SnO Anode 24
2.3 Ternary Tin Oxides 27
    2.3.1 M_2SnO_4; M=Mn, Mg, Co, and Zn 28
2.4 TMO 30
    2.4.1 Nickel Oxide 30
    2.4.2 Manganese Oxide 31
    2.4.3 Iron Oxide 32
    2.4.4 Cobalt Oxide 32
    2.4.5 Copper Oxide 32
2.5 TM_3O_4 33
    2.5.1 Co_3O_4 33
    2.5.2 Fe_3O_4 34
    2.5.3 Mn_3O_4 34
2.6 Summary 35
References 35

2.1 Introduction

One of the main and crucial components of a lithium-ion battery (LIB) is the anode, a negative electrode as equally significant as the cathode positive electrode. Anodes are typically characterized by delivering a high capacity with low operating potential, unlike cathodes that normally give a low capacity with high working potential. In order to design and develop ideal anode materials, there are important factors which should be taken into consideration, such as material that has a high specific surface, can offer more lithium insertion channels, low volume change during a Li-ion insertion/desertion process in order to get good cycling stability, high electronic and ionic conductivity, which leads to fast charging and discharging, low intercalation potential for Li to get high overall voltage for the full cell, low price, and being environmentally benign.

Generally, the anode materials are categorized into three groups according to the electrochemical energy storage mechanism and how the lithium ion enters into and leaves from the host structure during discharging and charging, and each group has its own advantages and challenges. The first group works by an intercalation/deintercalation-based reaction, the materials like $Li_4Ti_5O_{12}$ (LTO) including graphite are among those that normally deliver small specific capacity, but they are electrochemically stable upon cycling. In this group, lithium ions are electrochemically intercalating into the space between layers of the materials with insignificant volume changes (<4%) (Li et al. 2013, 356–63; Guan et al. 2016, e1501554). For example, the reversible lithium ion insertion and extraction from the anodes graphite and TiO_2 can be described by the following equations as given below:

$$xLi^+ + C_6 + xe^- \leftrightarrow Li_xC_6 \tag{2.1}$$

$$xLi^+ + TiO_2 + xe^- \leftrightarrow Li_xTiO_2. \tag{2.2}$$

The second mechanism is the alloying/de-alloying reaction-based; metal oxides, for instance, SnO_2 and SiO_2 are among them. This anode group consists of metals that can be alloyed with lithium such as Si, Sn, and their alloys (Wang et al. 2012, 544–52; Su et al. 2014, 1300882). In this category, the Li ions insert into the structure of anode material during the charge cycle, making an alloy with the anode. The reversible alloying reaction is shown in equation (2.3), where M is the anode material:

$$M + xLi^+ + xe^- \leftrightarrow Li_xM. \tag{2.3}$$

Mostly, the alloying/de-alloying reaction may result in the phase change, for instance, the alloying-based anode materials Si and Sn are accompanied with phase change (Yoo et al. 2011, 4324–28). Alloying reaction-based materials are most famous for their high theoretical capacity: for instance, 4200 mA h g^{-1} for Si in $Li_{4.4}Si$ and 992 mA h g^{-1} for Sn in $Li_{4.4}Sn$. However, the major disadvantage of these materials is their susceptibility to extremely large volume change during charge and discharge, which leads to cracking and pulverization of particles that result in rapid, severe capacity fading after a few cycles.

The third group consists of transition metal oxide (TMO) anodes (Co_3O_4, MnO_2, etc.), which serve an important function during a conversion reaction, such that they provide high theoretical capacity that can range from the low range of 350 mA h g^{-1} for Cu_2S to the high range of 1800 mA h g^{-1} for MnP_4 and practically with a high capacity of about >700 mA h g^{-1} with a voltage window of 0.05–3 V. Conversion reaction-based materials are based on the Faradaic reaction represented in equation (2.4):

$$MaXb + (b \cdot n) Li^+ + ae^- \leftrightarrow aM + bLinX, \tag{2.4}$$

where
   M = transition metal (such as Ti, Mn, Fe, Co, Ni, Cu, etc.)
   X = anion (such as O, N, F, and S)
   n = number of negative charge of X (Cao et al. 2017, 2213–42).

In this chapter, an overview of recent progress towards metal oxide-based anodes for LIB will be presented. Specifically, the different strategic approaches to curb the capacity-fading problem will be given more attention, and the current research trends will be reviewed and will be summarized. Future prospects in the design of advanced anodes and other functional materials by taking advantage of the merits of conversion reactions are highlighted.

2.2 Metal Oxide Anodes: TiO_2, $Li_4Ti_5O_{12}$, SnO_2, and SnO

2.2.1 TiO_2 Anode for LIB

The TiO_2-based anode materials have attracted a great interest owing to their natural abundance, intrinsic safety, and minimum environmental impact, as well as having excellent structural stability by showing low insignificant volume change (<4%) during Li$^+$ insertion/extraction, creating favourable empty sites for lithium insertion, etc. (Li et al. 2013, 356–63; Guan et al. 2016, e1501554).

TiO_2 also attracted attention since it is characterized by fast lithium ion diffusion; low cost; being environmentally friendly; and good safety. The ability to provide high operating voltage (~1.5 V), to have long cycle life, as well as enhanced safety makes a TiO_2 anode have very attractive properties. The main challenges regarding the practical use of TiO_2-based anodes are the aggregation tendencies of TiO_2 anode nanoparticles, the low capacity, as well as the low Li-ion diffusivity and low electronic/ionic conductivities, which significantly limit the rate performance (S. Guo et al. 2016, 11943–48; Song and Paik 2016, 14–31).

The electrochemical reaction due to insertion/extraction of Li^+ into TiO_2 can be described with the following equation:

$$x Li^+ + TiO_2 + xe^- \leftrightarrow Li_x TiO_2, \tag{2.5}$$

where x is between 0 and 1, depending on the crystal structure of TiO_2 (Huang et al. 2014, 201–8). In the case of bulk rutile, lithium capacity is very low, but it can be increased with rutile nanocrystals (Kebede, Zheng, and Ozoemena 2016, 55–91).

TiO_2 would result in significant improvements in battery safety as compared to a conventionally used graphite anode since TiO_2 has a higher operating voltage, which affects the formation of the solid-electrolyte interface (SEI) layer and prevents the formation of metallic lithium dendrite structures (Fröschl et al. 2012, 5313–60; Jiang and Zhang 2013, 97–122). The theoretical capacity of titania (~335 mA h g^{-1}) is comparable to that of graphite (~372 mA h g^{-1}), while it also exhibits a small volume expansion upon lithium insertion/extraction (~4% for anatase) (Jiang and Zhang 2013, 97–122).

There are various strategic approaches that can be employed to overcome the shortcomings of TiO_2 anode materials. For the purpose of improving their electronic conductivity and Li ion diffusion kinetics, the nano-architectural design of various morphologies of TiO_2 electrode materials have been utilized, such as zero dimensional (0D) nanoparticles, nanofibers or nanotubes (1D), nanobelts or nanosheets (2D), and 3D nanostructures (Tian et al. 2014, 6920–37). The mesoporous materials are the most preferred and efficient to use as anode materials (Yang et al. 2015, 8701–5). In the same way, some authors proposed different methods to overcome the poor TiO_2 anode conductivity, such as embedding noble metal nanoparticles on TiO_2 fibres (Nam et al. 2010, 2046–52) or nanostructuring TiO_2 and adding a C layer (Geng et al. 2015, 465–72).

L. Cheng et al. (2019, 417–425) synthesized flower-like mesoporous TiO_2, which were then uniformly loaded on the surface of a graphene aerogel (GA) to get composites by using a template-free synthesis method, which was able to improve the Li^+ diffusion kinetics and facilitate the penetration of the electrolyte. The cyclic voltammogram of the anode materials TiO_2/GA-3 for the first three cycles is given in Figure 2.1a. The peak at 1.79 V is due to the intercalation of Li^+ in the anatase TiO_2 lattice, which can be represented by the equation $x Li^+ + TiO_2 + xe^- \rightarrow Li_x TiO_2$. The strong peak appearing at around ~0.60 V is related to the formation of the SEI layer, due to the decomposition of the electrolyte. The peak located at around 0.01 V is attributed to the insertion of lithium in the GA, as well as the interfacial storage of lithium on the TiO_2/GA-3 nanostructures. During the first anodic sweep, an oxidation peak occurred at 2.02 V, which is attributed to the deintercalation of lithium from the anatase TiO_2 ($Li_x TiO_2 \rightarrow x Li^+ + TiO_2 + xe^-$). The synthesized TiO_2/GA composite anode material exhibited high capacity, long cycling stability, and good rate capability with a reversible capacity of 663.2 mA h g^{-1} being obtained at a current density of 100 mA g^{-1} after 250 cycles, and at a high current rate of 5 A g^{-1}, a reversible capacity of 215.5 mA h g^{-1} was achieved after 4000 cycles.

2.2.2 $Li_4Ti_5O_{12}$

The spinel LTO anode is categorized by its excellent safety and long lifetime that make it exceptionally suitable for application in high capacity, large scale such as for electric vehicles and smart grids. The conventional commercial graphite anode is threatened by power density and safety, as it develops lithium dendrite, which can lead to catching on fire (Ohzuku, Ueda, and Yamamoto 1995, 1431–35; Zaghib et al. 1999, 300–305; Yi et al. 2010, 1236–42).

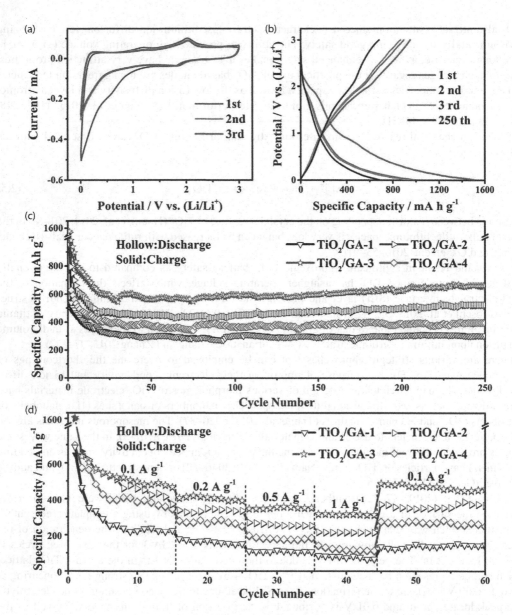

FIGURE 2.1 Electrochemical performance of TiO₂/GA-X (X ¼ 1, 2, 3, 4) electrodes in LIBs. (a) Cyclic voltammo-gram curves (CV) of TiO₂/GA-3 at a scan rate of 0.1 mV s⁻¹. (b) Galvanostatic discharge/charge curves of TiO₂/GA-3 at 100 mA g⁻¹. (c) Cycling performance of TiO₂/GA-X at 100 mA g⁻¹. (d) Rate capability of TiO₂/GA-X at different current densities. (Reprinted from *Electrochimica Acta*, 300, Cheng, L. et al., Template-free synthesis of mesoporous succulents-like TiO₂/graphene aerogel composites for lithium-ion batteries, 417–425. Copyright 2019, with permission from Elsevier.)

The LTO has a theoretical capacity of 175 mA h g⁻¹ by accommodating up to three Li ions via trans-forming from the spinel-LTO ($Li_4Ti_5O_{12}$) phase to the rock-salt-LTO ($Li_7Ti_5O_{12}$) phase during charging intercalation of the Li ions into the LTO host structure. The LTO exhibits insignificantly low volume change (only 2%–3%) with zero strain as compared to graphite volume changes of up to 10% during charging/discharging.

The characteristic charge-discharge curve for LTO gives a flat and extended voltage plateau at around 1.55 V vs Li+/Li (Flandrois and Simon 1999, 165–80), which can effectively reduce the reaction of an electrolyte on the surface of an electrode (less SEI layer) (Armand and Tarascon 2008, 652) and avoid the formation of lithium dendrites (Park et al. 2008, 14930–31). The flat and extended voltage plateau is typical to the phase transition between a Li-rich phase (rock-salt-LTO) and a Li-poor phase (spinel-LTO). The main frame for the lithium intercalation/deintercalation is [TiO_6], in the initial synthesized LTO spinel structure, the Li+ ions occupy the 8a sites. As the lithium ion insertion (discharge) proceeds, three lithium atoms from the 8a sites transfer to 16c sites, and the inserted three lithium ions move to the 16c sites via 8a sites (Aldon et al. 2004, 5721–25; Sorensen et al. 2006, 482–89; Wilkening et al. 2007, 1239–46).

Lin et al. (2017, 287–98) synthesized flower-like $Li_4Ti_5O_{12}$ hollow microspheres consisting of nanosheets and subsequently wrapped them by graphene. The graphene-wrapped $Li_4Ti_5O_{12}$@graphene has exhibited higher capacities and improved rate capabilities in the 0.01–3.0 V or 1.0–3.0 V potential range as presented in Figure 2.2. The $Li_4Ti_5O_{12}$@graphene composite shows a specific

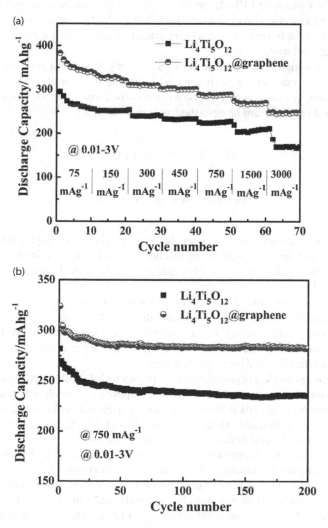

FIGURE 2.2 (a) Rate capabilities and (b) cyclic performances of the $Li_4Ti_5O_{12}$ and $Li_4Ti_5O_{12}$@graphene electrodes between 0.01 and 3 V. (Reprinted from *Electrochim. Acta*, 254, Lin, Z. et al., Graphene-Wrapped $Li_4Ti_5O_{12}$ Hollow Spheres Consisting of Nanosheets as Novel Anode Material for Lithium-Ion Batteries, 287–298. Copyright 2017, with permission from Elsevier.)

capacity of 272.7 mA h g^{-1} at 750 mA g^{-1} after 200 cycles in the potential range of 0.01 V to 3.0 V, while the pristine Li$_4$Ti$_5$O$_{12}$ only delivered a discharge capacity of 235.6 mA h g^{-1}. The improved electrochemical performance of Li$_4$Ti$_5$O$_{12}$@graphene should be attributed to a lower charge-transfer resistance, larger lithium ion diffusion coefficient, and lower activation energy. The electron transfer at the Li$_4$Ti$_5$O$_{12}$/graphene heterojunction interface, originating from a difference in the work function of two composites, reduces the localized work function of the composites, decreases the energy required for electrons to escape, and consequently, results in the improved electrochemical performance.

2.2.3 SnO$_2$ and SnO Anode

As the search for and development of new anode materials for LIBs with high specific capacity continue, tin oxide (SnO$_2$ and SnO)-based materials have attracted great attention as an anode candidate for the next generation high-performance LIBs due to their high specific capacity (782 mA h/g), low discharge potential (~0.5 V), and low cost (Bhaskar et al. 2014, 10762–71; Nam et al. 2015, 289–98; Zhou et al. 2017, 212–21). The benefit of having a low discharge potential anode is that it gives a high overall potential when used in a full cell configuration.

The SnO$_2$ anode has a high specific capacity of 782 mA h g^{-1}, which is about twice as large as that of graphite (372 mA h g^{-1}) due to the electrochemical alloying reaction of either 4.4 (Li$_{22}$Sn$_5$) or 4.25 (Li$_{17}$Sn$_4$) Li per Sn atom, and the reaction can be written in the form of the following equations.

The abovementioned reactions can be described as [equations (2.6), (2.7)]:

$$SnO_2 + 4Li^+ + 4e^- \rightarrow Li_2O + Sn \tag{2.6}$$

$$Sn + 4.4Li^+ + 4.4e^- \rightarrow 4Li_{4.4}Sn. \tag{2.7}$$

The SnO$_2$-based anode materials function with the same principle of alloying/de-alloying as Li ion inserting/leaving the SnO$_2$ host material during discharging/charging. As it is shown in equation (2.6), they usually suffer from large initial irreversible capacity induced by the formation of a SEI layer and the electrochemically inactive Li$_2$O.

The main challenge of using SnO$_2$ as a commercial anode is that it exhibits poor cyclability owing to large volume changes (~300%), which cause pulverization of electrodes associated during the lithium ion alloying/de-alloying process, as well as the intrinsically low conductivity. As the volume changes, SnO$_2$ anode materials are susceptible to agglomerate, crush, and suffer severe structural damage, resulting in low Coulombic efficiency and poor cycle performance.

There are two effective strategical approaches to curb the SnO$_2$ challenge and improve the electrochemical performance of SnO$_2$ anode materials, one of them is to composite SnO$_2$ with carbonaceous materials, i.e., graphene, carbon nanotubes, and carbon fibres which will be used as a buffer to accommodate the volume expansion during the Li$^+$ insertion and extraction process. In addition, the use of carbon also increases the intrinsic electronic conductivity and lithium ion migration inside SnO$_2$.

The other most commonly used approach is to improve the mobility of the lithium ions (Li$^+$), thereby the lithium diffusion kinetics by creating more surface area in contact with the electrolyte, which in turn improves the capacity and stability of the materials by scaling down the particle size of SnO$_2$ into nanoscale and nanoarchitecture such as nanobelts, nanorods, and nanosheets. The nano-sized particles can shorten the Li ion diffusion distances, provide more Li ion insertion sites, and better accommodate the absolute volume change. Specially, by combining both the strategies, it is possible to produce a high capacity and more stable SnO_2 anode for LIB applications (Palaniyandy, Kebede, et al. 2019, 1–9).

Recently, Zhao et al. (2019, 253–262) reported on the electrochemical performance associated with SnS/SnO_2 heterostructures sandwiched between spherical graphene. They synthesized ultrafine SnS/SnO_2 nanoparticles with heterostructures, which are sandwiched between multi-layers of graphene sheets (GS). Their composite with appropriate composition delivers the best lithium storage rate performance of 620 and 312.7 mA h g^{-1} at 1°C and 10°C, respectively. A very stable and high reversible specific capacity of 850 mA h g^{-1} is obtained after 200 cycles at 0.1°C as shown in Figure 2.3. The appropriate molar ratio of nano-heterostructures and the novel sandwich hollow spherical composite structure are attributed to the excellent electrochemical performances.

Zuo et al. (2018, 61–68) reported an SnO_2/graphene oxide (SnO_2/GO) composite and SnO_2, which were synthesized by the hydrothermal method to improve the cycling stability and rate performance of SnO_2-based anode materials for LIB applications. The composite exhibits excellent lithium ion storage capacity, cycling stability, and rate capability, as shown in Figure 2.4a and b, by providing

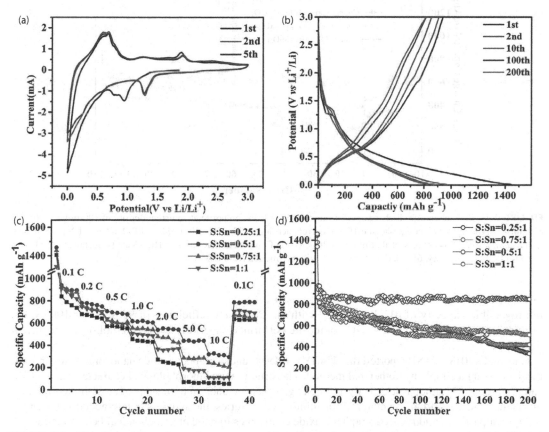

FIGURE 2.3 First five cyclic voltammograms curves (a) and typical galvanostatic voltage curves of SnS/SnO_2/SG-0.5 sample at 100 mA g^{-1} between 3.0 and 0.01 V vs Li/Li$^+$ (b); rate properties (c); and cycling performances (d) of SnS/SnO_2/SG composites with different S/Sn molar ratios. (Reprinted from *Electrochim. Acta*, 300, Zhao, B. et al., Composition-Dependent Lithium Storage Performances of SnS/SnO_2 Heterostructures Sandwiching Between Spherical Graphene, 253–262. Copyright 2019, with permission from Elsevier.)

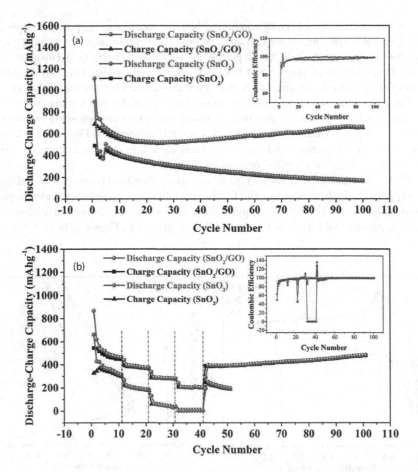

FIGURE 2.4 (a) The charge-discharge performance of SnO$_2$/GO composite at a current density of 100 mA g^{-1}; (b) rate capability of the SnO$_2$/GO composite at different current densities (100, 200, 500, 1000, and 100 mA g^{-1}). (Reprinted from *Electrochim. Acta*, 264, Zuo, S. et al., SnO$_2$/Graphene Oxide Composite Material with High Rate Performance Applied in Lithium Storage Capacity, 61–68. Copyright 2018, with permission from Elsevier.)

the reversible capacity of 612.2 mA h g^{-1} with the Coulombic efficiency of 98.8% after 100 cycles. Figure 2.4b shows that the SnO$_2$/GO composite performed superiorly to SnO$_2$ at all different current densities.

Gao et al. (2018, 72–81) reported that Tin(IV)oxide@reduced graphene oxide nanocomposites are synthesized using a simple hydrothermal method. The structural and morphological characterizations indicate that the SnO$_2$ nanoparticles fully and homogeneously anchor on both sides of cross-linked-reduced graphene oxide. As an anode material for lithium-ion batteries, the synergistic interaction between the SnO$_2$ nanoparticles and reduced graphene oxide contributes to good electrochemical behaviours, which enhance the cycling performance and rate capability. For a half cell, the SnO$_2$@reduced graphene oxide nanocomposites as an anode material exhibit a high reversible capacity of 1149 mA h g^{-1} at a current density of 0.2 A g^{-1} and a good capacity retention of 67.2% after 130 cycles as shown in Figure 2.5. For a full cell, it exhibits a capacity of 648 mA h g^{-1} at a current density of 0.2 A g^{-1} after 200 cycles, and the cycling retention of capacity reached 63.5%. The excellent storage capability and cycling performance of lithium-ion batteries make the composite a promising anode material in the practical application of lithium-ion batteries.

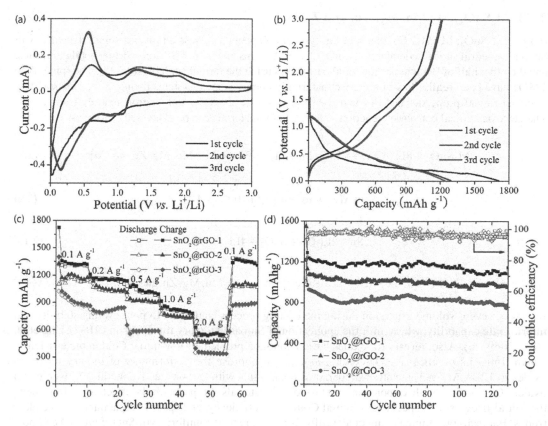

FIGURE 2.5 (a) Cyclic voltammetry, (b) galvanostatic discharge/charge profiles of the SnO_2@RGO-1 composite at the current density of 200 mA g⁻¹, (c) cycling performance of the SnO_2@RGO composites at the current density of 200 mA g⁻¹, and (d) rate performance of the SnO_2@RGO composites at different current densities. (Reprinted from *Electrochim. Acta*, 290, Gao, L. et al., Synthesis of Tin (IV) Oxide@ Reduced Graphene Oxide Nanocomposites with Superior Electrochemical Behaviors for Lithium-Ions Batteries, 72–81. Copyright 2019, with permission from Elsevier.)

2.3 Ternary Tin Oxides

In order to improve the electrochemical performance of binary tin oxides, ternary tin-based oxides have been considered as an attractive alternative anode material in which one or more electrochemically inactive/active matrix elements or oxides are incorporated for better performance. These matrices can buffer the volume changes to some extent during lithiation and de-lithiation reactions, thereby minimizing the capacity fading. Also, the ternary oxides have high electrical mobility and high electrical conductivity; therefore, it improves the Li cyclability during the electrochemical reactions of metal oxides. The mechanism of the ternary tin oxides during the Li+ cycling process is involved with lattice amorphization of the original mixed oxide and the pulverization of the crystal structure. As a result, metal (M) or metal oxides (MO) and Sn nanograins are generated and uniformly dispersed in amorphous Li_2O followed by forming the alloy of Li_xSn (Reddy, Subba Rao, and Chowdari 2013, 5364–5457). The most commonly investigated ternary tin-based oxides are M_2SnO_4; M=Mn, Co, Mg, and Zn, $MSnO_3$; M=Ca, Sr, Ba, Co, and Mg, $K_2M_2O_7$, and $M_2Sn_2O_7$; M=Y and Nd (Nithyadharseni et al. 2015, 1060–69, 2016, A540–45). This chapter explains only about M_2SnO_4; M=Mn, Co, Mg, and Zn materials, and its electrochemistry aspect.

2.3.1 M_2SnO_4; M=Mn, Mg, Co, and Zn

AB_2O_4 (M_2SnO_4; M=Mn, Co, Mg, and Zn) type of compounds possess an inverse spinel structure with Sn in tetrahedral sites O-coordination (SnO_6). Their A^{2+} ions and half of B^{3+} ions occupy octahedral sites, and the other half of B^{3+} ions occupy tetrahedral holes using the formula unit of $A(AB)O_4$. The spinel-type TMOs have been realised as an effective material in energy storage systems owing to their outstanding electrochemical properties (activity and stability), cost-effectiveness, and environmentally friendliness. The electrochemical conversion reaction of the M_2SnO_4 materials can be expressed as follows:

$$M_2SnO_4 + 8Li^+ + 8e^- \rightarrow 2M + Sn + Li_2O \left(\text{Where } M = Mn, Mg, Zn \text{ and } Co \right) \tag{2.8}$$

$$Sn + xLi^+ + xe^- \leftrightarrow Li_xSn \; (0 \leq x \leq 4.4) \tag{2.9}$$

$$Sn + 2Li_2O \leftrightarrow SnO_2 + 4Li^+ + 4e^- \tag{2.10}$$

$$M + Li_2O \leftrightarrow MO + 2Li^+ + 2e^- \left(\text{where } M = Mn, Mg, Zn, \text{ and } Co \right). \tag{2.11}$$

However, severe volume expansion during the cycling process could lead to poor cycling stability and inferior rate capability, which limit the application of Sn-based ternary tin oxides in LIBs (Liang et al. 2016, 3669–76). Also, ternary tin oxides have a common problem of low initial Coulombic efficiency, which limits its practical use. Therefore, in order to improve the performance of ternary tin-based oxides in LIBs, M_2SnO_4 are often prepared as composites with carbons, active metals, or even metal oxides. Composites with carbon- and metal-doped materials are particularly effective in improving the initial irreversible capacity loss, initial Coulombic efficiency, and specific capacitance of the electrodes. For example, Kuang Liang et al. (2016, 3669–76) reported uniform Mn_2SnO_4/Sn/C 3D composite cubes with a porous structure had been prepared via the hydrothermal method followed by further heat treatment. The Li cycling studies were performed at a constant current density of 500 mA g^{-1} and between 0.01 V and 3.0 V for Mn_2SnO_4/Sn/C composites, they exhibit high initial specific capacities of 2650 mA h g^{-1} (discharge) and 1579 mA h g^{-1} (charge) with the Coulombic efficiency of 59.6%. A high specific capacity of 908 mA g^{-1} was obtained after 100 cycles with the capacity retention of 50.5%. The enhanced electrochemical performance could be attributed to the 3D structure materials, which can improve the electrical conductivity of the electrode and result in fast kinetic properties of Li^+ (shown in Figure 2.6a and b), the porous structure offers a high surface area that can dramatically

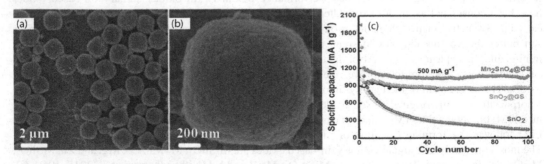

FIGURE 2.6 Scanning electron microscope (SEM) images of Mn_2SnO_4/Sn/carbon cubes composite at low magnification (a) and high magnification (b), and cyclic performance of Mn_2SnO_4@GS, SnO_2@GS, and SnO_2 at 500 mA g^{-1} (c). (Reprinted with permission from Liang, K. et al., *J. Phys. Chem. C*, 120, 3669–3676; Rehman, W.U. et al., *ACS Appl. Mater. Interfaces*, 10, 17963–17972. Copyright 2016; 2018, American Chemical Society.)

shorten the diffusion path length, and the uniformly distributed (Figure 2.6a and b) carbon network can effectively prevent the aggregation and also buffers the volume expansion of Li_xSn during the cycling process. Followed by this work, Wasif et al. reported bouquet-like Mn_2SnO_4@GS nanocomposites (Rehman et al. 2018, 17963–72) by using the hydrothermal method followed by annealing to investigate the lithium storage performance of the Mn_2SnO_4@GS material cycled at 0.01 V–3.0 V. The initial discharge-charge capacities of 1946–1195 mA h g^{-1} were observed with the highest Coulombic efficiency of 61.3%, which is apparently higher than that of pure SnO_2 (57%) and SnO_2@GS (59.1%). The specific capacity of 1042 mA h g^{-1} was obtained after 500 cycles and the highest capacity retention of 87% (shown in Figure 2.6c). Similar to Mn_2SnO_4 material, Mg_2SnO_4 has been reported as an electrochemically inactive matrix for the better performance of LIBs. For example, Tang et al. (2015, 167–72) utilized Mg_2SnO_4 nanoparticles prepared by a facile-precipitation method as a negative electrode in LIBs. The lithium ion intercalation/deintercalation of the Mg_2SnO_4 nanoparticles was enhanced due to the MgO spectator matrix, together with Li_2O, in which tin can be dissolved, could buffer the volume expansion, and contraction during cycling.

Among the reported spinel-type TMOs, Co_2SnO_4 and Zn_2SnO_4 have shown better electrochemical performance owing to the fact that their special properties of Co and Zn are electrochemically active with respect to lithium, as well as carbon coating improves the specific capacity of the material. For example, Co_2SnO_4/Co_3O_4 and Co_2SnO_4/SnO_2 are utilized as anodes for LIBs (An et al. 2014, 640–47; Mullaivananathan, Saravanan, and Kalaiselvi 2017, 1497–1502), there, Sn from SnO_2 is electrochemically active according to the equations (2.12) and (2.13), but the Co from Co_3O_4 is according to the following equations:

$$Co_3O_4 + 8Li \rightarrow 3Co + 4Li_2O \qquad (2.12)$$

$$CoO + 2Li \ll Co + Li_2O. \qquad (2.13)$$

These multi-phase reactions with Li^+ of the Co_2SnO_4 and doped oxides (Co_3O_4 and SnO_2) prove the excellent electrochemical performance of Co_2SnO_4 in LIBs. Apart from that, many researchers have proposed that the carbon-coated Co_2SnO_4 material exhibits significantly enhanced cyclability and superior rate capability via the increased exposed active sites between active materials and electrolytes, which can increase the kinetics of the Li ions (An et al. 2015, 2485–93; Chen et al. 2015, 203–13). For example, a Co_2SnO_4 hollow cube/graphene composite (Co_2SnO_4HC@RGO) (Zhang et al. 2014, 2728–34) was synthesized via the hydrothermal method assisted by pyrolysis-induced transformation. The Co_2SnO_4 hollow cubes are uniformly embedded in the graphene sheet which enhances the extra active sites, reduces the effective diffusion length for Li ions, and increases electrical conductivity which are beneficial for better electrochemical reversibility.

$ZnSn_2O_4$ material is one of the more extensively studied negative electrodes for LIBs owing to its outstanding properties, as 14.4 mol Li^+ per formula would take part/react with Zn_2SnO_4, Sn, and Zn, forming metallic Sn and Zn, as well as Li-Sn and Li-Zn alloys, during the discharge process. However, the large volume change upon the alloying/de-alloying processes, results in structural pulverization of the electrode, which causes many problems, including a capacity fading mechanism, poor rate capability, and cyclic performance. In order to overcome the drawbacks, the various synthesising strategies (1D, 2D, and 3D materials, hollow and sheet-like structures, nanowires) and carbon-coated $ZnSn_2O_4$ (Zhao et al. 2014, 128–32; Qin et al. 2015, 255–58, 2017, 124–30; Lim et al. 2016, 10691–99; Xia et al. 2018, 272–77) were used. For example, highly crystalline octahedron 3D-Zn_2SnO_4 fabricated (Santhoshkumar et al. 2018, 514–20) as a negative material for LIBs through hydrothermal synthesis showed enhanced capacity with stable structure and excellent rate capability performances. It could be noticed that this 3D structure suppressed the volume expansion during lithiation/de-lithiation and also the existence of a porosity electrode provides more room to accommodate the lithium ions.

2.4 TMO

TMO-type materials adopted the rock salt structure, in which the cations and anions are both octahedrally coordinated, as shown in Figure 2.7. In general, in LIB applications, TMOs have been attracting tremendous attention owing to their high theoretical capacity (718 mA h g^{-1} for NiO, 744 mA h g^{-1} for FeO, 716 mA h g^{-1} for CoO, 756 mA h g^{-1} for MnO, and 674 mA h g^{-1} for CuO), and their unique structural and mechanical properties could endow excellent stability, in order to alleviate the structure and volume changes of the electrodes generated during the intercalation/deintercalation process. Furthermore, these types of metal oxides possess high energy density and capacity, when they reacted with lithium, the conversion reaction is happening between the potential ranges of 0.01–3.0 V. The electrochemical reactions of the metal oxides as follow as,

$$MO + 2Li^+ + 2e^- = M + Li_2O \left(\text{where } M = Ni, Co, Mn, Fe, \text{and } Cu \right). \tag{2.14}$$

Despite all the efforts, however, one of the crucial drawbacks of TMOs for use as an alternative anode material for lithium-ion batteries are their low initial Coulombic efficiency.

2.4.1 Nickel Oxide

Among the above TMOs, NiO has received a considerable amount of attention for LIBs, due to its high specific capacity, easy synthesis, and low cost. However, the low intrinsic electronic conductivity of NiO limits its real capacity and rate capability. Meanwhile, the severe volume expansion results in pulverization of NiO electrodes during the charge-discharge process, which leads to fast capacity fade. Preparing porous/nanostructured materials with large specific surface area is an effective way to improve electrical contact and shorten ion diffusion length in order to reduce the resistivity. A lot of various structured NiO materials, such as the NiO nano-octahedron (C. Wang et al. 2017, 272–78), 1D porous (Zheng et al. 2018, 155–59), nanorods (X. Wang et al. 2017, 533–37), nanobelts (Oh et al. 2018, 889–99), nanofilms (Huang et al. 2015, 102–5), and metal oxide, carbon-doped NiO are fabricated, and they deliver high initial charge capacities in various ranges. For example, Liu et al. (2016, 29183) fabricated NiO nanowire growth on Ni foam (template free binder-less NWF) as a negative electrode fir LIBs, the schematic

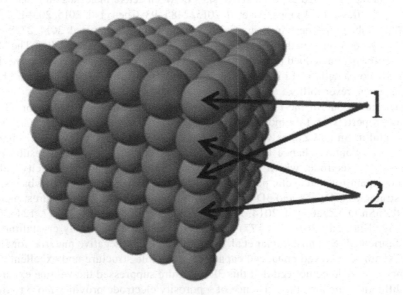

FIGURE 2.7 Schematic structure of metal oxides [MO, where M-metals (Fe, Co, Ni, Cu, and Mn) positioned in (1) and oxygen in (2)].

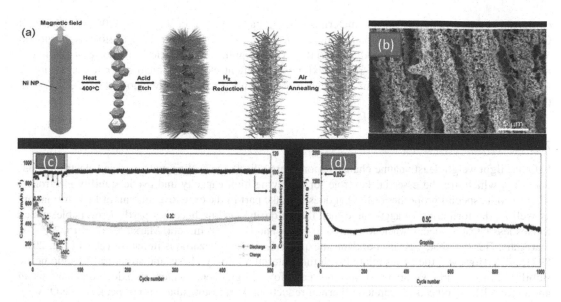

FIGURE 2.8 Schematic representation of NiO nanowire foam (NWF) preparation (a), SEM images of NiO NWF (b), and cyclic and rate performance of NiO NWF (c, d). (Reprinted by permission from Macmillan Publishers Ltd., *Sci. Rep.*, Liu, C., et al. 2016, 29183, copyright 2016.)

representation is illustrated in Figure 2.8a. Cycle life testing of the NiO NWF indicated about 32% capacity loss after 1000 cycles at 0.5°C (359 mA g^{-1}) over a potential range between 0.02 V and 3.0 V (shown in Figure 2.8c and d). The superior capacity with stable cycling and rate performance can be attributed to the intimate electrical contact between NiO active materials and the conductive metallic Ni support, and the porous framework (shown in Figure 2.8b) providing access for the electrolyte resulting in short ionic diffusion length.

Z. Wang et al. (2017, 2909–15) fabricated ternary composites, wherein NiO-SnO$_2$ particles were distributed on graphene nanosheets via the hydrothermal approach. The initial Coulombic efficiency was from 41% to 66% when the NiO-SnO$_2$ particles were on graphene nanosheets, when cycled between 0.01 and 3.0 V. However, rapid capacity fading was observed from the initial cycles to the end of the 50th cycle. Later on, Ding et al. (2018, 479–85) constructed carbon-encapsulated NiO nanoparticles on a porous-reduced graphene oxide (pRGO) matrix (NiO@C@pRGO) through the hydrothermal approach, followed by thermal treatment employed as a negative electrode for LIBs. High capacity retention of 86% was achieved after 1000 cycles at 2000 mA g^{-1} over a potential window of 0.05–3.0 V. This excellent electrochemical behaviour can be attributed to the dual-carbon functionalization, which buffers the volume changes, improves the electronic conductivity, and enhances the kinetic properties of Li ions.

2.4.2 Manganese Oxide

MnO is deemed to be a superior choice of materials because of its outstanding properties, such as low conversion potential (<0.8 V), high density (5.43 g cm^{-3}), low abundance in nature, eco-friendly, and low electromotive force (1.032 V vs Li/Li$^+$). However, the inferior electrical conductivity and the drastic volume and structural changes related to the formation of nanometre-sized Mn grains which are <5 nm embedded into the Li$_2$O matrix during the repeated charge-discharge cycling process give rise to the pulverization of electrode materials and gradual aggregation of active materials. This results in the instability of the SEI and fast capacity fading, which prevent the subsequent widespread applications. Therefore, in order to overcome the drawbacks, various strategies have been proposed to prepare nanoscale materials (Han et al. 2017, 377–86), porous nano-structured (W. Zhang et al. 2016, 16936–45; Zheng et al. 2017, 474–80), metal-doped (Kong et al. 2018, 419–26), and construct composite with conductive carbon materials (Petnikota, Srikanth, et al. 2015, 3205–13; Xiao and Cao 2015, 12840–49). For example, Kong et al. (2018, 419–26) constructed uniformly Ni-doped MnO porous microspheres as anode materials for LIB applications. Cycle life testing of the material indicated that Ni/MnO microspheres exhibited excellent electrochemical reversible capacity of 537.5 mA h g^{-1} that was observed for 1000 cycles at 1000 mA g^{-1}, potential window of 0.01–3.0 V, with the capacity retention of ~ 62%.

2.4.3 Iron Oxide

FeO has light weight least volume changes upon charge/discharge cycling and enhanced electronic conductivity, which are the essential key role for achieving high capacity and cyclic stability electrodes. Thanks to its special properties of FeO, it does not take part in the excessive amount of Li$_2$O formation, as well as the formation of aggregated metallic Fe during cycling becomes partly irreversible, which causes less capacity decay when compared to Fe$_3$O$_4$ and Fe$_2$O$_3$ (Petnikota, Marka, et al. 2015, 253–63) materials. Besides that, FeO is abundant, inexpensive, and non-hazardous in nature (D. Li et al. 2016, 89715–20). However, the issues related with the material are its synthesis and structural/stoichiometric stability. For example, Petnikota et al. reported exfoliated graphene oxide/iron oxide composite as an anode for LIBs, synthesised graphene-thermal reduction. As a result, small minor peaks of Fe$_2$O$_3$ were observed, in the X-ray pattern. They reported the reversible capacity of 857 mA h g^{-1} at the 60th cycle (cycled at a current rate of 50 mA g^{-1} over a potential window between 0.005 and 3.0 V), with the capacity retention of 84%.

2.4.4 Cobalt Oxide

The CoO material is the same as MnO, suffering from poor conductivity, large volume expansion, and particle aggregation during the lithiation/de-lithiation process, thus, resulting in poor electrochemical performance (cycling and rate capability performance). To overcome these drawbacks, new strategies have been proposed by various researchers. That is, adding the carbon hybrid composite materials (Kang et al. 2016, 15920–25) with different types of morphologies, such as nanotubes (L. Zhang et al. 2016, 278–83), nanowires (K. Cao et al. 2015, 1082–89), and the addition of anion (N) to the structure (F. Li et al. 2017, 524–32). N-doping can generate extrinsic defects, and also the N-doped carbon matrix exhibits enhanced lithium ion diffusion and improved electronic conductivity. Recently, Cao et al. assembled a CoO/reduced graphene oxide (CoO/RGO) composite for LIBs (L. Cao et al. 2018, 96–105). The composite delivers a highest capacity of 1309 mA h g^{-1} after 100 cycles at 0.1 A g^{-1}, cycle and rate performance. F. Li et al. (2017, 524–32) fabricated a sea urchin-like nitrogen-doped CoO/Co carbon matrix composite as a negative electrode by the hydrothermal method. It was reported that the co-effect of the NC matrix and Co-doping accelerates the diffusion of lithium ions and electrons in the system, which is the main contribution to improve the excellent cycling stability (811 mA h g^{-1} at 5000 mA g^{-1} after 1000 cycles) and rate performance.

2.4.5 Copper Oxide

Copper oxide is drawing great focus due to its merits such as inexpensiveness, environmentally friendly nature, good safety, and low toxicity. Nevertheless, similar with other TMOs, CuO is also suffering from low electrical conductivity, which is unfavourable for charge transfer. In addition, CuO-based electrodes usually experience severe volume expansion and dispersion (about 174%) of Cu particles in the Li$_2$O matrix upon discharge-charge processes, which leads to severe mechanical strain and rapid capacity decay (Kundu et al. 2017, 1595–604). Therefore, it is necessary to implement new strategies to overcome these challenges. For example, Wang et al. (2014, 1243–50) fabricated nanostructured CuO anode materials with controllable morphologies such as leaf-like CuO, oatmeal-like CuO, and hollow-spherical CuO by changing the ligand agents. When these materials were used in LIBs, the hollow-spherical CuO exhibits more enhanced reversible capacity than the other materials. Wang and his co-workers developed yucca fern-shaped CuO nanowire growth on Cu foam as a negative anode for LIB applications. The stable reversible capacity of 151 mA h g^{-1} was achieved after 100 cycles, cycled at 1000 mA g^{-1}. This enhanced electrochemistry performance could be due to the 3D CuO NWs network where a porous architecture simultaneously reduces the ion diffusion distances, promotes the electrolyte permeation, and electronic conductivity.

The TMO's performances are good when it's used as anode materials in LIBs, however, these metal oxide anode materials only retained around 70%–80% of their initial capacity after 1000 charge/discharge cycles, and low initial Coulombic efficiency performance.

2.5 TM$_3$O$_4$

TM$_3$O$_4$ (where M is Mn, Co, and Fe) is a special kind of material based on high theoretical capacity and low cost. In terms of active electrode materials, the possible use of TMOs as anode materials for lithium batteries has drawn extensive research attention. The Li cycling properties of TM$_3$O$_4$ (TM-Co and Fe) materials by conversion reaction were first reported by the group of Tarascon. However, TM$_3$O$_4$-type materials suffering from severe volume expansion during alloying-de-alloying processes leads to instability of the materials. The electrochemical reaction of TM$_3$O$_4$-type materials could be observed by the following equations:

$$TM_3O_4 + 8Li^+ + 8e^- = 4\,Li_2O + 3TM\,(TM = Mn, Co, and\,Fe)$$

2.5.1 Co$_3$O$_4$

Co$_3$O$_4$ adopts the inverse spinel structure, and Co$_3$O$_4$, being as a mixed valence compound, its formula is sometimes written as Co$^{3+}_t$ [Co^{2+}, $^{3+}$]O$_4$, where 't' refers to the tetrahedral site and 'o' to the octahedral site occupancy (Figure 2.9) and sometimes as CoO•Co$_2$O$_3$. Co$_3$O$_4$ is a promising anode material due to its high theoretical capacity (890 mA h g^{-1}), eco-friendliness, high corrosion stability, excellent catalytic property, and low cost (Z. Li, Yu, and Paik 2016, 41–46; Fan et al. 2017, 2046–53). However, this material suffers from poor conductivity, SEI growing between

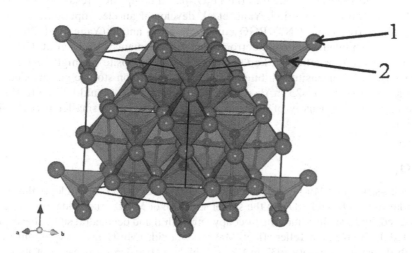

FIGURE 2.9 Crystallographic presentation of Co$_3$O$_4$ spinel oxides with outline unit cell in which (1) represents 'O', (2) stands for 'Co' and 'CoO$_4$-tetrahedra' and 'CoO$_6$-octahedra' are shown. (With kind permission from Springer science + Business media: Nanomaterials in Advanced Batteries and Supercapacitors, "Metal Oxides and Lithium Alloys as Anode Materials for Lithium-Ion Batteries," 2016, 55–91, Kebede, M. et al. Copyright 2016, Springer.)

the electrode-electrolyte surface, and a large amount of volume changes during the redox reaction that could destroy the integrity of the anode, which lead to poor rate capability and rapid capacity fading (L. Li et al. 2017, 7960–65). Also, it is interesting, but challenging to design and fabricate Co_3O_4 electrode materials with a stable structure. Hence, there are different approaches to circumvent these issues, that is, to design or fabricate nanocrystallization, hierarchical porous structure, and coating/supporting TMO with various carbon materials and metals (G. Huang et al. 2015, 1592–99; Liu et al. 2015, 878–84; Yin et al. 2016, 410–19). For example, Lee et al. (2015, 3861–65) reported a 3D multi-walled carbon nanotube (MWCNT) layer-by-layer structure composed of Co_3O_4 nanoplates as a anode for LIB applications. This structure exhibits a high-rate and long-term cycling performance compared to a conventional structure, which is confirmed by mesoporous peapod-like Co_3O_4@carbon nanotube arrays materials (Gu et al. 2015, 7060–64).

L. Guo et al. (2016, 234–42) fabricated Co_3O_4 nanoparticles that were loaded on a nitrogen-doped porous carbon spheres by the hydrothermal method. The cycling performance and reversible capacity of the materials was improved. Yan et al. (2017, 495–501) synthesized Co_3O_4/Co nanoparticles enclosed in graphitic carbon as anode material, it exhibits a reversible capacity of 843 mA h g^{-1} after 60 cycles with capacity retention of 89%, which is due to the graphitic carbon and metallic cobalt.

2.5.2 Fe_3O_4

Fe_3O_4 is a naturally available mineral magnetite, and it adopts an inverse spinal structure similar to Co_3O_4. The single phase of Fe_3O_4 contains both Fe^{2+} and Fe^{3+} ions, which is sometimes written as $FeO \cdot Fe_2O_3$. Fe_3O_4 has been considered to be one of the most promising electrode materials due to its high theoretical capacity of 928 mA h g^{-1}, natural abundance, environmental friendliness, and low processing cost. Unfortunately, Fe_3O_4 has low electric conductivity and severe volume expansion during alloying-de-alloying processes, which leads to serious agglomeration and pulverization, which results in large irreversible capacity loss, poor cycling stability, and inferior rate capability. Moreover, multiple phase transitions occurred during lithiation that are strongly dependent on the electrochemical environment and Li$^+$ diffusion. To address these drawbacks, great efforts have been devoted, based on Fe_3O_4 nanocomposites, and various metal oxide/foreign ions and various nanostructured materials have been fabricated to achieve stable cyclability and high conductivity (Luo et al. 2017, 151–61; Wu et al. 2017, 74–84; Q. Zhao et al. 2018, 233–40). For example, the introduction of conductive carbonaceous materials, such as carbon nanotubes, nanofibers (X. Qin et al. 2015, 347–56), and reduced graphene oxide into the Fe_3O_4-based composites result in both the enhanced capacity and great capacity retention. Yang et al. developed an electrophoretic deposition route to fabricate a finder-free Fe_3O_4/CNTs/RGO composite as an anode (Yang et al. 2016, 26730–39), and the composite exhibited excellent electrochemical performances. S. Li et al. (2016, 17343–51) fabricated a bio-inspired nanofibrous Fe_3O4-TiO_2-carbon composite for high performance anode material for LIBs. This composite exhibited a superior lithium ion storage performance. It showed a stable reversible capacity of 525 mAh g^{-1} after 100 cycles. This could be due to the composite in which the high loading density of active material, as well as the smaller particle sizes of the Fe_3O_4 material.

2.5.3 Mn_3O_4

The hausmannite-structured Mn_3O_4 mineral is found in nature. Mn_3O_4 also adopts the inverse spinel structure similar to Co_3O_4 and Fe_3O_4. In the spinel structure of Mn_3O_4, the oxide ions are cubic, which are closely packed, and the Mn^{2+} and Mn^{3+} occupy tetrahedral and octahedral sites. However, the structure is contorted due to the Jahn-Teller effect. Manganese oxide (Mn_3O_4) is a very promising candidate due to its high theoretical capacity (937 mA h g^{-1}), high energy density because of its low operating potential of manganese (~1.2 V), good thermal stability, and environmental benignity. In particular, as an anode for LIBs, Mn_3O_4 is low cost and non-toxic (Palaniyandy, Nkosi, et al. 2019, 79–92). However, Mn_3O_4 has poor electrical conductivity ($10 - 7 \sim 10^{-8}$ S cm^{-1}) compared with other metal oxides, such

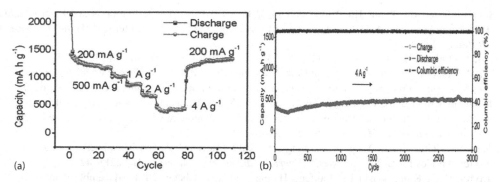

FIGURE 2.10 (a) Rate capability studies at different current densities of 200 mA g⁻¹, 500 mA g⁻¹, 1 A g⁻¹, 2 A g⁻¹, and 4 A g⁻¹ and (b) long cyclability test at 4 A g⁻¹ of Mn₃O₄/C nanosphere anode. (Reprinted from *J. Electroanal. Chem.*, 833, Palaniyandy, N. et al., "Conversion of Electrolytic MnO₂ to Mn₃O₄ Nanowires for High-Performance Anode Materials for Lithium-Ion Batteries," 65. Copyright 2018, with permission from Elsevier.)

as Fe_3O_4 and Co_3O_4, which leads to kinetically limited rate capability, unstable SEI, and large volume change that render rapid capacity loss during charge/discharge processes. To tackle this issue, effective strategies have been regarded, such as to design or synthesize the unique structure of carbon composites, various nanostructures and metal and foreign cations/anions doped (Chen et al. 2017, 4655–62), Mn_3O_4. For example, Liu et al. self-assembled Mn_3O_4/C nanospheres as anode material for LIBs, which exhibits a high reversible capacity (1237 mA h g⁻¹ at 200 mA g⁻¹) with an extremely long cycle life and excellent rate capability, which is shown in Figure 2.10. Palaniyandy et al. (2018, 65–75) reported fluorine-doped Mn_3O_4 nanosphere as an anode for LIBs, this exhibits more enhanced electrochemical performance than the pristine Mn_3O_4.

2.6 Summary

Having excellent prospects in getting high capacity and high energy density electrode materials for LIB, different anode materials are attracting a great deal of attention. The three kinds of anodes: intercalation, alloying, and conversion type of anodes are so crucial in establishing a suitable matching anode for respective cathodes of lithium-ion batteries. The intercalation type anodes (C, LTO, TiO_2, etc.) are offering low, reasonable capacity, and they have a stable cyclability. The alloying type of anodes Si- (4211 mA h g⁻¹), Sn (911 mA h g⁻¹)-based anodes deliver remarkably high capacity, but they face severe capacity-fading challenges due to volume expansion. The conversion type metal oxide anodes, such as Fe_3O_4, Co_3O_4, Mn_3O_4, etc., can provide high capacities (>700), and they are somehow better in terms of stability as compared to alloying anodes like Si and Sn.

In terms of curbing the capacity-fading challenges, especially for anode materials providing high theoretical capacities, which have lithiation mechanisms due to alloying and conversion, two commonly acceptable strategies have been implemented. They are nano-architecturing of the anode materials, as nanostructured materials are capable of better endurance at the huge volume expansion and coating the materials with carbonaceous materials enhances the conductivity and buffer the volume expansion as well.

REFERENCES

Aldon, L., P. Kubiak, M. Womes, J. C. Jumas, J. Olivier-Fourcade, J. L. Tirado, J. I. Corredor, and C. Pérez Vicente. 2004. "Chemical and Electrochemical Li-Insertion into the $Li_4Ti_5O_{12}$ Spinel." *Chemistry of Materials* 16 (26): 5721–25.

An, Bonan, Qiang Ru, Shejun Hu, Xiong Song, and Chang Chen. 2015. "Enhanced Electrochemical Performance of Nanomilling Co_2SnO_4/C Materials for Lithium Ion Batteries." *Ionics* 21 (9): 2485–93.

An, Bonan, Qiang Ru, Shejun Hu, Xiong Song, and Juan Li. 2014. "Facile Synthesis and Electrochemical Performance of Co_2SnO_4/Co_3O_4 Nanocomposite for Lithium-Ion Batteries." *Materials Research Bulletin* 60: 640–47.

Armand, Michel, and J.-M. Tarascon. 2008. "Building Better Batteries." *Nature* 451 (7179): 652.

Bhaskar, Akkisetty, Melepurath Deepa, and Tata Narasinga Rao. 2014. "Size-Controlled SnO_2 Hollow Spheres via a Template Free Approach as Anodes for Lithium Ion Batteries." *Nanoscale* 6 (18): 10762–71.

Cao, Kangzhe, Lifang Jiao, Yongchang Liu, Huiqiao Liu, Yijing Wang, and Huatang Yuan. 2015. "Ultra-high Capacity Lithium-Ion Batteries with Hierarchical CoO Nanowire Clusters as Binder Free Electrodes." *Advanced Functional Materials* 25 (7): 1082–89.

Cao, Kangzhe, Ting Jin, Li Yang, and Lifang Jiao. 2017. "Recent Progress in Conversion Reaction Metal Oxide Anodes for Li-Ion Batteries." *Materials Chemistry Frontiers* 1 (11): 2213–42.

Cao, Liyun, Qian Kang, Jiayin Li, Jianfeng Huang, and Yayi Cheng. 2018. "Assembly Control of CoO/Reduced Graphene Oxide Composites for Their Enhanced Lithium Storage Behavior." *Applied Surface Science* 455: 96–105.

Chen, Chang, Qiang Ru, Shejun Hu, Bonan An, Xiong Song, and Xianhua Hou. 2015. "Co_2SnO_4 Nanocrystals Anchored on Graphene Sheets as High-Performance Electrodes for Lithium-Ion Batteries." *Electrochimica Acta* 151: 203–13.

Chen, Jiayuan, Xiaofeng Wu, Yan Gong, Pengfei Wang, Wenhui Li, Qiangqiang Tan, and Yunfa Chen. 2017. "Synthesis of Mn_3O_4/N-Doped Graphene Hybrid and Its Improved Electrochemical Performance for Lithium-Ion Batteries." *Ceramics International* 43 (5): 4655–62.

Cheng, Lingli, Dandan Qiao Pandeng Zhao, Yongchao He, Wangfei Sun, Hongchuan Yu, Zheng Jiao. 2019. "Template-Free Synthesis of Mesoporous Succulents-Like TiO_2/Graphene Aerogel Composites for Lithium-Ion Batteries." *Electrochimica Acta* 300 (2019): 417–425

Ding, Chunyan, Weiwei Zhou, Xiangyuan Wang, Bin Shi, Dong Wang, Pengyu Zhou, and Guangwu Wen. 2018. "Hybrid Aerogel-Derived Carbon/Porous Reduced Graphene Oxide Dual-Functionalized NiO for High-Performance Lithium Storage." *Chemical Engineering Journal* 332: 479–85.

Fan, Lei, Weidong Zhang, Shoupu Zhu, and Yingying Lu. 2017. "Enhanced Lithium Storage Capability in Li-Ion Batteries Using Porous 3D Co_3O_4 Nanofiber Anodes." *Industrial & Engineering Chemistry Research* 56 (8): 2046–53.

Flandrois, S., and B. Simon. 1999. "Carbon Materials for Lithium-Ion Rechargeable Batteries." *Carbon* 37 (2): 165–80.

Fröschl, T., U. Hörmann, P. Kubiak, G. Kučerová, M. Pfanzelt, Clemens K. Weiss, R. J. Behm, N. Hüsing, U. Kaiser, and Katharina Landfester. 2012. "High Surface Area Crystalline Titanium Dioxide: Potential and Limits in Electrochemical Energy Storage and Catalysis." *Chemical Society Reviews* 41 (15): 5313–60.

Gao, Lvlv, Cuiping Gu, Haibo Ren, Xinjie Song, and Jiarui Huang. 2018. "Synthesis of Tin (IV) Oxide@Reduced Graphene Oxide Nanocomposites with Superior Electrochemical Behaviors for Lithium-Ions Batteries." *Electrochimica Acta* 290: 72–81.

Geng, Hongbo, Xueqin Cao, Yu Zhang, Kaiming Geng, Genlong Qu, Minghua Tang, Junwei Zheng, Yonggang Yang, and Hongwei Gu. 2015. "Hollow Nanospheres Composed of Titanium Dioxide Nanocrystals Modified with Carbon and Gold for High Performance Lithium Ion Batteries." *Journal of Power Sources* 294: 465–72.

Gu, Dong, Wei Li, Fei Wang, Hans Bongard, Bernd Spliethoff, Wolfgang Schmidt, Claudia Weidenthaler, Yongyao Xia, Dongyuan Zhao, and Ferdi Schüth. 2015. "Controllable Synthesis of Mesoporous Peapod-like Co_3O_4@ Carbon Nanotube Arrays for High-Performance Lithium-Ion Batteries." *Angewandte Chemie International Edition* 54 (24): 7060–64.

Guan, Bu Yuan, Le Yu, Ju Li, and Xiong Wen David Lou. 2016. "A Universal Cooperative Assembly-Directed Method for Coating of Mesoporous TiO_2 Nanoshells with Enhanced Lithium Storage Properties." *Science Advances* 2 (3): e1501554.

Guo, Liangui, Yu Ding, Caiqin Qin, Wei Li, Jun Du, Zhengbin Fu, Wulin Song, and Feng Wang. 2016. "Nitrogen-Doped Porous Carbon Spheres Anchored with Co_3O_4 Nanoparticles as High-Performance Anode Materials for Lithium-Ion Batteries." *Electrochimica Acta* 187: 234–42.

Guo, Sheng-qi, Meng-meng Zhen, Lu Liu, and Zhi-hao Yuan. 2016. "Facile Preparation and Lithium Storage Properties of TiO_2@ Graphene Composite Electrodes with Low Carbon Content." *Chemistry–A European Journal* 22 (34): 11943–48.

Han, Cheng-Gong, Chunyu Zhu, Yoshitaka Aoki, Hiroki Habazaki, and Tomohiro Akiyama. 2017. "MnO/N–C Anode Materials for Lithium-Ion Batteries Prepared by Cotton-Templated Combustion Synthesis." *Green Energy & Environment* 2 (4): 377–86.

Huang, Gang, Feifei Zhang, Xinchuan Du, Yuling Qin, Dongming Yin, and Limin Wang. 2015. "Metal Organic Frameworks Route to In Situ Insertion of Multiwalled Carbon Nanotubes in Co_3O_4 Polyhedra as Anode Materials for Lithium-Ion Batteries." *ACS Nano* 9 (2): 1592–99.

Huang, Haijian, Ying Huang, Mingyue Wang, Xuefang Chen, Yang Zhao, Ke Wang, and Haiwei Wu. 2014. "Preparation of Hollow Zn_2SnO_4 Boxes@ C/Graphene Ternary Composites with a Triple Buffering Structure and Their Electrochemical Performance for Lithium-Ion Batteries." *Electrochimica Acta* 147: 201–8.

Huang, X. H., P. Zhang, J. B. Wu, Y. Lin, and R. Q. Guo. 2015. "Porous NiO/Graphene Hybrid Film as Anode for Lithium Ion Batteries." *Materials Letters* 153: 102–5.

Jiang, Chunhai, and Jinsong Zhang. 2013. "Nanoengineering Titania for High Rate Lithium Storage: A Review." *Journal of Materials Science & Technology* 29 (2): 97–122.

Kang, Qian, Liyun Cao, Jiayin Li, Jianfeng Huang, Zhanwei Xu, Yayi Cheng, Xin Wang, Jinying Bai, and Qianying Li. 2016. "Super P Enhanced CoO Anode for Lithium-Ion Battery with Superior Electrochemical Performance." *Ceramics International* 42 (14): 15920–25.

Kebede, Mesfin, Haitao Zheng, and Kenneth I. Ozoemena. 2016. "Metal Oxides and Lithium Alloys as Anode Materials for Lithium-Ion Batteries." *Nanomaterials in Advanced Batteries and Supercapacitors*. Springer.

Kong, Xiangzhong, Yaping Wang, Jiande Lin, Shuquan Liang, Anqiang Pan, and Guozhong Cao. 2018. "Twin-Nanoplate Assembled Hierarchical Ni/MnO Porous Microspheres as Advanced Anode Materials for Lithium-Ion Batteries." *Electrochimica Acta* 259: 419–26.

Kundu, Manab, Gopalu Karunakaran, Evgeny Kolesnikov, Mikhail V. Gorshenkov, and Denis Kuznetsov. 2017. "Negative Electrode Comprised of Nanostructured CuO for Advanced Lithium Ion Batteries." *Journal of Cluster Science* 28 (3): 1595–1604.

Lee, Tae Il, Jong-Pil Jegal, Ji-Hyeon Park, Won Jin Choi, Jeong-O Lee, Kwang-Bum Kim, and Jae-Min Myoung. 2015. "Three-Dimensional Layer-by-Layer Anode Structure Based on Co_3O_4 Nanoplates Strongly Tied by Capillary-Like Multiwall Carbon Nanotubes for Use in High-Performance Lithium-Ion Batteries." *ACS Applied Materials & Interfaces* 7 (7): 3861–65.

Li, Di, Kangli Wang, Hongwei Tao, Xiaohong Hu, Shijie Cheng, and Kai Jiang. 2016. "Facile Synthesis of an Fe_3O_4/FeO/Fe/C Composite as a High-Performance Anode for Lithium-Ion Batteries." *RSC Advances* 6 (92): 89715–20.

Li, Fanyan, Manman Ren, Weiliang Liu, Guangda Li, Mei Li, Liwei Su, Cuiling Gao, Jinpei Hei, and Hongxia Yang. 2017. "Sea Urchin-Like CoO/Co/N-Doped Carbon Matrix Hybrid Composites with Superior High-Rate Performance for Lithium-Ion Batteries." *Journal of Alloys and Compounds* 701: 524–32.

Li, Li, Zichao Zhang, Sijia Ren, Bingke Zhang, Shuhua Yang, and Bingqiang Cao. 2017. "Construction of Hollow Co_3O_4 Cubes as a High-Performance Anode for Lithium Ion Batteries." *New Journal of Chemistry* 41 (16): 7960–65.

Li, Shun, Mengya Wang, Yan Luo, and Jianguo Huang. 2016. "Bio-Inspired Hierarchical Nanofibrous Fe_3O_4–TiO_2–Carbon Composite as a High-Performance Anode Material for Lithium-Ion Batteries." *ACS Applied Materials & Interfaces* 8 (27): 17343–51.

Li, Yanjuan, Xiao Yan, Wenfu Yan, Xiaoyong Lai, Nan Li, Yue Chi, Yingjin Wei, and Xiaotian Li. 2013. "Hierarchical Tubular Structure Constructed by Mesoporous TiO_2 Nanosheets: Controlled Synthesis and Applications in Photocatalysis and Lithium Ion Batteries." *Chemical Engineering Journal* 232: 356–63.

Li, Zhangpeng, Xin-Yao Yu, and Ungyu Paik. 2016. "Facile Preparation of Porous Co_3O_4 Nanosheets for High-Performance Lithium Ion Batteries and Oxygen Evolution Reaction." *Journal of Power Sources* 310: 41–46.

Liang, Kuang, Tuck–Yun Cheang, Tao Wen, Xiao Xie, Xiao Zhou, Zhi–Wei Zhao, Cong–Cong Shen, Nan Jiang, and An–Wu Xu. 2016. "Facile Preparation of Porous Mn_2SnO_4/Sn/C Composite Cubes as High Performance Anode Material for Lithium-Ion Batteries." *The Journal of Physical Chemistry C* 120 (7): 3669–76.

Lim, Young Rok, Chan Su Jung, Hyung Soon Im, Kidong Park, Jeunghee Park, Won Il Cho, and Eun Hee Cha. 2016. "Zn_2GeO_4 and Zn_2SnO_4 Nanowires for High-Capacity Lithium-and Sodium-Ion Batteries." *Journal of Materials Chemistry A* 4 (27): 10691–99.

Lin, Zhiya, Yanmin Yang, Jiamen Jin, Luya Wei, Wei Chen, Yingbin Lin, and Zhigao Huang. 2017. "Graphene-Wrapped Li$_4$Ti$_5$O$_{12}$ Hollow Spheres Consisting of Nanosheets as Novel Anode Material for Lithium-Ion Batteries." *Electrochimica Acta* 254: 287–98.

Liu, Chueh, Changling Li, Kazi Ahmed, Zafer Mutlu, Cengiz S. Ozkan, and Mihrimah Ozkan. 2016. "Template Free and Binderless NiO Nanowire Foam for Li-Ion Battery Anodes with Long Cycle Life and Ultrahigh Rate Capability." *Scientific Reports* 6: 29183.

Liu, Yanguo, Zhiying Cheng, Hongyu Sun, Hamidreza Arandiyan, Jinpeng Li, and Mashkoor Ahmad. 2015. "Mesoporous Co$_3$O$_4$ Sheets/3D Graphene Networks Nanohybrids for High-Performance Sodium-Ion Battery Anode." *Journal of Power Sources* 273: 878–84.

Luo, Honglin, Dehui Ji, Zhiwei Yang, Yuan Huang, Guangyao Xiong, Yong Zhu, Ruisong Guo, and Yizao Wan. 2017. "An Ultralight and Highly Compressible Anode for Li-Ion Batteries Constructed from Nitrogen-Doped Carbon Enwrapped Fe$_3$O$_4$ Nanoparticles Confined in a Porous 3D Nitrogen-Doped Graphene Network." *Chemical Engineering Journal* 326: 151–61.

Mullaivananathan, V., K. R. Saravanan, and N. Kalaiselvi. 2017. "Synthetically Controlled, Carbon-Coated Co$_2$SnO$_4$/SnO$_2$ Composite Anode for Lithium-Ion Batteries." *JOM* 69 (9): 1497–1502.

Nam, Sang Hoon, Hee-Sang Shim, Youn-Su Kim, Mushtaq Ahmad Dar, Jong Guk Kim, and Won Bae Kim. 2010. "Ag or Au Nanoparticle-Embedded One-Dimensional Composite TiO$_2$ Nanofibers Prepared via Electrospinning for Use in Lithium-Ion Batteries." *ACS Applied Materials & Interfaces* 2 (7): 2046–52.

Nam, Seunghoon, Seung Jae Yang, Sangheon Lee, Jaewon Kim, Joonhyeon Kang, Jun Young Oh, Chong Rae Park, Taeho Moon, Kyu Tae Lee, and Byungwoo Park. 2015. "Wrapping SnO$_2$ with Porosity-Tuned Graphene as a Strategy for High-Rate Performance in Lithium Battery Anodes." *Carbon* 85: 289–98.

Nithyadharseni, P., M. V. Reddy, Kenneth I. Ozoemena, Fabian I. Ezema, R. Geetha Balakrishna, and B. V. R. Chowdari. 2016. "Electrochemical Performance of BaSnO$_3$ Anode Material for Lithium-Ion Battery Prepared by Molten Salt Method." *Journal of the Electrochemical Society* 163 (3): A540–45.

Nithyadharseni, P., M. V. Reddy, Kenneth I. Ozoemena, R. Geetha Balakrishna, and B. V. R. Chowdari. 2015. "Low Temperature Molten Salt Synthesis of Y$_2$Sn$_2$O$_7$ Anode Material for Lithium Ion Batteries." *Electrochimica Acta* 182: 1060–69.

Oh, Se Hwan, Jin-Sung Park, Min Su Jo, Yun Chan Kang, and Jung Sang Cho. 2018. "Design and Synthesis of Tube-in-Tube Structured NiO Nanobelts with Superior Electrochemical Properties for Lithium-Ion Storage." *Chemical Engineering Journal* 347: 889–99.

Ohzuku, Tsutomu, Atsushi Ueda, and Norihiro Yamamoto. 1995. "Zero-strain Insertion Material of Li [Li1/3Ti5/3] O 4 for Rechargeable Lithium Cells." *Journal of the Electrochemical Society* 142 (5): 1431–35.

Palaniyandy, Nithyadharseni, Funeka P. Nkosi, Kumar Raju, and Kenneth I. Ozoemena. 2018. "Fluorinated Mn$_3$O$_4$ nanospheres for lithium-ion batteries: Low-cost synthesis with enhanced capacity, cyclability and charge-transport." *Materials Chemistry and Physics* 209: 65–75.

Palaniyandy, Nithyadharseni, Funeka P. Nkosi, Kumar Raju, and Kenneth I. Ozoemena. 2019. "Conversion of Electrolytic MnO$_2$ to Mn$_3$O$_4$ Nanowires for High-Performance Anode Materials for Lithium-Ion Batteries." *Journal of Electroanalytical Chemistry* 833: 79–92.

Palaniyandy, Nithyadharseni, Mesfin A. Kebede, Kenneth I. Ozoemena, and Mkhulu K. Mathe. 2019. "Rapidly Microwave-Synthesized SnO$_2$ Nanorods Anchored on Onion-Like Carbons (OLCs) as Anode Material for Lithium-Ion Batteries." *Electrocatalysis*, Springer, 1–9.

Park, Kyu-Sung, Anass Benayad, Dae-Joon Kang, and Seok-Gwang Doo. 2008. "Nitridation-Driven Conductive Li$_4$Ti$_5$O$_{12}$ for Lithium Ion Batteries." *Journal of the American Chemical Society* 130 (45): 14930–31.

Petnikota, Shaikshavali, Sandeep Kumar Marka, Arkaprabha Banerjee, M. V. Reddy, V. V. S. S. Srikanth, and B. V. R. Chowdari. 2015. "Graphenothermal Reduction Synthesis of 'Exfoliated Graphene Oxide/Iron (II) Oxide' Composite for Anode Application in Lithium Ion Batteries." *Journal of Power Sources* 293: 253–63.

Petnikota, Shaikshavali, Vadali V. S. S. Srikanth, P. Nithyadharseni, M. V. Reddy, S. Adams, and B. V. R. Chowdari. 2015. "Sustainable Graphenothermal Reduction Chemistry to Obtain MnO Nanonetwork Supported Exfoliated Graphene Oxide Composite and Its Electrochemical Characteristics." *ACS Sustainable Chemistry & Engineering* 3 (12): 3205–13.

Qin, Liping, Shuquan Liang, Anqiang Pan, and Xiaoping Tan. 2015. "Facile Solvothermal Synthesis of Zn$_2$SnO$_4$ Nanoparticles as Anode Materials for Lithium-Ion Batteries." *Materials Letters* 141: 255–58.

Qin, Liping, Shuquan Liang, Xiaoping Tan, and Anqiang Pan. 2017. "Zn_2SnO_4/Graphene Composites as Anode Materials for High Performance Lithium-Ion Batteries." *Journal of Alloys and Compounds* 692: 124–30.

Qin, Xianying, Haoran Zhang, Junxiong Wu, Xiaodong Chu, Yan-Bing He, Cuiping Han, Cui Miao, Shuan Wang, Baohua Li, and Feiyu Kang. 2015. "Fe_3O_4 Nanoparticles Encapsulated in Electrospun Porous Carbon Fibers with a Compact Shell as High-Performance Anode for Lithium Ion Batteries." *Carbon* 87: 347–56.

Reddy, M. V., G. V. Subba Rao, and B. V. R. Chowdari. 2013. "Metal Oxides and Oxysalts as Anode Materials for Li Ion Batteries." *Chemical Reviews* 113 (7): 5364–5457.

Rehman, Wasif Ur, Youlong Xu, Xiaofei Sun, Inam Ullah, Yuan Zhang, and Long Li. 2018. "Bouquet-like Mn_2SnO_4 Nanocomposite Engineered with Graphene Sheets as an Advanced Lithium-Ion Battery Anode." *ACS Applied Materials & Interfaces* 10 (21): 17963–72.

Santhoshkumar, P., K. Prasanna, Yong Nam Jo, Suk Hyun Kang, Youn Cheol Joe, and Chang Woo Lee. 2018. "Synthesis of Highly Crystalline Octahedron 3D-Zn_2SnO_4 as an Advanced High-Performance Anode Material for Lithium Ion Batteries." *Applied Surface Science* 449: 514–20.

Song, Taeseup, and Ungyu Paik. 2016. "TiO_2 as an Active or Supplemental Material for Lithium Batteries." *Journal of Materials Chemistry A* 4 (1): 14–31.

Sorensen, Erin M., Scott J. Barry, Ha-Kyun Jung, James M. Rondinelli, John T. Vaughey, and Kenneth R. Poeppelmeier. 2006. "Three-Dimensionally Ordered Macroporous $Li_4Ti_5O_{12}$: Effect of Wall Structure on Electrochemical Properties." *Chemistry of Materials* 18 (2): 482–89.

Su, Xin, Qingliu Wu, Juchuan Li, Xingcheng Xiao, Amber Lott, Wenquan Lu, Brian W. Sheldon, and Ji Wu. 2014. "Silicon-Based Nanomaterials for Lithium-Ion Batteries: A Review." *Advanced Energy Materials* 4 (1): 1300882.

Tang, Hao, Cuixia Cheng, Gaige Yu, Haowen Liu, and Weiqing Chen. 2015. "Structure and Electrochemical Properties of Mg_2SnO_4 Nanoparticles Synthesized by a Facile Co-Precipitation Method." *Materials Chemistry and Physics* 159: 167–72.

Tian, Jian, Zhenhuan Zhao, Anil Kumar, Robert I. Boughton, and Hong Liu. 2014. "Recent Progress in Design, Synthesis, and Applications of One-Dimensional TiO_2 Nanostructured Surface Heterostructures: A Review." *Chemical Society Reviews* 43 (20): 6920–37.

Wang, Bin, Bin Luo, Xianglong Li, and Linjie Zhi. 2012. "The Dimensionality of Sn Anodes in Li-Ion Batteries." *Materials Today* 15 (12): 544–52.

Wang, Chen, Qing Li, Fangfang Wang, Guofeng Xia, Ruiqing Liu, Deyu Li, Ning Li, Jacob S. Spendelow, and Gang Wu. 2014. "Morphology-Dependent Performance of CuO Anodes via Facile and Controllable Synthesis for Lithium-Ion Batteries." *ACS Applied Materials & Interfaces* 6 (2): 1243–50.

Wang, Chengzhi, Yongjie Zhao, Dezhi Su, Caihua Ding, Lin Wang, Dong Yan, Jingbo Li, and Haibo Jin. 2017. "Synthesis of NiO Nano Octahedron Aggregates as High-Performance Anode Materials for Lithium Ion Batteries." *Electrochimica Acta* 231: 272–78.

Wang, Xinghui, Leimeng Sun, Xiaolei Sun, Xiuwan Li, and Deyan He. 2017. "Size-Controllable Porous NiO Electrodes for High-Performance Lithium Ion Battery Anodes." *Materials Research Bulletin* 96: 533–37.

Wang, Zhao, Jiechen Mu, Yong Li, Jin Chen, Lipeng Zhang, Degang Li, and Pingping Zhao. 2017. "Preparation and Lithium Storage Properties of NiO-SnO_2/Graphene Nanosheet Ternary Composites." *Journal of Alloys and Compounds* 695: 2909–15.

Wilkening, Martin, Roger Amade, Wojciech Iwaniak, and Paul Heitjans. 2007. "Ultraslow Li Diffusion in Spinel-Type Structured Li 4 Ti_5O_{12}—A Comparison of Results from Solid State NMR and Impedance Spectroscopy." *Physical Chemistry Chemical Physics* 9 (10): 1239–46.

Wu, Qianhui, Rongfang Zhao, Wenjie Liu, Xiue Zhang, Xiao Shen, Wenlong Li, Guowang Diao, and Ming Chen. 2017. "In-Depth Nanocrystallization Enhanced Li-Ions Batteries Performance with Nitrogen-Doped Carbon Coated Fe_3O_4 Yolk–Shell Nanocapsules." *Journal of Power Sources* 344: 74–84.

Xia, Ji, Ran Tian, Yiping Guo, Qi Du, Wen Dong, Runjiang Guo, Xiuwu Fu, Lin Guan, and Hezhou Liu. 2018. "Zn_2SnO_4-Carbon Cloth Freestanding Flexible Anodes for High-Performance Lithium-Ion Batteries." *Materials & Design* 156: 272–77.

Xiao, Ying, and Minhua Cao. 2015. "Carbon-Anchored MnO Nanosheets as an Anode for High-Rate and Long-Life Lithium-Ion Batteries." *ACS Applied Materials & Interfaces* 7 (23): 12840–49.

Yan, Zhiliang, Qiyang Hu, Guochun Yan, Hangkong Li, Kaimin Shih, Zhewei Yang, Xinhai Li, Zhixing Wang, and Jiexi Wang. 2017. "Co_3O_4/Co Nanoparticles Enclosed Graphitic Carbon as Anode Material for High Performance Li-Ion Batteries." *Chemical Engineering Journal* 321: 495–501.

Yang, Shuliang, Changyan Cao, Peipei Huang, Li Peng, Yongbin Sun, Fang Wei, and Weiguo Song. 2015. "Sandwich-Like Porous TiO_2/Reduced Graphene Oxide (RGO) for High-Performance Lithium-Ion Batteries." *Journal of Materials Chemistry A* 3 (16): 8701–5.

Yang, Yang, Jiaqi Li, Dingqiong Chen, and Jinbao Zhao. 2016. "A Facile Electrophoretic Deposition Route to the Fe_3O_4/CNTs/RGO Composite Electrode as a Binder-Free Anode for Lithium Ion Battery." *ACS Applied Materials & Interfaces* 8 (40): 26730–39.

Yi, Ting-Feng, Li-Juan Jiang, J. Shu, Cai-Bo Yue, Rong-Sun Zhu, and Hong-Bin Qiao. 2010. "Recent Development and Application of $Li_4Ti_5O_{12}$ as Anode Material of Lithium Ion Battery." *Journal of Physics and Chemistry of Solids* 71 (9): 1236–42.

Yin, Dongming, Gang Huang, Qujiang Sun, Qian Li, Xuxu Wang, Dongxia Yuan, Chunli Wang, and Limin Wang. 2016. "RGO/Co_3O_4 Composites Prepared Using GO-MOFs as Precursor for Advanced Lithium-Ion Batteries and Supercapacitors Electrodes." *Electrochimica Acta* 215: 410–19.

Yoo, Hana, Jung-In Lee, Hyunjung Kim, Jung-Pil Lee, Jaephil Cho, and Soojin Park. 2011. "Helical Silicon/ Silicon Oxide Core–Shell Anodes Grown onto the Surface of Bulk Silicon." *Nano Letters* 11 (10): 4324–28.

Zaghib, K., M. Simoneau, M. Armand, and M. Gauthier. 1999. "Electrochemical Study of $Li_4Ti_5O_{12}$ as Negative Electrode for Li-Ion Polymer Rechargeable Batteries." *Journal of Power Sources* 81: 300–305.

Zhang, Jingjing, Jianwen Liang, Yongchun Zhu, Denghu Wei, Long Fan, and Yitai Qian. 2014. "Synthesis of Co_2SnO_4 Hollow Cubes Encapsulated in Graphene as High Capacity Anode Materials for Lithium-Ion Batteries." *Journal of Materials Chemistry A* 2 (8): 2728–34.

Zhang, Linsen, Zhitao Wang, Huan Wang, Kun Yang, Lizhen Wang, Xiaofeng Li, Yong Zhang, and Huichao Dong. 2016. "Preparation and Electrochemical Performances of CoO/3D Graphene Composite as Anode for Lithium-Ion Batteries." *Journal of Alloys and Compounds* 656: 278–83.

Zhang, Wei, Jinzhi Sheng, Jie Zhang, Ting He, Lin Hu, Rui Wang, Liqiang Mai, and Shichun Mu. 2016. "Hierarchical Three-Dimensional MnO Nanorods/Carbon Anodes for Ultralong-Life Lithium-Ion Batteries." *Journal of Materials Chemistry A* 4 (43): 16936–45.

Zhao, Bing, Hua Zhuang, Yaqing Yang, Yanyan Wang, Haihua Tao, Zhixuan Wang, Jinlong Jiang, Zhiwen Chen, Shoushuang Huang, and Yong Jiang. 2019. "Composition-Dependent Lithium Storage Performances of SnS/SnO_2 Heterostructures Sandwiching Between Spherical Graphene." *Electrochimica Acta* 300 (2019): 253–262.

Zhao, Qingshan, Jialiang Liu, Yixian Wang, Wei Tian, Jingyan Liu, Jiazhen Zang, Hui Ning, Chaohe Yang, and Mingbo Wu. 2018. "Novel In-Situ Redox Synthesis of Fe_3O_4/RGO Composites with Superior Electrochemical Performance for Lithium-Ion Batteries." *Electrochimica Acta* 262: 233–40.

Zhao, Yang, Ying Huang, Xu Sun, Haijian Huang, Ke Wang, Meng Zong, and Qiufen Wang. 2014. "Hollow Zn_2SnO_4 Boxes Wrapped with Flexible Graphene as Anode Materials for Lithium Batteries." *Electrochimica Acta* 120: 128–32.

Zheng, Fangcai, Zhichen Yin, Hongyu Xia, Guoliang Bai, and Yuanguang Zhang. 2017. "Porous MnO@ C Nanocomposite Derived from Metal-Organic Frameworks as Anode Materials for Long-Life Lithium-Ion Batteries." *Chemical Engineering Journal* 327: 474–80.

Zheng, Qingyuan, Yingying Liu, Hongtao Guo, Xiaoyan Hua, Shaojun Shi, and Mingming Zuo. 2018. "Synthesis of Hierarchical 1D NiO Assisted by Microwave as Anode Material for Lithium-Ion Batteries." *Materials Research Bulletin* 98: 155–59.

Zhou, Dan, Xiaogang Li, Li-Zhen Fan, and Yonghong Deng. 2017. "Three-Dimensional Porous Graphene-Encapsulated CNT@SnO_2 Composite for High-Performance Lithium and Sodium Storage." *Electrochimica Acta* 230: 212–21.

Zuo, Shiyong, Deren Li, Zhiguo Wu, Yunqiang Sun, Qihai Lu, Fengyi Wang, Renfu Zhuo, De Yan, Jun Wang, and Pengxun Yan. 2018. "SnO_2/Graphene Oxide Composite Material with High Rate Performance Applied in Lithium Storage Capacity." *Electrochimica Acta* 264: 61–68.

3

Sodium-Ion Battery Anode Materials and Its Future Prospects and Challenges

Nithyadharseni Palaniyandy and Mesfin A. Kebede

CONTENTS

3.1   Introduction 41
3.2   Carbon-Based Materials 42
    3.2.1   Graphene 42
    3.2.2   Hard Carbon 43
3.3   Transition Metal Oxide-Based Materials 45
    3.3.1   Iron Oxide and Its Derivatives 45
    3.3.2   Copper Oxide and Its Derivatives 45
    3.3.3   Cobalt Oxide and Its Derivatives 46
3.4   Alloy-Based Materials 48
3.5   Organic Materials-Based Anodes 51
3.6   Phosphorus-Based Anodes 52
3.7   Conclusions 53
References 53

3.1 Introduction

So far lithium ion batteries (LIBs) have been considered to be prominent energy storage devices owing to its high specific energy and power densities. However, the potential scarcity of lithium resources makes it expensive and its low safety make it necessary to look for new alternative electrochemical batteries. Sodium is less expensive and abundant, as it is the fourth most abundant element in the earth's crust and is uniformly distributed around the world. Sodium-ion batteries (SIBs) adopt the same technology and electrochemistry principle as LIBs, however, the charge carriers are Na^+ instead of Li^+. Recently, SIBs have continued to receive tremendous exposure due to their natural abundance (earth's crust and sea water) [2.6 wt. %, compared to lithium (0.06%)], cost effectiveness, and similar chemical and electrochemical properties to the LIBs. However, the drawbacks such as higher redox potential (−2.71 V vs Standard hydrogen electrode (SHE)), higher atomic mass (23 g mol^{-1}), and the large ionic radii (1.02 Å) of sodium metal leading to a lower energy density of electrode material, face a challenge for the choice of the electrode materials for the SIBs. Therefore, it's necessary to use suitable nanostructured anode/cathode materials in order to occupy the sodium ions during the insertion and de-insertion process. There are many reports available on nanostructured SIBs electrode materials elsewhere (Nkosi et al. 2017; Palaniyandy et al. 2019). However, it is very challenging to discover appropriate electrode materials to accommodate Na^+ ions, to realize the fast insertion and de-insertion due to the larger ionic radii of Na^+ ions than Li^+ ions, and to overcome the sluggish diffusion kinetics of Na^+ ions.

Vaalma et al. investigated the cost effect of replacing lithium with sodium in the positive electrode and the electrolyte. Since the other components of the battery remain the same, a reduction in cost is possible for shifting to Na-containing materials. Moreover, an additive cost reduction is achieved by replacing the expensive copper current collector of the LIB negative electrode with aluminum since the latter does not alloy with Na at low potentials, in contrast to LIBs. Thus, theoretically replacing Li with Na and Cu with Al results in a cost reduction of ~12.5% based on the calculation of raw material cost. However, there are lot of challenges to achieve a SIB with similar energy density to commercial LIBs (Vaalma et al. 2018).

The sodium-ion storage of anode material is one of the major issues in developing the commercial SIBs. Up to now, various anode materials with different storage mechanisms such as intercalations, conversion, and alloying have been explored. Materials employed in the intercalation process usually possess high cycle stability and a low sodium storage capacity. Consistently, the anode material in the other two mechanisms of conversion and alloying usually indicates a high initial sodium ion storage capacity. However, these types of anode materials suffered from huge volume changes which led to losing contact between the particles and the current collector, resulting in damage to the electrodes. Although different strategies, such as those that involve the nanostructure and the carbon coat have been adopted to prevent the pulverization, poor cycle stability is still a great challenge for the conversion and alloying anode materials. In this scenario, so far, various anode materials such as metal alloys, metal oxides, metal sulfides, and carbonaceous materials have been explored for SIBs. In this chapter, a discussion of various materials and their progress on the development of negative electrode materials is provided below.

3.2 Carbon-Based Materials

To date, graphite and hard carbon materials have been considered as a promising anode material for LIBs, owing to their low potential, high capacity, abundance, and low cost. For the same reasons, carbon-based anode materials are also considered to be a most promising choice for SIBs. However, graphite cannot be utilized as a negative electrode for SIBs due to the thermodynamic instability of stage-1 compounds such as NaC_6 or NaC_8 and the smaller interlayer distance (~0.34 nm) result in poor sodium storage capacity. The theoretical calculations proposed that a minimum interlayer distance of 0.37 nm is required for sodium ion insertion in the carbon. In this scenario, the recognized disordered carbons, namely, graphitizable- (soft carbon), non-graphitizable- (hard carbon), and graphene-based materials can accommodate sodium ions to a great extent. Herein, two types of carbon-based materials, such as graphene- and hard carbon-based materials are discussed elaborately.

3.2.1 Graphene

Graphene is a specific type of carbon material, and it was discovered by Geim in 2004. Graphene consist of the sp^2-hybridized carbon atoms arranged in the hexagonal lattice. Also, graphene has high electric conductivity and is active for the Na^+ ion storage due to its two-dimensional structure with large specific surface area. However, the specific capacity of graphene materials is low. New strategies such as doping heteroatoms (S, N, P, and B) into carbon anodes have been employed to enhance specific capacity of the sodium ion storage. In addition, the incorporation of heteroatoms into the carbon structure could increase the conductivity of graphitic materials, resulting in improved rate performance. So far, many reviews have been published on graphene-based nanomaterials for SIBs (Lu et al. 2018; Zhang et al. 2019). For instance, nitrogen-rich graphene hollow microspheres, N-doped graphene composites, three dimensional (3D)-nitrogen graphene foam, and graphene-based nitrogen-doped carbon sandwich nanosheets (N-GCs) anode materials have been developed via a simple calcination of the graphene oxide (GO)-coated polystyrene (PS@GO) microspheres in the presence of a melamine and an in-situ polymerization method that was followed by pyrolysis for sodium ion storage devices (Xu et al. 2015; Li et al. 2016; Chang et al. 2018; Yue et al. 2018).

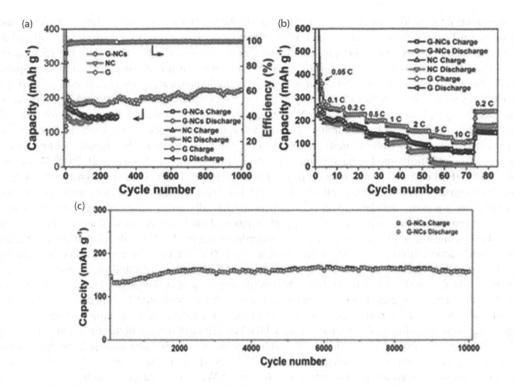

FIGURE 3.1 (a) Cycling performance of G-NCs, NC, and G at 500 mA g⁻¹, (b) rate capability of G-NCs, NC, and G between 0.01 and 3 V, and (c) long-cycle performance of G-NCs at 5000 mA g⁻¹. (Li, D. et al., *J. Mater. Chem. A*, 4, 8630–8635, 2016. Reproduced by permission of The Royal Society of Chemistry.)

The nitrogen-doped graphene materials exhibit excellent reversible capacity, rate performances, and outstanding cycle stability. For example, N-GCs exhibit excellent rate performance with a reversible capacity of 110 mA h g⁻¹, even at a high current density of 10,000 mA g⁻¹, and outstanding cycle stability (a capacity retention of 154 mA h g⁻¹ after 10,000 cycles at 5000 mA g⁻¹), shown in Figure 3.1 (Li, Zhang et al. 2016). Quan and his co-workers demonstrated S-doped graphene (Quan et al. 2018) as an anode and delivers a high reversible capacity of 380 mA h g⁻¹ over 300 cycles at 100 mA g⁻¹.

Later on, Xu et al. (2018) developed nitrogen and sulfur co-doped graphene nanosheets that were synthesized through a solvothermal method using carbonization of poly (2,5-dimercapto-1,3,4-thia-diazole (PDMcT) polymerized on the surface of GO represented as PDMcT/RGO. It exhibited high capacities (240 mA h g⁻¹ at 500 mA g⁻¹), improved rate performance (144 mA h g⁻¹ at 10 A g⁻¹), and good cycling stability (153 mA h g⁻¹ over 5000 cycles at 5000 mA g⁻¹). It is concluded that the enhanced performances could be attributed to the enlarged interlayer spacing and electronic conductivity from the heteroatoms which facilitate the sodium ion's kinetics during the insertion and de-insertion.

3.2.2 Hard Carbon

So far, hard carbon is the best carbon anode material for SIBs due to its large interlayer distance, therefore, recently, many researches have been concentrated on hard carbon material. Hard carbon is also disordered sp² non-graphitizable carbon and cannot be graphitized even at high temperatures and has two domains, that is, carbon layers (graphene-like) and micropores (nano-sized pores) in bulk (Jian et al. 2016; Li, Jian et al. 2017). Owing to its excellent properties of high capacity, lower redox potential, natural abundance, the expanded graphene interlayers with a stable structure, and low cost, hard carbon is very attractive in high-performance SIBs. Also, hard carbon was utilized as electrodes in the first commercial Li-ion cells and delivered comparable reversible capacity, however, it had low volumetric density due to defective stacking and low density compared to graphite (El Moctar et al. 2018). Stevens and Dahn were the first to demonstrate glucose-derived hard carbons at 1000°C as an anode for SIBs in 2000, they delivered a reversible capacity of 300 mA h g^{-1} (Hameed et al. 2019). Many researches proposed a Na-ion storage mechanism named "house of cards model," also known as insertion-absorption in hard carbon, wherein Na-ion insertion in graphene sheets could occur in two ways, i.e., Na-ion intercalation between graphene sheets in the sloping voltage region and Na-ions adsorption onto the nanoporous sites in the plateau region (Irisarri et al. 2015; Wang, Jin et al. 2017). Hard carbon can be prepared from different carbohydrates, such as polymers and sugars, recently, biomass-derived card carbon has been extensively investigated for large scale production because of its low cost, abundance, and environmentally friendly nature. The carbonaceous precursors and carbonization temperatures influence the performance of hard carbons, as well as the optimization of synthesis conditions (Luo et al. 2016). For example, Hong et al. (2014) developed 3D-connected porous hard carbon through a H$_3$PO$_4$-treated biomass at 700°C, which delivered a reversible capacity of 181 mAh g^{-1} at 200 mA g^{-1} over 220 cycles. Sun et al. (2015) demonstrated biomass-derived hard carbon at 800°C–1400°C, which exhibits a reversible capacity of 430 mA h g^{-1} at 30 mA g^{-1} over 200 charge-discharge cycles. Later on, Wang, Liu et al. (2018) fabricated N-doped-graphitized hard carbon (N-GHC) for high performance LIBs and SIBs, via a simple carbonization and activation of urea-soaked self-crosslinked Co-alginate. The N-GHC in SIBs delivered a highest reversible capacity of 178 mA h g^{-1} at 100 mA g^{-1}, with a stable cycling performance (300 charge-discharge cycles) and excellent rate performances as shown in Figure 3.2. Li et al. (2018) demonstrated a highly defective hard carbon through microwave treatment (6 s at 90 W) using a carbon which was made from pyrolysis of cellulose at 650°C for high performance SIBs. For comparison, the high temperature (1100°C) hard carbon was also prepared. The reversible capacity of microwaved hard carbon at 650°C is 308 mA h g^{-1}, which is apparently higher than that of other carbons. The improved electrochemistry could be due to the microwave treatment, in which the structural vacancies of hard carbon were maintained to a great extent.

Unfortunately, in all the above-discussed results, hard carbon suffers with a low first cycle Coulombic efficiency and fast capacity fading. The poor performance may be caused by the more prominent formation of a solid electrolyte interphase layer on the large surface area. It is noted that the low first cycle Coulombic efficiency of the hard carbon is the critical reason to reduce the irreversible active Na$^+$ ions, which will be provided by the cathode decreases in the full-cell capacity (Luo et al. 2015; Izanzar et al. 2018).

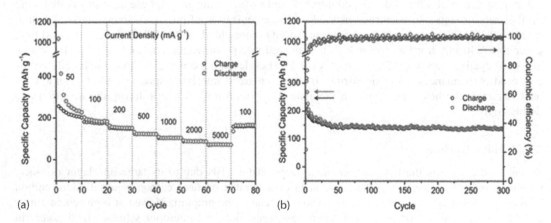

FIGURE 3.2   Rate performance (a) and cycling performance at 100 mA g^{-1} of N-GHC in SIBs (b). (Reprinted with permission from Wang, N. et al., *Sci. Rep.*, 8, 9934, 2018. Copyright 2016 Springer Nature.)

3.3 Transition Metal Oxide-Based Materials

Generally, transition metal oxides (TMOs-FeO, CoO, CuO, NiO, MnO, and ZnO) are considered to be potential electrode materials due to their high specific capacity which is associated with a conversion reaction mechanism; however, some TMO's can adopt an alloying/de-alloying or insertion/extraction combined with a conversion mechanism (Hasa et al. 2015; Wang et al. 2019). In 2002, Alcantara et al. (2002) demonstrated conversion-based $NiCo_2O_4$ spinel oxide for the first time, in which Na_2O and metals are formed during cycling, as follows as $NiCo_2O_4 + 8Na \rightarrow Ni + 2Co + 4Na_2O$, demonstrating a reversible capacity of ~200 mA h g^{-1}. Followed by this report, recently, many TMOs have emerged, such as iron oxide, cobalt oxide, tin (di) oxide, copper oxide, molybdenum oxide, nickel oxide, and manganese oxide. Nevertheless, due to the large volume expansion of TMOs accelerated to damaged electrodes, this led to the loss of electrical contact, and subsequently, rapid capacity fading (Hwang et al. 2017; Wu et al. 2018). To tackle these issues, many strategies including nanostructured materials and carbon composites have been proposed recently by many researchers. Herein, iron oxide, copper oxide, and cobalt oxide and its derivatives are discussed briefly.

3.3.1 Iron Oxide and Its Derivatives

Iron oxide is one of the most promising negative materials for SIBs, due to its two different oxidation states of Fe^{2+} and Fe^{3+}, named as FeO, Fe_2O_3, and Fe_3O_4. Recently, many research works have been focused on Fe_2O_3 and Fe_3O_4 anode materials, owing to their high specific capacities (1007 mA h g^{-1} for Fe_2O_3 and 900 mA h g^{-1} for Fe_3O_4); however, FeO is not electrochemically active with Na-ion. In addition, iron oxide materials have excellent chemical stability, richness in natural sources, low cost, and non-toxicity (Jiang et al. 2014; Liu, Jia et al. 2016; Modafferi et al. 2019). Unfortunately, its practical exploitation is impeded by its rapid capacity fading initiated by poor electrical conductivity and huge volume expansion during the electrochemical process, this leads to the loss of electrical disconnection between the anode powders that results in short lifespan and poor rate capability (Huang et al. 2014; Li, Xu et al. 2016). Many strategies have been focused to mitigate these issues, such as the fabrication of the nanostructured particles, carbon coating, and doped heteroatoms could improve the electronic conductivity of the materials. In 2010, the electrochemical performances of nanocrystalline Fe_3O_4 and α-Fe_2O_3 negative electrode materials for SIBs were first demonstrated by Komaba et al. (2010), which delivered a reversible capacity of 170 mA h g^{-1} at 100 mA g^{-1}. Komaba et al. suggested that the electrochemistry of Fe_3O_4 is highly improved by controlling the size of the particles. Recently, Fe_2O_3 embedded with a nitrogen-doped carbon matrix has been studied for the high specific capacity materials (Liu et al. 2015; Zhang, Wang et al. 2015; Guo et al. 2018; Meng et al. 2018). Liu, Zhang et al. (2016) reported a $MnFe_2O_4$@C (MFO@C) nanofibers anode for high-performance SIBs. MFO@C delivered a high specific capacity of 504 mA h g^{-1} at 100 mA g^{-1} and excellent rate capabilities of 305 mA h g^{-1} at 10,000 mA g^{-1}, as shown in Figure 3.3.

3.3.2 Copper Oxide and Its Derivatives

Cu-based oxides, such as CuO and Cu_2O, have received great attention among researchers due to their high capacity (674 mA h g^{-1}), abundance in nature, eco-friendliness, and low cost. In addition, during electrochemistry, CuO conversion is only partially reversible, but oxygen becomes redox active as well, and Na_2O_2 forms as an additional intermediate. On the contrary, oxygen is inactive in Li-CuO cells (Chen et al. 2016; Wang, Liu et al. 2016). Klein et al. (2015) demonstrated that during the first discharge, CuO is converted to Cu_2O and to Cu at 0.4 V and 0.01 V, when CuO is employed as an anode for SIBs, as with LIBs, however, the redox reaction voltage is not same. However, copper oxide (CuO and Cu_2O) suffers with huge volume variation (173%), poor electronic conductivity, and structural collapse leading to rapid capacity fading. To tackle these issues, great efforts have been made such as morphology controlled nanostructures and carbon coating composites. Recently, many researches have been focused on CuO nanosheets, 3D structures, micro-nanostructured spheres, and nanoparticles to improve the electrochemistry (Lu et al. 2015; Chen et al. 2016; Fan et al. 2017). For example, Rath et al. (2018) reported morphology-dependent various CuO nanostructures such

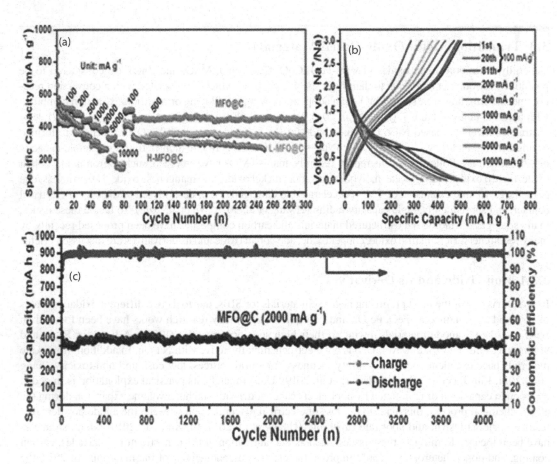

FIGURE 3.3 (a) Rate capability and cycling performance of MFO@C, L-MFO@C, and H-MFO@C materials between 0.01 and 3.0 V and (b) galvanostatic charge-discharge profiles of MFO@C at different current rates. (c) Long-term cycling stability of MFO@C electrode at 2000 mA g⁻¹. (Reprinted with permission from Liu, Y. et al., *Nano Lett.*, 16, 3321–3328. Copyright 2016 American Chemical Society.)

as nanoflakes, nanoellipsoids, and nanorods as anodes for SIBs, through a hydrothermal approach, as shown in Figure 3.4. Among them, nanorods exhibit an initial specific capacity of 600 mA h g⁻¹ and excellent rate capability of 206 mA h g⁻¹ at 1000 mA g⁻¹, with a capacity retention of 73% over 150 cycles. However, the rational design and syntheses of porous structures remain a challenge. Therefore, research has focused on metal-organic framework-based materials (Zhang, Qin et al. 2015; Kim, Kim, Cho et al. 2016; Li, Yan et al. 2017). Kim et al. developed CuO/Cu_2O in a graphitized porous carbon matrix derived by Cu-based metal-organic framework for LIBs and SIBs (Kim, Kim, Cho et al. 2016). The electrochemistry of SIBs showed a high specific capacity of 303 mA h g⁻¹ at 50 mA g⁻¹, with an excellent cycling performance of almost 100% capacity retention over 200 charge-discharge cycles, as illustrated in Figure 3.5.

3.3.3 Cobalt Oxide and Its Derivatives

As with FeO, CoO is also electrochemically inactive with Na-ion. Thus, Co_3O_4 and its derivatives are regarded as a promising anode for SIBs, owing to their special characteristics of different oxidation states, low cost, and abundance in nature (Longoni et al. 2016). Moreover, Co_3O_4 adopts an inverse spinel structure and also has been widely studied as an anode for SIBs because of its high specific capacity of 890 mA h g⁻¹, as with LIBs. However, its practical application is hindered because its low

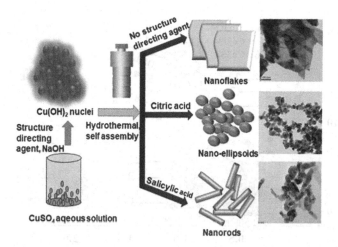

FIGURE 3.4 Schematic illustration of CuO nanostructures such as nanoflakes, nanoellipsoids, and nanorods as anode for SIBs. (Reprinted with permission from Rath, P.C. et al., *ACS Sustain. Chem. Eng.*, 6, 10876–10885. Copyright 2018 American Chemical Society.)

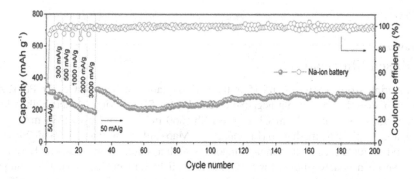

FIGURE 3.5 Cycling stability of CuO/Cu$_2$O-GPC material at different current densities for SIBs. (Reprinted with permission from Kim, H. et al., *Adv. Funct. Mater.*, 26, 5042–5050. Copyright 2016 American Chemical Society.)

intrinsic electronic conductivity and huge volume variation during the sodiation/desodiation process resulted in poor cycle life. Strategies to mitigate these issues include fabricating nanostructures with different morphologies and a heteroatom-doped carbon matrix (Wang, Wang et al. 2016; Kang et al. 2017; Chen, Chen et al. 2018; Xu, Xia et al. 2018). Nevertheless, Xu et al. (2017) reported a sodiation/lithiation mechanism process in the porous cobalt oxide material using operando synchrotron X-ray techniques. It was noticed that the porous cobalt oxide material possessed low sodiation activity compared to the lithiation activity in LIBs, which is shown in Figure 3.6. This could be due to the less pore structure changes, oxidation state, and local structure changes, as well as crystal structure evolution during its insertion/extraction process. Kim and his co-workers also performed the detailed conversion and reconversion reaction of Co$_3$O$_4$ in SIBs and its mechanism (Kim, Kim, Kim et al. 2016). Therefore, in recent years, metal oxide-doped (MnO$_2$ and ZnO) (Wang, Cao et al. 2017; Chen, Deng et al. 2018), ternary metal oxides, such as Ni- and Cu-doped cobalt oxides can be widely studied due to their similar atomic radii of Mn, Ni, and Co. The introduction of Ni into MnCo$_2$O$_4$ can partly replace Co atoms to obtain a Mn–Co–Ni ternary oxide without a distinct crystal structure change. Besides, it is expected to offer a synergistic effect on redox reactions, including contributions from cobalt, nickel, and manganese ions, compared to the corresponding single component oxides (Wu et al. 2016; Wang, Huang et al. 2017).

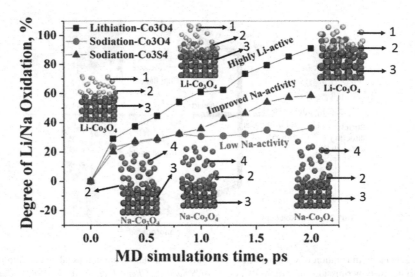

FIGURE 3.6 Illustrates the degree of oxidation of as a function of lithiated and sodiated electrodes of Co_3O_4 material, in which 1, 2, 3, and 4 denote as lithium, oxygen, cobalt and sodium atoms. (Reprinted with permission from Xu, G.-L. et al., *Nano Lett.*, 17, 953–962, 2017. Copyright 2017 American Chemical Society.)

3.4 Alloy-Based Materials

Alloy-based materials from groups 14 and 15 in the periodic table, such as Si, Ge, Sn, Pb, P, As, Sb, and Bi metals, are considered as promising negative electrodes for SIBs, owing to their high volumetric and gravimetric capacities ($Na_{15}Sn_4$ (847 mA h g^{-1}), Na_3Sb (660 mA h g^{-1}), Na_3Bi (385 mAh g^{-1}), $Na_{15}Pb_4$ (485 mAh g^{-1}), and Na_3P (2560 mAh g^{-1})) (Li et al. 2015; Mao et al. 2016; Wang, Li et al. 2018). Moreover, the groups 14 and 15 metals, as well as their composites such as oxides, selenides, and binary and ternary alloy anodes have been widely studied for SIBs due to their high reversible capacities, compared to the carbon-based anode materials (Tang et al. 2017). However, the large size of Na^+ ions (55%) still causes huge volume changes during the charge/discharge reactions, ultimately leading to formation of cracks or pulverization of the electrode (loss of electrical contact between the active materials) and capacity fade (Wang, Li et al. 2018). In addition, during continuous cycling, the thick solid electrolyte interphase layers are formed between the electrode and electrolyte surfaces, which significantly block charge transfer and lead to quick capacity fading. Several strategies to mitigate the capacity fade include fabrication of nanostructured materials, choice of binders, and electrolyte additives (Hameed et al. 2019). The general alloying reaction for groups 14 and 15 metals, during electrochemical sodiation, was illustrated as below reaction (3.1).

$$\text{Alloy anodes}\left(\text{Si, Ge, Sn, Pb, Sb, or P}\right) + x\text{Na}^+ + xe^- \rightarrow \text{Na}x^- \text{ TM alloys} \tag{3.1}$$

Though, Si has been delivered a high reversible capacity of ~3600 mA h g^{-1} for LIBs, it delivers a low reversible capacity for SIBs, due to its inferior electrical conductivity. Also, Ge reacts with Na, forming an alloying of Na-Ge at 0.15 to 0.6 V, which can deliver a reversible capacity of 350 mA h g^{-1}. P has a high reversible capacity (2596 mA h g^{-1}), however, it suffers low electrical conductivity which causes a low specific capacity and rate capability (which will be discussed in the P section).

Amongst elements, Sn is the most widely used anode material for SIBs, due to its low reaction potential, the low cost and abundant source of tin, and high theoretical capacity of 847 mA h g^{-1}. Also Sn undergoes multiple sodiation steps, forming four different intermetallic compounds, such as $NaSn_5$, $NaSn$, Na_9Sn_4, and $Na_{15}Sn_4$ (Lao et al. 2017). In addition, the variety of Sn-based compounds with different physicochemical properties, such as SnO, SnO_2, SnS_2, and SnP, have been looked at as anodes for SIBs. However, Sn metal suffers from huge capacity decay due to the huge volume expansion (420%) and slow reaction kinetics (Ying and Han 2017). To tackle these issues, tremendous efforts have been made to prepare novel electrodes such as, employing carbon matrix and binary and ternary Sn-based composite electrodes (Xie et al. 2017). For example, Ruan et al. (2017) designed a carbon-encapsulated Sn@N-doped carbon nanotube composite through a hydrothermal approach. The N-doped carbon nanotubes could increase the cycling performance of the Sn-based materials. Electrochemical results showed that the carbon encapsulated Sn@N-doped carbon nanotube could attain a reversible capacity of 398 mA h g⁻¹ at 100 mA g⁻¹ and a capacity retention of 67% was observed over 150 cycles, as shown in Figure 3.7. Another example relates to a Sn@CNT nanopillar grown on carbon paper (Xie et al. 2015) for sodium storage devices. The as-prepared Sn@CNT-CP

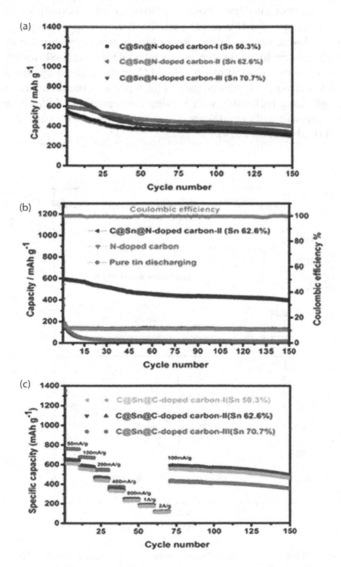

FIGURE 3.7 Cycling performances of C@Sn@N-doped carbon composites (a), comparison cycling performance of C@Sn@N-doped, N-doped carbon and cure Sn particles (b), rate capability of C@Sn@N-doped carbon composites (c). (Reprinted with permission from Ruan, B. et al., *ACS Appl. Mater. Interfaces*, 9, 37682–37693, 2017. Copyright 2016 American Chemical Society.)

electrode delivered a reversible capacity of 550 mA h cm^{-2} after 100 cycles. A nanoporous CuSn (Zhang et al. 2017) binary alloy has been fabricated by a de-alloying method. The as-prepared nanoporous CuSn alloy delivered a reversible capacity of 233 mAh g^{-1} over 100 charge-discharge cycles. Cheng et al. (2017) demonstrated a SnOx-Sn@ few-layered graphene composite negative electrode via oxygen plasma-assisted milling for SIBs and delivered a reversible capacity 448 mA h g^{-1} in the cycle and excellent cycling performance, with 83% of the capacity retained after 250 cycles.

On the other hand, Sb-based alloys have demonstrated impressive performance due to their alloying reaction with three sodium (Na$_3$Sb), which delivered a reversible capacity of 660 mA h g^{-1}, similar to LIBs. Sb-based intermetallic anodes that undergo an alloying or conversion reaction would suffer major structural changes during the charge and discharge process, that is, during the end of the sodiation process, an amorphous NaxSb was transformed into the hexagonal Na$_3$Sb phase as evidenced by X-ray diffraction patterns during cycling (Hameed et al. 2019). Upon desodiation, the crystalline Na$_3$Sb was found to transform into amorphous Sb. However, Sb-based alloys are often limited by the cracking and pulverization of active anode particles during long-term cycling processes. Many strategies have been employed, such as fabrication of binary or ternary alloys to improve the electrochemistry. He et al. (2015) developed SnSb nanocrystals for SIBs, as SnSb widely studied for LIBs, delivered a reversible capacity of >350 mA h g^{-1} at 1 C, with excellent cycling and rate capability as shown in Figure 3.8. Farbod et al. demonstrated film-type tin-germanium-antimony ternary alloys through a magnetic sputtering technique, with varying concentration of Sn, Sb, and Ge. It is noted that, Sn$_{50}$Ge$_{25}$Sb$_{25}$ delivers an excellent rate capability, displaying a stable capacity of 381 mA h g^{-1} at 8500 mA g^{-1} (~10 C) (Farbod et al. 2014).

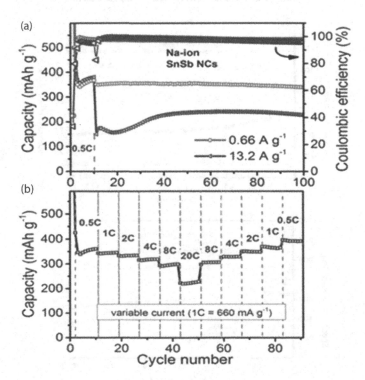

FIGURE 3.8 Cycling stability tests at 1°C and 20°C rates (a) and rate capability tests (b) for SIB anodes comprising SnSb NCs as energy storage material. (He, M., *Nanoscale*, 7, 455–459, 2015. Reproduced by permission of The Royal Society of Chemistry.)

3.5 Organic Materials-Based Anodes

There has been a huge effort made in searching for suitable organic materials to serve as an anode for SIB that possess good electrochemical activity, natural abundance, and are environmentally-friendly. In recent times, many organic carboxylate-based materials have been introduced as the anode materials for SIBs, which exhibited a long discharge plateau below 0.6 V vs Na/Na+ (Abouimrane et al. 2012). Specially, organic materials are attractive since they are composed of naturally abundant elements (C, H, O, N, S), such that they are more favorable than inorganic materials due to the latter are limited resources, non-degradable, and require tedious synthesis and purification procedures. Though organic electrode materials are considered as promising alternatives to inorganic electrodes, only few organic materials have been applied in SIBs so far.

Especially, the organic material sodium terephthalate $Na_2C_8H_4O_4$ (Na_2TP) is gaining extensive attention and has shown reliable electrochemical properties, as it is capable of delivering high sodium storage capacity, good cycling performance, and rate capability. Na_2TP was first reported by Yong-Sheng Hu's group, and their sample demonstrated a high reversible capacity of 250 mA h g^{-1}, with good cyclability (Zhao et al. 2012). The electrochemical performance of the organic compound sodium terephthalate $Na_4C_8H_4O_4$ (Na_2TP) was investigated by Chen and co-workers and used as both an anode and cathode for SIBs (Wang et al. 2014). Guoxiu Wang's group reported an organic-based composite in order to improve the intrinsic conductivity of Na_2TP, a sodium terephthalate@ graphene ($Na_2TP@GE$) hybrid, synthesized via a freeze-drying technique, and this hybrid material demonstrated a high reversible capacity of 268.9 mA h g^{-1} and continued cyclability (with a capacity retention of 77.3% over 500 cycles as shown in Figure 3.9a and b) (Wang, Kretschmer et al. 2016). Furthermore, Yong Lei's group reported several conjugated organic carboxylates, which demonstrated an effective way to improve the electrochemical performances of organic material in the molecular design (Wang et al. 2015).

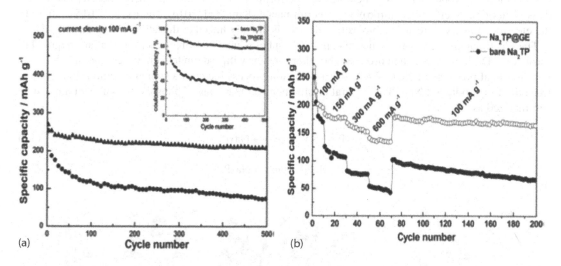

FIGURE 3.9 (a) Cycling performances of bare Na_2TP (circle) and $Na_2TP@GE$ hybrid (triangle) for 500 cycles (corresponding Coulombic efficiencies shown in the inset) and (b) rate performances of bare Na_2TP and $Na_2TP@GE$ hybrid at various current densities. (Wang, Y. et al., *RSC Adv.*, 6, 57098–57102, 2016. Reprinted by permission of The Royal Society of Chemistry.)

3.6 Phosphorus-Based Anodes

P is a non-metallic chemical element, one of the most abundant elements (ranks 11th), and accounts for about 0.11% of the earth's crust. Phosphorous can be found in three common allotropes, which are the white P, the red P, and the black P, and their structures are represented in Figure 3.10 (Ni et al. 2018). Amongst them, the black P is thermodynamically the most stable allotrope and can be prepared when white P (white P is the most reactive, which forms a tetrahedral shape) is heated under high pressure (12,000 atmospheres). The black P has, interestingly, different properties from other allotropes, as it is a conductor of electricity, and much like graphite, with both being black and peeling and having puckered sheets of linked atoms.

Phosphorus, by having an extremely high theoretical capacity of about 2596 mAh g^{-1}, suitably low operating potential (0.4 V vs Na/Na$^+$), being an abundant resource, and environmental benignity, has aroused a substantial amount of attention as a promising anode material for SIBs. With the capability to offer a high specific capacity, thereby high energy density despite cycling stability challenges, phosphorous is still one of the most promising candidates as an anode for SIB. SIBs are more suitable as an energy storage device for scalable and affordable stationary applications like smart grids and to backup renewable energy, such as solar and wind energy (Slater et al. 2013; Yabuuchi, Kubota et al. 2014; Hwang et al. 2017). During the electrochemical reaction taking place by virtue of having a high specific capacity, many sodium ions obviously leaving and entering the phosphorous anode structure, it apparently starts to experience large volume expansion (~400%) during repetitive cycling, which leads to severe capacity fading, and it also has the challenge of the poor electronic conductivity (~1 × 10^{-14} S cm^{-1}), which leads to low electrochemical activity.

Phosphorus is capable of storing three Na ions at reasonable potentials with a high theoretical specific capacity of 2596 mA h g^{-1}, which considerably exceeds most other SIB anodes currently available, such as Si (954 mA h g^{-1}), Sn (847 mA h g^{-1}), and Sb (660 mA h g^{-1}), which makes it so attractive (Sangster 2010; Kim et al. 2013; Yabuuchi, Matsuura et al. 2014; Walter et al. 2015; Dahbi et al. 2016; Li, Carter et al. 2017; Liu et al. 2017). It is observed that elements, which are known to deliver extremely high theoretical capacity in LIBs, are not automatically giving high theoretical capacity for SIBs. Since the size of a Na$^+$ ion is larger by 55% relative to a Li$^+$ ion, the LIB electrodes are not necessarily suitable to SIBs. Thus, it is necessary to search for suitable electrode materials which have high theoretical capacity and develop electrode materials which are well performing in terms of both energy density and power density in order to effectively improve SIB performance (Kim et al. 2014; Xiang et al. 2015). After all, the electrochemistry of rechargeable battery devices is directly linked to the active electrode materials.

The electrochemical sodiation mechanism of phosphorus involves three electrons that are required to form Na$_3$P. During the sodiation process, phosphorus reacts with sodium to form the compounds of NaxP, with the final products of Na$_3$P. While the desodiation process includes a stepwise sodium ion removal from the fully sodiatied Na$_3$P. The general electrochemical reactions of phosphorus with sodium can be summarized as follows:

$$P + x\text{Na}^+ + x e^- \leftrightarrow \text{Na}x\text{P} \tag{3.2}$$

$$P + 3\text{Na}^+ + 3e^- \rightarrow \text{Na}_3\text{P}. \tag{3.3}$$

FIGURE 3.10 The phosphorus allotropes structures for (a) white P, (b) red P, and (c) black P. (Reprinted with permission from Ni, J. et al., *ACS Energy Lett.*, 3, 1137–1144, 2016. Copyright 2016, American Chemical Society.)

Owing to the poor electronic conductivity nature of phosphorus ($\sim 1 \times 10^{-14}$ S cm^{-1}), its practical capacity is so low as compared to the theoretical capacity. However, one of the effective strategies to enhance the electrochemical performance of most phosphorus-based electrodes is by forming different types of phosphorus carbon composites and metal phosphides (MPs) or their composites. In addition to volume change expansion, the other issue of P anode is facing the low conductivity issue. Particularly, the conductivity of red P improves by creating composites of carbon, such as carbon wrapping, doping carbon into bulk P, and compositing and hybridizing by strong interaction between carbon and P and weak interaction between carbon and P. The most common wrapping red P can be realized by mixing with carbon nanoparticles via various techniques. Unfortunately, there will be capacity trade-off, as the resulting P/C composite only affords a modest capacity of about 1000 mA h g^{-1} based on the mass of the composite, and the rate capacity is yet satisfactory.

On the other hand, when phosphorous P combines with metals to form MPs, it works better than the use of phosphorous P alone as an anode for SIBs. Hence, currently more and more MPs-based materials on SIBs with a high prospect of delivering high capacity and electrochemical stability have attracted much attention and gained huge efforts to studying them. However, still the challenges of huge volumetric expansion of the structure during charging/discharging and relatively poor electrical conductivity have greatly limited its practical application. The main problems for MPs in SIBs application are to achieve long-term cycle stability and rapid charge/discharge capabilities (Wu et al. 2018). Some of the strategies used to solve these challenges are regulating the MPs nanostructure to relieve the pulverization of electrode materials and the development of MPs composite with carbon could enhance electrode stability in a long cycle test through a synergistic effect of composite materials (Qian et al. 2014).

3.7 Conclusions

Recently, many researchers have been focusing their research on SIBs due to their great advantages, compared to LIBs. So far, tremendous research has been done on anode materials with regard to various structural designs in order to improve the electrochemical performances of SIBs. In this chapter, we have summarized different types of anode materials for SIBs according to their reaction mechanism, such as carbon, transition-based metal oxides, alloy-based, and organic- and phosphorus-based metals. The highest problems of each type of anode material have been pointed out, such as low specific capacity and poor rate capability for the insertion-type materials and huge volume expansion for the conversion and alloying reaction materials during charge/discharge processes. Specific structural designs have been illustrated for each type of anode material with unique structures and outstanding electrochemical properties. The transition-metal-containing binary compounds have high theoretical capacities through conversion reactions with sodium, offering the potential to develop high energy density SIBs. On the other hand, conversion reactions can couple with alloying reactions in the compounds (e.g., Sn and Sb) to give rise to ultrahigh capacity. The abundant species of conversion-type compounds provide opportunities for seeking appropriate anode candidates with multiple advantages, such as reasonable potential, low cost, and environmental friendliness, although they suffer from many challenging difficulties, which need to be overcome before practical applications can be realized.

REFERENCES

Abouimrane, Ali, Wei Weng, Hussameldin Eltayeb, Yanjie Cui, Jens Niklas, Oleg Poluektov, and Khalil Amine. 2012. "Sodium Insertion in Carboxylate Based Materials and Their Application in 3.6 V Full Sodium Cells." *Energy & Environmental Science* 5 (11): 9632–38.

Alcántara, R., M. Jaraba, P. Lavela, and J. L. Tirado. 2002. "NiCo$_2$O$_4$ Spinel: First Report on a Transition Metal Oxide for the Negative Electrode of Sodium-Ion Batteries." *Chemistry of Materials* 14 (7): 2847–48.

Chang, Bin, Jing Chen, Mingan Zhou, Xiaojie Zhang, Prof. Wei Wei, Bin Dai, Sheng Han, and Yanshan Huang. 2018. "Three-Dimensional Graphene-Based Na-doped Carbon Composites as High-Performance Anode Materials for Sodium-Ion Batteries." *Chemistry–An Asian Journal* 13 (24): 3859–64.

Chen, Chengcheng, Yanying Dong, Songyue Li, Zhuohan Jiang, Yijing Wang, Lifang Jiao, and Huatang Yuan. 2016. "Rapid Synthesis of Three-Dimensional Network Structure CuO as Binder-Free Anode for High-Rate Sodium Ion Battery." *Journal of Power Sources* 320: 20–27.

Chen, Huanhui, Libo Deng, Shan Luo, Xiangzhong Ren, Yongliang Li, Lingna Sun, Peixin Zhang, Guoqiang Chen, and Yuan Gao. 2018. "Flexible Three-Dimensional Heterostructured $ZnO-Co_3O_4$ on Carbon Cloth as Free-Standing Anode with Outstanding Li/Na Storage Performance." *Journal of the Electrochemical Society* 165 (16): A3932–42.

Chen, Xiaobo, Xi Chen, Junjie Ni, Yadong Pan, Zhenjie Jia, Guoce Zhuang, and Zhihai Zhang. 2018. "Co_3O_4/Carbon Allotrope Composites as Anode Material for Sodium-Ion Batteries." *Journal of Electroanalytical Chemistry* 830/831: 116–21.

Cheng, Deliang, Jiangwen Liu, Xiang Li, Renzong Hu, Meiqing Zeng, Lichun Yang, and Min Zhu. 2017. "A Highly Stable (SnOx-Sn)@ Few Layered Graphene Composite Anode of Sodium-Ion Batteries Synthesized by Oxygen Plasma Assisted Milling." *Journal of Power Sources* 350: 1–8.

Dahbi, Mouad, Naoaki Yabuuchi, Mika Fukunishi, Kei Kubota, Kuniko Chihara, Kazuyasu Tokiwa, Xuefang Yu, Hiroshi Ushiyama, Koichi Yamashita, and Jin-Young Son. 2016. "Black Phosphorus as a High-Capacity, High-Capability Negative Electrode for Sodium-Ion Batteries: Investigation of the Electrode/Electrolyte Interface." *Chemistry of Materials* 28 (6): 1625–35.

El Moctar, Ismaila, Qiao Ni, Ying Bai, Feng Wu, and Chuan Wu. 2018. "Hard Carbon Anode Materials for Sodium-Ion Batteries." *Functional Materials Letters* 11.

Fan, Mouping, Haiying Yu, and Yu Chen. 2017. "High-Capacity Sodium Ion Battery Anodes Based on CuO Nanosheets and Carboxymethyl Cellulose Binder." *Materials Technology* 32 (10): 598–605.

Farbod, Behdokht, Kai Cui, W. Peter Kalisvaart, Martin Kupsta, Beniamin Zahiri, Alireza Kohandehghan, Elmira Memarzadeh Lotfabad, Zhi Li, Erik J. Luber, and David Mitlin. 2014. "Anodes for Sodium Ion Batteries Based on Tin-Germanium-Antimony Alloys." *ACS Nano* 8 (5): 4415–29.

Guo, Tianxiao, Hanxiao Liao, Peng Ge, Yu Zhang, Yiqi Tian, Wanwan Hong, Zidan Shi, Chunsheng Shao, Hongshuai Hou, and Xiaobo Ji. 2018. "Fe_2O_3 Embedded in the Nitrogen-Doped Carbon Matrix with Strong C-O-Fe Oxygen-Bridge Bonds for Enhanced Sodium Storages." *Materials Chemistry and Physics* 216: 58–63.

Hameed, A. Shahul, Kei Kubota, and Shinichi Komaba. 2019. "From Lithium to Sodium and Potassium Batteries". In *Future Lithium-Ion Batteries*, pp. 181–219, Cambridge, UK: RSC Publishing.

Hasa, Ivana, Roberta Verrelli, and Jusef Hassoun. 2015. "Transition Metal Oxide-Carbon Composites as Conversion Anodes for Sodium-Ion Battery." *Electrochimica Acta* 173: 613–18.

He, Meng, Marc Walter, Kostiantyn V. Kravchyk, Rolf Erni, Roland Widmer, and Maksym V. Kovalenko. 2015. "Monodisperse SnSb Nanocrystals for Li-Ion and Na-Ion Battery Anodes: Synergy and Dissonance Between Sn and Sb." *Nanoscale* 7 (2): 455–459.

Hong, Kun-lei, Long Qie, Rui Zeng, Zi-qi Yi, Wei Zhang, Duo Wang, Wei Yin, Chao Wu, Qing-jie Fan, Wu-xing Zhang, and Yun-hui Huang. 2014. "Biomass Derived Hard Carbon Used as a High Performance Anode Material for Sodium Ion Batteries." *Journal of Materials Chemistry A* 2 (32): 12733–38.

Huang, Bo, Kaiping Tai, Mingou Zhang, Yiran Xiao, and Shen J. Dillon. 2014. "Comparative Study of Li and Na Electrochemical Reactions with Iron Oxide Nanowires." *Electrochimica Acta* 118: 143–49.

Hwang, Jang-Yeon, Seung-Taek Myung, and Yang-Kook Sun. 2017. "Sodium-Ion Batteries: Present and Future." *Chemical Society Reviews* 46 (12): 3529–614.

Irisarri, E., A. Ponrouch, and M. R. Palacin. 2015. "Hard Carbon Negative Electrode Materials for Sodium-Ion Batteries." *Journal of The Electrochemical Society* 162 (14): A2476–482.

Izanzar, Ilyasse, Mouad Dahbi, Manami Kiso, Siham Doubaji, Shinichi Komaba, and Ismael Saadoune. 2018. "Hard Carbons Issued from Date Palm as Efficient Anode Materials for Sodium-Ion Batteries." *Carbon* 137: 165–73.

Jian, Zelang, Zhenyu Xing, Clement Bommier, Zhifei Li, and Xiulei Ji. 2016. "Hard Carbon Microspheres: Potassium-Ion Anode Versus Sodium-Ion Anode." *Advanced Energy Materials* 6 (3): 1501874.

Jiang, Yinzhu, Meijuan Hu, Dan Zhang, Tianzhi Yuan, Wenping Sun, Ben Xu, and Mi Yan. 2014. "Transition Metal Oxides for High Performance Sodium Ion Battery Anodes." *Nano Energy* 5: 60–66.

Kang, Wenpei, Yu Zhang, Lili Fan, Liangliang Zhang, Fangna Dai, Rongming Wang, and Daofeng Sun. 2017. "Metal-Organic Framework Derived Porous Hollow Co_3O_4/Na-C Polyhedron Composite with Excellent Energy Storage Capability." *ACS Applied Materials & Interfaces* 9 (12): 10602–609.

Kim, A.-Young, Min Kyu Kim, Keumnam Cho, Jae-Young Woo, Yongho Lee, Sung-Hwan Han, Dongjin Byun, Wonchang Choi, and Joong Kee Lee. 2016. "One-Step Catalytic Synthesis of CuO/Cu_2O in a Graphitized Porous C Matrix Derived from the Cu-Based Metal-Organic Framework for Li- and Na-Ion Batteries." *ACS Applied Materials & Interfaces* 8 (30): 19514–23.

Kim, Haegyeom, Hyunchul Kim, Hyungsub Kim, Jinsoo Kim, Gabin Yoon, Kyungmi Lim, Won-Sub Yoon, and Kisuk Kang. 2016. "Understanding Origin of Voltage Hysteresis in Conversion Reaction for Na Rechargeable Batteries: The Case of Cobalt Oxides." *Advanced Functional Materials* 26 (28): 5042–50.

Kim, Youngjin, Kwang-Ho Ha, Seung M. Oh, and Kyu Tae Lee. 2014. "High-Capacity Anode Materials for Sodium-Ion Batteries." *Chemistry &"A European Journal* 20 (38): 11980–92.

Kim, Youngjin, Yuwon Park, Aram Choi, Nam-Soon Choi, Jeongsoo Kim, Junesoo Lee, Ji Heon Ryu, Seung M. Oh, and Kyu Tae Lee. 2013. "An Amorphous Red Phosphorus/Carbon Composite as a Promising Anode Material for Sodium Ion Batteries." *Advanced Materials* 25 (22): 3045–49.

Klein, Franziska, Ricardo Pinedo, Philipp Hering, Angelika Polity, Jürgen Janek, and Philipp Adelhelm. 2015. "Reaction Mechanism and Surface Film Formation of Conversion Materials for Lithium-and Sodium-Ion Batteries: An XPS Case Study on Sputtered Copper Oxide (CuO) Thin Film Model Electrodes." *The Journal of Physical Chemistry C* 120 (3): 1400–14.

Komaba, Shinichi, Takashi Mikumo, Naoaki Yabuuchi, Atsushi Ogata, Hiromi Yoshida, and Yasuhiro Yamada. 2010. "Electrochemical Insertion of Li and Na Ions into Nanocrystalline Fe_3O_4 and alpha-Fe_2O_3 for Rechargeable Batteries." *Journal of The Electrochemical Society* 157 (1): A60–65.

Lao, Mengmeng, Yu Zhang, Wenbin Luo, Qingyu Yan, Wenping Sun, and Shi Xue Dou. 2017. "Alloy-Based Anode Materials Toward Advanced Sodium-Ion Batteries." *Advanced Materials* 29 (48): 1700622.

Li, Dongdong, Lei Zhang, Hongbin Chen, Jun Wang, Liang-Xin Ding, Suqing Wang, Peter J. Ashman, and Haihui Wang. 2016. "Graphene-Based Nitrogen-Doped Carbon Sandwich Nanosheets: A New Capacitive Process Controlled Anode Material for High-Performance Sodium-Ion Batteries." *Journal of Materials Chemistry A* 4 (22): 8630–35.

Li, Dongsheng, Dong Yan, Xiaojie Zhang, Jiabao Li, Ting Lu, and Likun Pan. 2017. "Porous CuO/Reduced Graphene Oxide Composites Synthesized from Metal-Organic Frameworks as Anodes for High-Performance Sodium-Ion Batteries." *Journal of Colloid and Interface Science* 497: 350–58.

Li, Henan, Li Xu, Hansinee Sitinamaluwa, Kimal Wasalathilake, and Cheng Yan. 2016. "Coating Fe_2O_3 with Graphene Oxide for High-Performance Sodium-Ion Battery Anode." *Composites Communications* 1: 48–53.

Li, Mengya, Rachel Carter, Landon Oakes, Anna Douglas, Nitin Muralidharan, and Cary L. Pint. 2017. "Role of Carbon Defects in the Reversible Alloying States of Red Phosphorus Composite Anodes for Efficient Sodium Ion Batteries." *Journal of Materials Chemistry A* 5 (11): 5266–72.

Li, Zhi, Jia Ding, and David Mitlin. 2015. "Tin and Tin Compounds for Sodium Ion Battery Anodes: Phase Transformations and Performance." *Accounts of Chemical Research* 48 (6): 1657–65.

Li, Zhifei, Yicong Chen, Zelang Jian, Heng Jiang, Joshua James Razink, William F. Stickle, Joerg C. Neuefeind, and Xiulei Ji. 2018. "Defective Hard Carbon Anode for Na-Ion Batteries." *Chemistry of Materials* 30 (14): 4536–42.

Li, Zhifei, Zelang Jian, Xingfeng Wang, Ismael A. Rodríguez-Pérez, Clement Bommier, and Xiulei Ji. 2017. "Hard Carbon Anodes of Sodium-Ion Batteries: Undervalued Rate Capability." *Chemical Communications* 53 (17): 2610–13.

Liu, Huan, Mengqiu Jia, Qizhen Zhu, Bin Cao, Renjie Chen, Yu Wang, Feng Wu, and Bin Xu. 2016. "3D-0D Graphene-Fe_3O_4 Quantum Dot Hybrids as High-Performance Anode Materials for Sodium-Ion Batteries." *ACS Applied Materials & Interfaces* 8 (40): 26878–85.

Liu, Shuai, Jinkui Feng, Xiufang Bian, Jie Liu, Hui Xu, and Yongling An. 2017. "A Controlled Red Phosphorus@ Ni-P core@ Shell Nanostructure as an Ultralong Cycle-Life and Superior High-Rate Anode for Sodium-Ion Batteries." *Energy & Environmental Science* 10 (5): 1222–33.

Liu, Xinjuan, Taiqiang Chen, Haipeng Chu, Lengyuan Niu, Zhuo Sun, Likun Pan, and Chang Q. Sun. 2015. "Fe_2O_3-Reduced Graphene Oxide Composites Synthesized via Microwave-Assisted Method for Sodium Ion Batteries." *Electrochimica Acta* 166: 12–16.

Liu, Yongchang, Ning Zhang, Chuanming Yu, Lifang Jiao, and Jun Chen. 2016. "$MnFe_2O_4$@C Nanofibers as High-Performance Anode for Sodium-Ion Batteries." *Nano Letters* 16 (5): 3321–28.

Longoni, Gianluca, Michele Fiore, Joo-Hyung Kim, Young Hwa Jung, Do Kyung Kim, Claudio M. Mari, and Riccardo Ruffo. 2016. "Co_3O_4 Negative Electrode Material for Rechargeable Sodium Ion Batteries: An Investigation of Conversion Reaction Mechanism and Morphology-Performances Correlations." *Journal of Power Sources* 332: 42–50.

Lu, Yanying, Ning Zhang, Qing Zhao, Jing Liang, and Jun Chen. 2015. "Micro-Nanostructured CuO/C Spheres as High-Performance Anode Materials for Na-Ion Batteries." *Nanoscale* 7 (6): 2770–76.

Lu, Yong, Yanying Lu, Zhiqiang Niu, and Jun Chen. 2018. "Graphene-Based Nanomaterials for Sodium-Ion Batteries." *Advanced Energy Materials* 8 (17): 1702469.

Luo, Wei, Clement Bommier, Zelang Jian, Xin Li, Rich Carter, Sean Vail, Yuhao Lu, Jong-Jan Lee, and Xiulei Ji. 2015. "Low-Surface-Area Hard Carbon Anode for Na-Ion Batteries via Graphene Oxide as a Dehydration Agent." *ACS Applied Materials & Interfaces* 7 (4): 2626–31.

Luo, Wei, Fei Shen, Clement Bommier, Hongli Zhu, Xiulei Ji, and Liangbing Hu. 2016. "Na-Ion Battery Anodes: Materials and Electrochemistry." *Accounts of Chemical Research* 49 (2): 231–40.

Mao, Jianfeng, Xiulin Fan, Chao Luo, and Chunsheng Wang. 2016. "Building Self-Healing Alloy Architecture for Stable Sodium-ion Battery Anodes: A Case Study of Tin Anode Materials." *ACS Applied Materials & Interfaces* 8 (11): 7147–55.

Meng, Shuo, Dong-Lin Zhao, Lu-Lu Wu, Ze-Wen Ding, Xing-Wang Cheng, and Tao Hu. 2018. "Fe_2O_3/ Nitrogen-Doped Graphene Nanosheet Nanocomposites as Anode Materials for Sodium-Ion Batteries with Enhanced Electrochemical Performance." *Journal of Alloys and Compounds* 737: 130–35.

Modafferi, Vincenza, Saveria Santangelo, Michele Fiore, Enza Fazio, Claudia Triolo, Salvatore Patanè, Riccardo Ruffo, and Maria G. Musolino. 2019. "Transition Metal Oxides on Reduced Graphene Oxide Nanocomposites: Evaluation of Physicochemical Properties." *Journal of Nanomaterials*, Article ID 1703218, 1–9.

Ni, Jiangfeng, Liang Li, and Jun Lu. 2018. "Phosphorus: An Anode of Choice for Sodium-Ion Batteries." *ACS Energy Letters* 3 (5): 1137–44.

Nkosi, Funeka P., Kumar Raju, Nithyadharseni Palaniyandy, M. V. Reddy, Caren Billing, and Kenneth I. Ozoemena. 2017. "Insights into the Synergistic Roles of Microwave and Fluorination Treatments Towards Enhancing the Cycling Stability of P2-Type Na0.67[Mg0.28Mn0.72]O_2 Cathode Material for Sodium-Ion Batteries." *Journal of The Electrochemical Society* 164 (13): A3362–70.

Palaniyandy, Nithyadharseni, Remegia Mmalewane Modibedi, and Mkhulu K. Mathe. 2019. "Preparation and Characterization of α-MnO_2 Nanoplatelets/Onion like Carbon (OLC) Composite: High Capacity and Long-Cycle Life Rechargeable Sodium Ion Batteries." *Meeting Abstracts* MA2019-03 (2): 170.

Qian, Jiangfeng, Ya Xiong, Yuliang Cao, Xinping Ai, and Hanxi Yang. 2014. "Synergistic Na-storage Reactions in Sn_4P_3 as a High-Capacity, Cycle-Stable Anode of Na-Ion Batteries." *Nano Letters* 14 (4): 1865–69.

Quan, Bo, Aihua Jin, Seung-Ho Yu, Seok Mun Kang, Juwon Jeong, Hactor D. Abruna, Longyi Jin, Yuanzhe Piao, and Yung-Eun Sung. 2018. "Solvothermal-Derived S-Doped Graphene as an Anode Material for Sodium-Ion Batteries." *Advanced Science* 5 (5): 1700880.

Rath, Purna Chandra, Jagabandhu Patra, Diganta Saikia, Mrinalini Mishra, Chuan-Ming Tseng, Jeng-Kuei Chang, and Hsien-Ming Kao. 2018. "Comparative Study on the Morphology-Dependent Performance of Various CuO Nanostructures as Anode Materials for Sodium-Ion Batteries." *ACS Sustainable Chemistry & Engineering* 6 (8): 10876–85.

Ruan, Boyang, Hai-peng Guo, Yuyang Hou, Qiannan Liu, Yuanfu Deng, Guohua Chen, Shu-lei Chou, Hua-kun Liu, and Jia-zhao Wang. 2017. "Carbon-Encapsulated Sn@ N-doped Carbon Nanotubes as Anode Materials for Application in SIBs." *ACS Applied Materials & Interfaces* 9 (43): 37682–93.

Sangster, James M. 2010. "Na-P (Sodium-Phosphorus) System." *Journal of Phase Equilibria and Diffusion* 31 (1): 62–67.

Slater, Michael D., Donghan Kim, Eungje Lee, and Christopher S. Johnson. 2013. "Sodium-Ion Batteries." *Advanced Functional Materials* 23 (8): 947–58.

Sun, Ning, Huan Liu, and Bin Xu. 2015. "Facile Synthesis of High Performance Hard Carbon Anode Materials for Sodium Ion Batteries." *Journal of Materials Chemistry A* 3 (41): 20560–66.

Tang, Qiming, Yanhui Cui, Junwei Wu, Deyang Qu, Andrew P. Baker, Yiheng Ma, Xiaona Song, and Yanchen Liu. 2017. "Ternary Tin Selenium Sulfide (SnSe0.5S0.5) Nano Alloy as the High-Performance Anode for Lithium-Ion and Sodium-Ion Batteries." *Nano Energy* 41: 377–86.

Vaalma, Christoph, Daniel Buchholz, Marcel Weil, and Stefano Passerini. 2018. "A Cost and Resource Analysis of Sodium-Ion Batteries." *Nature Reviews Materials* 3: 18013.

Walter, Marc, Rolf Erni, and Maksym V. Kovalenko. 2015. "Inexpensive Antimony Nanocrystals and Their Composites with Red Phosphorus as High-Performance Anode Materials for Na-Ion Batteries." *Scientific Reports* 5: 8418.

Wang, Chengliang, Yang Xu, Yaoguo Fang, Min Zhou, Liying Liang, Sukhdeep Singh, Huaping Zhao, Andreas Schober, and Yong Lei. 2015. "Extended TT-Conjugated System for Fast-Charge and-Discharge Sodium-Ion Batteries." *Journal of the American Chemical Society* 137 (8): 3124–30.

Wang, Kun, Yu Jin, Shixiong Sun, Yangyang Huang, Jian Peng, Jiahuan Luo, Qin Zhang, Yuegang Qiu, Chun Fang, and Jiantao Han. 2017. "Low-Cost and High-Performance Hard Carbon Anode Materials for Sodium-ion Batteries." *ACS Omega* 2 (4): 1687–95.

Wang, Lei, Zengxi Wei, Minglei Mao, Hongxia Wang, Yutao Li, and Jianmin Ma. 2019. "Metal Oxide/Graphene Composite Anode Materials for Sodium-Ion Batteries." *Energy Storage Materials* 16: 434–54.

Wang, Ning, Qinglei Liu, Boya Sun, Jiajun Gu, Boxuan Yu, Wang Zhang, and Di Zhang. 2018. "N-Doped Catalytic Graphitized Hard Carbon for High-performance Lithium/Sodium-Ion Batteries." *Scientific Reports* 8 (1): 9934.

Wang, Shiwen, Lijiang Wang, Zhiqiang Zhu, Zhe Hu, Qing Zhao, and Jun Chen. 2014. "All Organic Sodium-Ion Batteries with $Na_4C_8H_2O_6$." *Angewandte Chemie International Edition* 53 (23): 5892–96.

Wang, Wanlin, Weijie Li, Shun Wang, Zongcheng Miao, Hua Kun Liu, and Shulei Chou. 2018. "Structural Design of Anode Materials for Sodium-Ion Batteries." *Journal of Materials Chemistry A* 6 (15): 6183–205.

Wang, Xiaojun, Kangzhe Cao, Yijing Wang, and Lifang Jiao. 2017. "Controllable N-doped $CuCo_2O_4$@C Film as a Self-Supported Anode for Ultrastable Sodium-Ion Batteries." *Small* 13 (29): 1700873.

Wang, Xiaojun, Yongchang Liu, Yijing Wang, and Lifang Jiao. 2016. "CuO Quantum Dots Embedded in Carbon Nanofibers as Binder-Free Anode for Sodium Ion Batteries with Enhanced Properties." *Small* 12 (35): 4865–72.

Wang, Yanrong, Hui Huang, Qingshui Xie, Yuanpeng Wang, and Baihua Qu. 2017. Rational Design of Graphene-Encapsulated $NiCo_2O_4$ Core-shell Nanostructures as an Anode Material for Sodium-Ion Batteries." *Journal of Alloys and Compounds* 705: 314–19.

Wang, Ying, Caiyun Wang, Yijing Wang, Huakun Liu, and Zhenguo Huang. 2016. "Superior Sodium-Ion Storage Performance of Co_3O_4@nitrogen-doped Carbon: Derived from a Metal-Organic Framework." *Journal of Materials Chemistry A* 4 (15): 5428–35.

Wang, Ying, Katja Kretschmer, Jinqiang Zhang, Anjon Kumar Mondal, Xin Guo, and Guoxiu Wang. 2016. "Organic Sodium Terephthalate@ Graphene Hybrid Anode Materials for Sodium-Ion Batteries." *RSC Advances* 6 (62): 57098–102.

Wu, Chao, Shi-Xue Dou, and Yan Yu. 2018. "The State and Challenges of Anode Materials Based on Conversion Reactions for Sodium Storage." *Small* 14 (22): 1703671.

Wu, Lijun, Junwei Lang, Peng Zhang, Xu Zhang, Ruisheng Guo, and Xingbin Yan. 2016. "Mesoporous Ni-Doped $MnCo_2O_4$ Hollow Nanotubes as an Anode Material for Sodium Ion Batteries with Ultralong Life and Pseudocapacitive Mechanism." *Journal of Materials Chemistry A* 4 (47): 18392–400.

Xiang, Xingde, Kai Zhang, and Jun Chen. 2015. "Recent Advances and Prospects of Cathode Materials for Sodium-Ion Batteries." *Advanced Materials* 27 (36): 5343–64.

Xie, Hezhen, W. Peter Kalisvaart, Brian C. Olsen, Erik J. Luber, David Mitlin, and Jillian M. Buriak. 2017. "Sn-Bi-Sb Alloys as Anode Materials for Sodium Ion Batteries." *Journal of Materials Chemistry A* 5 (20): 9661–70.

Xie, Xiuqiang, Katja Kretschmer, Jinqiang Zhang, Bing Sun, Dawei Su, and Guoxiu Wang. 2015. "Sn@ CNT Nanopillars Grown Perpendicularly on Carbon Paper: A Novel Free-Standing Anode for Sodium Ion Batteries." *Nano Energy* 13: 208–217.

Xu, Gui-Liang, Tian Sheng, Lina Chong, Tianyuan Ma, Cheng-Jun Sun, Xiaobing Zuo, Di-Jia Liu, Yang Ren, Xiaoyi Zhang, Yuzi Liu, Steve M. Heald, Shi-Gang Sun, Zonghai Chen, and Khalil Amine. 2017. "Insights into the Distinct Lithiation/Sodiation of Porous Cobalt Oxide by in Operando Synchrotron X-Ray Techniques and Ab Initio Molecular Dynamics Simulations." *Nano Letters* 17 (2): 953–62.

Xu, Jiantie, Min Wang, Nilantha P. Wickramaratne, Mietek Jaroniec, Shixue Dou, and Liming Dai. 2015. "High-Performance Sodium Ion Batteries Based on a 3D Anode from Nitrogen-Doped Graphene Foams." *Advanced Materials* 27 (12): 2042–48.

Xu, Meng, Qiuying Xia, Jili Yue, Xiaohui Zhu, Qiubo Guo, Junwu Zhu, and Hui Xia. 2018. "Rambutan-like Hybrid Hollow Spheres of Carbon Confined Co_3O_4 Nanoparticles as Advanced Anode Materials for Sodium-Ion Batteries." *Advanced Functional Materials* 29 (6): 1807377.

Xu, Xiangdong, Hongliang Zeng, Dezhi Han, Ke Qiao, Wei Xing, Mark J. Rood, and Zifeng Yan. 2018. "Nitrogen and Sulfur Co-Doped Graphene Nanosheets to Improve Anode Materials for Sodium-Ion Batteries." *ACS Applied Materials & Interfaces* 10 (43): 37172–180.

Yabuuchi, Naoaki, Kei Kubota, Mouad Dahbi, and Shinichi Komaba. 2014. "Research Development on Sodium-Ion Batteries." *Chemical Reviews* 114 (23): 11636–82.

Yabuuchi, Naoaki, Yuta Matsuura, Toru Ishikawa, Satoru Kuze, Jin-Young Son, Yi-Tao Cui, Hiroshi Oji, and Shinichi Komaba. 2014. "Phosphorus Electrodes in Sodium Cells: Small Volume Expansion by Sodiation and the Surface-Stabilization Mechanism in Aprotic Solvent." *ChemElectroChem* 1 (3): 580–89.

Ying, Hangjun, and Wei-Qiang Han. 2017. "Metallic Sn-Based Anode Materials: Application in High-Performance Lithium-Ion and Sodium-Ion Batteries." *Advanced Science* 4 (11): 1700298.

Yue, Xiyan, Ning Huang, Zhongqing Jiang, Xiaoning Tian, Zhongde Wang, Xiaogang Hao, and Zhong-Jie Jiang. 2018. "Nitrogen-Rich Graphene Hollow Microspheres as Anode Materials for Sodium-Ion Batteries with Super-high Cycling and Rate Performance." *Carbon* 130: 574–83.

Zhang, Ruie, Zhifeng Wang, Wenqing Ma, Wei Yu, Shanshan Lu, and Xizheng Liu. 2017. "Improved Sodium-Ion Storage Properties by Fabricating Nanoporous CuSn Alloy Architecture." *RSC Advances* 7 (47): 29458–63.

Zhang, Xiaojie, Wei Qin, Dongsheng Li, Dong Yan, Bingwen Hu, Zhuo Sun, and Likun Pan. 2015. "Metal-Organic Framework Derived Porous CuO/Cu_2O Composite Hollow Octahedrons as High Performance Anode Materials for Sodium Ion Batteries." *Chemical Communications* 51 (91): 16413–16.

Zhang, Yan, Xinhui Xia, Bo Liu, Shengjue Deng, Dong Xie, Qi Liu, Yadong Wang, Jianbo Wu, Xiuli Wang, and Jiangping Tu. 2019. "Multiscale Graphene-Based Materials for Applications in Sodium Ion Batteries." *Advanced Energy Materials* 9 (8): 1803342.

Zhang, Zhi-Jia, Yun-Xiao Wang, Shu-Lei Chou, Hui-Jun Li, Hua-Kun Liu, and Jia-Zhao Wang. 2015. "Rapid Synthesis of α-Fe_2O_3/rGO Nanocomposites by Microwave Autoclave as Superior Anodes for Sodium-Ion Batteries." *Journal of Power Sources* 280: 107–113.

Zhao, Liang, Junmei Zhao, Yong-Sheng Hu, Hong Li, Zhibin Zhou, Michel Armand, and Liquan Chen. 2012. "Disodium Terephthalate ($Na_2C_8H_4O_4$) as High Performance Anode Material for Low-Cost Room-Temperature Sodium-Ion Battery." *Advanced Energy Materials* 2 (8): 962–65.

4

Cathode Materials for Sodium-Ion-Based Energy Storage Batteries

Assumpta C. Nwanya, Mesfin A. Kebede, Fabian I. Ezema, and M. Maaza

CONTENTS

4.1 Introduction 59
4.2 Operating Principle of Na-Ion Battery 60
    4.2.1 Calculation of Capacity of Battery 61
4.3 Electrode Materials Used in SIBs 62
    4.3.1 Layered Oxides 62
        4.3.1.1 Single Metal Layered Oxides 64
        4.3.1.2 Multiple Metal Layered Oxides 70
    4.3.2 Phosphates Polyanions 72
        4.3.2.1 Single Phosphates 72
    4.3.3 Other Cathodes Materials 74
4.4 Conclusion and Perspective 74
Acknowledgements 75
References 75

4.1 Introduction

The quest for mobile energy storage has revolutionized research in energy storage, especially in rechargeable batteries. Presently, lithium-ion batteries dominate the market. They are widely used in products ranging from consumer electronics to electric vehicles. The high cost and scarcity of lithium occasioned by high demand are driving research to develop alternatives to lithium-ion batteries. These alternatives became expedient to meet future needs in energy storage especially in expanding applications including batteries to store power from sources such as solar and wind energy for use on the power grid. The sodium-ion battery (SIB) is taking a centre stage in battery research because sodium resources are abundant in nature and is less expensive. Secondly, the electrochemical properties of SIBs are similar to that of a lithium-ion battery (LIB). Actually, studies in SIBs and LIBs started at almost the same time; however, investigation of SIBs was significantly reduced due to the success achieved in the commercialization of LIBs in the 1990s (Wang et al. 2018).

The major drawback to SIBs is that sodium is slightly heavier than lithium, making the insertion and de-insertion of sodium ion a bit more sluggish than lithium ion. The anodes in most LIBs are made primarily from thin layers of graphitic carbon. Lithium ions can fit easily and are reversible between the layers, but the larger-sized sodium ions cannot pass through these layers reversibly. This makes the insertion/de-insertion of Na ions in a host material difficult and often results in severe structural degradation (Wang et al. 2018). Hence, SIBs may not be able to compete with LIBs in terms of energy density. Nevertheless, they can be well suited for storing off-peak grid power, as well as fluctuating renewable energies, such as wind and solar energy, where the weight and footprint requirement is not critical. In addition, the low cost of sodium (which is pivotal for stationary energy storage systems on a large

TABLE 4.1

Physical Characteristics of Sodium Ion in Comparison to Lithium, Potassium, and Magnesium Ions

Properties	Li$^+$	Na$^+$	K$^+$	Mg$^+$
Relative atomic mass	6.94	23.00	39.10	24.31
Mass-to-electron ratio	6.94	23.00	39.10	12.16
Ionic radius/Å	0.76	1.02	1.38	0.72
E° (vs standard hydrogen electrode [SHE])/V	−3.04	−2.71	−2.93	−1.55
Melting point/°C	180.5	97.7	63.4	650.0
Theoretical capacity of metal electrodes/mAh g^{-1}	3861	1166	685	2205
Material abundance (ppm)	20	28,300	25,900	20,900

Sources: Wang et al., *Electrochem. Energy Rev.*, 1, 200–237, 1998; Ellis and Nazar, *Curr. Opin. Solid St. M.*, 16, 168–177, 2012; Fleischer, *Recent Estimates of the Abundances of the Elements in the Earth's Crust*, United States Department of the Interior, Geological Survey Circular 285, 3, 1963; Yabuuchi et al., *Chem. Rev.*, 114, 11636–11682, 2014.

scale) and abundant nature (sodium is ranked the sixth most abundant element in the earth's crust) (Fleischer 1963) is likely to outweigh this concern, as sodium is the second lightest and smallest alkali metal next to lithium as Table 4.1 (Yabuuchi et al. 2014) shows. Moreover, it is feasible to replace Cu with Al current collectors since alloying reactions do not occur between Na and Al, and this will further reduce the costs and weight concerns of SIBs (Guo et al. 2016). The increasing interest in SIBs has resulted in several companies such as Faradion in the UK and Aquion and Valence Technologies in the USA to designing prototypes and commercializing SIBs (Brazil 2014). Other similar sodium-based storage systems, such as sodium-sulphur (Na-S) (Breeze 2014; Xu et al. 2018), sodium air (Na-O) (Liu, Liu, and Luo 2016; Liang et al. 2018), as well as zebra cells (Steinbock and Dustmann 2001; Skyllas-Kazacos 2010) have also been the subject of intense research.

This chapter gives a short overview of sodium-ion-based batteries; the electrochemistry and some of the successfully used cathode materials.

4.2 Operating Principle of Na-Ion Battery

The components, structures, and energy storage principles of SIBs are basically the same with LIBs, only that sodium ions are used to replace lithium ions as shown schematically in Figure 4.1. Batteries are typically comprised of two electrodes; anode (negative electrode) and cathode (positive electrode), a porous separator, and electrolyte. Typically, SIBs consist of a cathode that is most often a layered sodium metal oxide (NaMO) and a carbon-based anode. This is similar to the layered lithium metal oxide (LiMO) cathode and carbon anode that are typically used in LIBs. The separator, which must be porous, is used as a barrier that provides electrode separation to prevent short circuiting and promote isotropic ion transfer. The electrolytes, which are obtained by dissolving the electrolyte salts in polar solvents, act as ionic conductors.

Energy introduced into the battery during charging causes sodium ion de-insertion from the NaMO structure, oxidation reactions take place at the cathode, a Na ion migrates through the electrolyte that is ionically conductive to the anode, while an electron also moves through the external circuit to the anode (equation 4.1). At the anode, an electron and Na ion are inserted into the interlayer spacing of the anode forming a stable sodium atom (equation 4.2) (Tang et al. 2015).

$$Na \rightarrow Na^+ + e^- \tag{4.1}$$

$$Na^+ + e^- \rightarrow Na \tag{4.2}$$

During discharge, the reverse reactions take place; sodium ion insertion occurs, reduction reactions take place at the cathode, and a Na ion moves from the anode through the electrolyte to the cathode releasing the stored energy. A good SIB should be low cost, have good cycling stability, be safe, and have high specific capacity, energy, and power. In this regard, the cathode material plays a significant role in

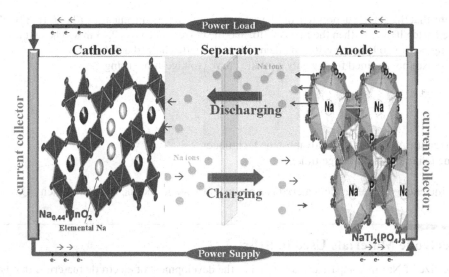

FIGURE 4.1 Schematic diagram of the operation of a sodium ion battery.

determining the operation voltage and specific capacity of the battery. Therefore, good cathode materials should exhibit high specific energy, very good cycling stability, as well as rate capability.

4.2.1 Calculation of Capacity of Battery

The theoretical capacity of the battery is related to its molecular weight (*Mw*) and is calculated using Faraday's law.

$$Q_{\text{theoretical}} = \left(\frac{n\text{F}}{3.6\text{Mw}} \right) \text{mAhg}^{-1} \tag{4.3}$$

where
   n is the electron transfer number or the number of charge carrier during charge/discharge
   F is the Faraday's constant
   Mw is the molecular weight of the active material of the electrode.

For example:

$$\text{Na}_{0.44}\text{MnO}_2 \rightarrow \left(\text{Na}_{1-0.44} \right)\text{MnO}_2 + 0.44\text{Na}^+ + 1e^- \tag{4.4}$$

The molecular weight of $\text{Na}_{0.44}\text{MnO}_2$ is $(23 \times 0.44 + 55 + 32) = 97.12$

$$Q_{\text{theoretical}} = \left(\frac{0.44 \times 96485.33}{3.6 \times 97.12} \right) = 121.42\,\text{mAhg}^{-1}$$

The practical specific capacity depends on the C-rate used. The term "C-rate" is used to represent a constant current and gives the expected capacity of the working electrode as a ratio of the desired number of hours to fully charge or discharge that capacity. The C-rate is a measure of the rate at which a battery is completely charged or discharged, relative to its nominal capacity. A C-rate of 1 C means that the current necessary for complete charge or discharge in 1 hour is applied.

For example, cycling at 1 C-rate for an active mass of 0.2 g of $\text{Na}_{0.44}\text{MnO}_2$ will imply:

$$1\text{C} \rightarrow 0.2\text{g} \times 121.42\,\text{mAg}^{-1}$$

$$\Rightarrow 24.284\,\text{mA}$$

This means that the cell is supposed to discharge completely at this current after 1 hour. If the cell is to be discharged after 10 hours, then the rate is C/10, and the current becomes 2.43 mA. The cell might finish discharging before 10 hours, but is still taken to have been discharged at C/10 rate.

Capacity can be obtained from a galvanostatic curve (voltage-time) using:

$$Q = i \times t \tag{4.5}$$

where
   i = current (mA) (constant in galvanostatic technique)
   t = time (charge and discharge time).

The specific energy of a cathode material is a product of voltage and the specific capacity Q.

4.3 Electrode Materials Used in SIBs

The large size of Na+ in comparison to Li+ makes the development of electrode materials that have substantial interlayer and interstitial spaces to hold Na ions paramount, in order to obtain a very good battery performance (Wang et al. 2018). In this regard, many materials have been studied for applications as cathodes and anodes for SIBs. Figure 4.2 (Ellis and Nazar 2012) summarizes the cathodes and anodes that have been used in SIBs, indicating their specific capacities and operating voltages. However, our focus for the chapter is on the cathode materials.

Different types of materials with crystal structures that have substantial interlayer and interstitial spacings for sodium ion migration and storage have been studied for cathodes in SIBs. These include the following.

4.3.1 Layered Oxides

Layered oxides are transition metal oxides (with a 3d electron configuration) of the type A_xMO_2 (A = Na, M = Co, Mn, Fe or Ni) $(0.5 \leq x \leq 1)$ (Tang et al. 2015; Das and Goplan 2019). Studies on layered Na_xMO_2 started more than four decades ago seeking potential materials in SIBs, but have increased recently because of its high energy density potential. The Na ion can intercalate/de-intercalate in the interstitial sites between the MO_6 octahedral in the Na_xMO_2 forming transition metal layers. Ideally,

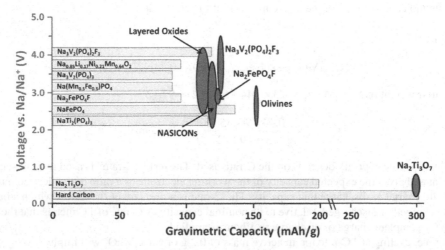

FIGURE 4.2 Key cathodes and anodes that have been used in SIBs indicating their specific capacities and operating voltages. (Theoretical capacities of the various materials at their various potentials are shown with ovals, while achieved capacities are shown with bars). (Reprinted from *Current Opinion in Solid State and Materials Science*, 16, Ellis, B.L. and Nazar, L.F., Sodium and Sodium-Ion Energy Storage Batteries, 168–177. Copyright 2012, with permission from Elsevier.)

electrochemical cycling of layered $NaMO_2$ materials should be seen as topotactic. That is, the crystal structure of the oxide materials containing the Na ion should not change as it intercalates/de-intercalates. However, in reality, the structure remains unchanged only in the sense that it retains its layered formation.

Na_xMO_2 layered oxides undergo various reversible and irreversible phase transformations to form ordered structures due to the rearrangement of Na and vacancies that remain as a Na ion de-intercalates. This is seen as steps and plateaus on cycling voltage curves.

Two main groups of layered oxides were identified by Delmas et al. (1981) for SIBs based on the Na ion environment. The first is the octahedral (O) type, in which the Na ions occupy octahedral coordination sites in the unit cell, while the second is the prismatic (P), in which the Na ions occupy prismatic coordination sites in the unit cell (Delmas et al. 1981). For $0.5 \leq x \leq 0.8$, P2-type (γ) Na_xMO_2 is formed, while for $0.85 \leq x \leq 1$, O3-type (α) Na_xMO_2 is formed (Das and Goplan 2019). The numerals (2 and 3) indicate the number of distinguishable units of sodium layers in the crystal. Hence, O3-type Na_xMO_2 indicates octahedral sites and oxygen stacking which repeats with every three sodium layers corresponding to an $R3m$ crystal space group. Hence, an O3-configuration consists of AB CA BC layers, P3-has AB BC CA layers, O2 has AB AC layers, while P2 has AB BA layers with various O stacking, as illustrated in Figure 4.3 (Radin et al. 2017). Na ion de-intercalation from O3-type and P2-type Na_xMO_2 leads to phase changes. O3-type and P2-type materials transform into P3-type and O2-type materials, respectively, by the sliding of MO_2 slabs without the splitting of bonds (Yabuuchi et al. 2014; Guo et al. 2016). However, a P3-type can also be formed from solid state reactions in the absence of electrochemical Na ion extraction reactions (Tsuchiya and Yabuuchi 2016), while an O2-type can also be formed by ion exchange reactions (Paulsen et al. 2002). It is difficult for an O3/P3-type to transition to P2-type layered oxides at a normal operating temperature. However, such transitions can only be achieved at an elevated temperature because it involves the splitting of M-O bonds (Smirnova et al. 2006; Yabuuchi et al. 2014).

A loss of symmetry for a phase through lattice distortion or sodium ion extraction (to a monoclinic phase) is indicated with a 'prime' symbol in all the structures. Hence, an O3 becomes O'3, for example, in O'3-$NaMnO_2$, which has a monoclinic distortion in the MnO_6 octahedra containing Mn^{3+} ions because of the Jahn-Teller distortion (Ma et al. 2011).

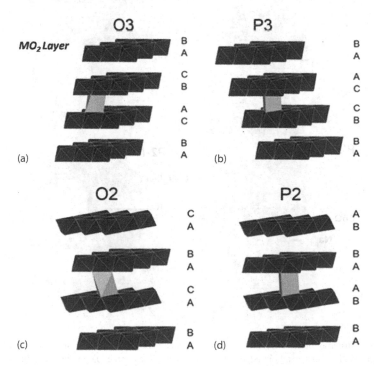

FIGURE 4.3 Crystal structures for Na-containing layered oxides (Na_xMO_2) (a) O3, (b) P3, (c) O2, and (d) P2. (Reprinted with permission from Radin, M.D. et al., *Nano Lett.*, 17, 7789–7795, 2017. Copyright 2017 American Chemical Society.)

4.3.1.1 Single Metal Layered Oxides

Na_xCoO_2 is one the single layered metal oxides earlier studied for SIBs because of the success achieved by $LiCoO_2$ (Xiang et al. 2015). However, the electrochemical performance of Na_xCoO_2 is poor due to the larger Na ion size yielding moderate discharge capacities of 70–107 mAh g^{-1} (Xia and Dahn 2012; Rai et al. 2014). Nonetheless, the diffusion constant D of Na ion (~ 0.5–1.5 × 10^{-10} cm^{2}s^{-1}) in Na_xCoO_2 is found to be higher than that of Li ion (<1.0 × 10^{-11} cm^{2}s^{-1}) in $LiCoO_2$ (Ding et al. 2013). A solid state reaction method is primarily used to obtain the various phases of Na_xCoO_2 by varying the ratio of Na and Co in the precursors in stoichiometric quantities (Xia and Dahn 2012; Ding et al. 2013; Lei et al. 2014; Shibata et al. 2015). The O3- and P2-type phases of Na_xCoO_2 are formed at x values between 0.83–1.0 and 0.67–0.80, respectively (Xia and Dahn 2012; Lei et al. 2014). Studies show that the diffusion constant of O3-type Na_2CoO_2 is higher than that of P2-type Na_2CoO_2 due probably to the sequential structural transition from O3–O′3–P′3–P3–P′3 with Na ion de-intercalation (Lei et al. 2014; Shibata et al. 2015). Figure 4.4 shows the structures of the crystal of the various phases of Na_xCoO_2, while Table 4.2 gives the compositions, phase types, and the specific capacity obtained for Na_xCoO_2. It is evidenced that the P2 phase is normally obtained at a higher temperature, while the other phases are obtained at lower temperatures structures.

Na_xMnO_2 is another good single metal oxide used as a cathode in SIBs. It has a lower cost and higher theoretical capacity (243 mAh g^{-1}) than Na_xCoO_2. The lower cost of manganese-based materials is as a result of the high abundance of elemental manganese on the surface of the earth. Various phases of Na_xMnO_2 have shown to be electrochemically active for SIBs. Na_xMnO_2 (x = 0.2, 0.4, 0.44, 1) and $Na_{0.7}MnO_{2+y}$ (0 ≤ y ≤ 0.25) were initially reported by Parant et al. (1971). Based on the sodium content and reaction temperatures, either tunnel, layered, or a mix of tunnel and layered structure of Na_xMnO_2 can be obtained.

A tunnel structure is made up of sheets of MnO_6 octahedron and edge-sharing MnO_5 square pyramidal sites connected either by edge or corner sharing (Xu et al. 2014; Wu et al. 2017). Two types of tunnels

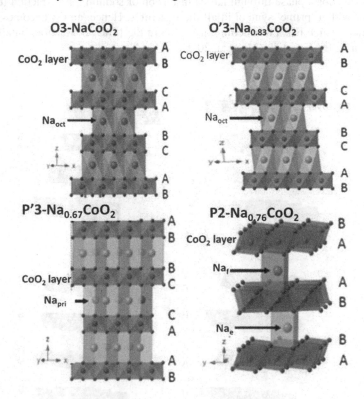

FIGURE 4.4 Crystal structure of the different phases of the Na_xCoO_2. Symbols A, B, and C represent layers of oxygen with different stacking. All Na ions reside in edge-sharing octahedral sites in the O3 and O′3 structures, while the Na ions have prismatic coordination in the P′3 structure with one side edge-sharing and another face sharing with a Co octahedron. (Reprinted with permission from Lei, Y. et al., *Chem. Mater.*, 26, 5288–5296. Copyright 2014 American Chemical Society.)

TABLE 4.2

Compositions, Phase Types, and the Specific Capacity Obtained for Na_xCoO_2

Compound	Phase Type	Synthesis Temperature (°C)	Obtained Specific Capacity (mAh.g^{-1})	References
$Na_{0.67}CoO_2$	P2	850	146.8	Hwang et al. (2017)
$Na_{0.7}CoO_2$	P2	850	–	Berthelot et al. (2010)
$Na_{0.74}CoO_2$	P2	850	134.0	Matsui et al. (2015)
$Na_{0.67}CoO_2$	P2	800	105.0	Shibata et al. (2015)
$Na_{0.67}CoO_2$	P2	900	146.0	Kang et al. (2018)
$NaCoO_2$	O3(single crystal)	1050	–	Takahashi et al. (2003)
$NaCoO_2$	O3	900	140.0	Yoshida et al. (2013)
$Na_{0.99}CoO_2$	O3	550	118	Shibata et al. (2015)
$Na_{0.62}CoO_2$	P'3	550	–	Carlier et al. (2009)
$Na_{0.67}CoO_2$	P'3	530	–	Ono et al. (2002)
$Na_{0.67}CoO_2$	O'3			Delmas et al. (1981)

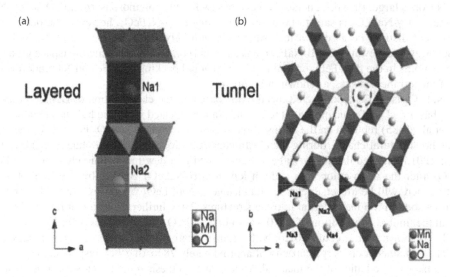

FIGURE 4.5 Demonstrative crystal structures of layered (a) and tunnel compounds (b) in the Na_xMnO_2 system. (Reprinted with permission from Wu, Z.-G. et al., *ACS Appl. Mater. Interfaces*, 9, 21267–21275. Copyright 2017 American Chemical Society.)

can be formed; the larger S-shaped double tunnel and a single six-sided tunnel, both occupied by Na ions (Cao et al. 2011; Xu et al. 2014; Wu et al. 2017), as shown in Figure 4.5. The larger S-shaped tunnel can tolerate structural stress during cycling, therefore it offers better cycling stability (Cao et al. 2011; Wu et al. 2017). More Na ions (Na2 and Na3 in Figure 4.5) are able to occupy the larger S-shaped tunnel and are more mobile, thereby enabling high diffusion channels for Na ion transport (Wu et al. 2017). A tunnel-type structure is normally formed at $0.22 \leq x \leq 0.44$, a mix of tunnel and layered structures is sometimes obtained at $0.44 \leq x \leq 0.66$, while a fully layered structure is normally obtained at $0.66 \leq x \leq 1$ (Kuratani et al. 2007; Han et al. 2014; Liu et al. 2017; Ferrara et al. 2018).

The layered Na_xMnO_2 structure can be classified into two main phases: P2 or O3 phase (Xu et al. 2014) based on the synthesis protocols, such as the annealing temperature and the atomic ratio of sodium and manganese. Mendiboure et al. (1985) reported an O3 layered $NaMnO_2$ structure with a monoclinic distortion in the structure (O3'-type) because of the Jahn-Teller distortion of the Mn^{3+} ion. The range for reversible change shown by an O'3-type was very narrow ($x < 0.2$ in $Na_{1-x}MnO_2$), however, Ma et al. (2011) showed that the O'3-type $Na_{1-x}MnO_2$ could have a reversible range that is much wider ($x < 0.8$), though, with poor cycling stability. A P'2 phase type of Na_xMnO_2 having an orthorhombic crystal lattice (space group *Cmcm*) (Stoyanova et al. 2010) is one of the polymorphs of layered Na_xMnO_2 oxides that has also been shown to be thermodynamically stable. P2-type Na_xMnO_2 ($0.6 < x < 0.7$) may have a much larger Na-ion storage capacity. This is because the trigonal prismatic site occupied by Na ion in P2 phase is larger than the octahedral site in O3 phase. This makes Na ion transport in the P2 phase faster than that in the O3 phase. It is more difficult for the P2 phase to transition to other phases because that involves the breaking of M–O bonds, thus, the P2 phase is expected to exhibit better cyclic performance due to insignificant structural change. However, the biggest challenge for a P2-type Na_xMnO_2 is the deficiency of Na in their crystal structure, which is also a peculiar problem with most P-type cathodes. The high reversible capacities mostly reported for P2-type cathode materials more often than not were obtained in a half cell configuration using Na metal anodes, which also serves as an additional Na source. In a practical SIB cell, the charge storage capacity obtained from the Na ions will solely depend on the sodium ions obtainable from the cathode side of the cell. This will consequently stop the use of anode materials, such as carbon-based materials and phosphorous that has no sodium content.

Na_xMnO_2 is mostly synthesized by solid state reactions (Sauvage et al. 2007; Xu et al. 2014; G. Ma et al. 2016; T. Ma et al. 2017), though other synthesis routes such as hydrothermal (Kuratani et al. 2007; Zhang et al. 2009; Su et al. 2013), solution combustion (G. Ma et al. 2016), etc. have been used (Table 4.3).

$NaFeO_2$ is amongst the layered single metal oxide that have been shown to be electrochemically active for SIBs. Na ion is larger than Fe^{3+}, hence, $NaFeO_2$ has two stable polymorphs: α-$NaFeO_2$ and β-$NaFeO_2$. The structure of γ-$NaFeO_2$ is said to be almost the same as β-$NaFeO_2$, however, the detailed crystal structure is not distinct because of a reversible transition between β-$NaFeO_2$ and γ-$NaFeO_2$ (Takeda et al. 1980; Grey and Hewat 1990). β-$NaFeO_2$ has an orthorhombic crystal structure (space group: *Pna2*). The oxygen atoms in the β-$NaFeO_2$ crystal form a hexagonal packing with Fe and Na randomly occupying half of the tetrahedral sites, as shown in Figure 4.6.

The α-$NaFeO_2$ is an O3-type layered metal oxide based on the classification of Delmas et al. (1981). α-$NaFeO_2$ has a rock salt structure with the Fe and Na ions located in octahedral sites (Ma et al. 2018). Kikkawa et al. (1985) reported on the de-insertion of a Na ion from α-$NaFeO_2$ in 1985 for the first time. There was no structural change during the electrochemical process. However, Takeda et al. (1994) and Zhao et al. (2013) showed that the α-$NaFeO_2$ transformed to a new monoclinic phase of $Na_{0.5}FeO_2$ and $Na_{0.58}FeO_2$ after the de-insertion of a sodium ion from α-$NaFeO_2$. Studies by Yabuuchi et al. (2012) indicate that there will be irreversible structural change in $Na_{1-x}FeO_2$ when it is charged beyond $x > 0.5$. This disturbs sodium ion intercalation into the structures. They further showed that reversible cycling is achieved in the limited composition of $x = 0 - 0.45$ in $Na_{1-x}FeO_2$. The quasi-reversibility of α-$NaFeO_2$ in a SIB cell was shown by Lee et al. (2015), with a transformation into a new O″3 layered phase (denoted as O″3). They obtained a discharge capacity of approximately 78 mAh g^{-1} corresponding to about 0.3 Na intercalated back to the lattice of the material. Most of the synthesis route for α-$NaFeO_2$ is by solid state reaction, though few other methods such as hydrothermal have been used. Table 4.4 gives a summary of some of the phases of $NaFeO_2$ that have been synthesized indicating the synthesis route, as well as the capacities obtained.

$NaCrO_2$ is another O3-type layered oxide that has been actively studied for use in SIBs. This is because of its high theoretical capacity of about 250 mAh g^{-1} and large interslab distance that allows easy insertion into and extraction from the host structure. However, in practical applications, only about 110 mAh g^{-1} ($Na_{1-x}CrO_2$, $0 \leq x \leq 0.5$) is obtained (Xia and Dahn 2012; Kubota et al. 2015; Yu et al. 2015). Braconnier et al. (1982) showed the possibility of Na^+ insertion/de-insertion into the electrode, although they could only cycle 0.15 Na reversibly, which was not satisfactory as a battery material. Studies by Kubota et al. (2015) showed that irreversible phase transitions take place when $Na_{1-x}CrO_2$ is charged by de-sodiation beyond 3.8 V and $x > 0.5$. This phase transition occurs due to the movement of chromium ions from the octahedral sites in CrO_2 slabs to the tetrahedral sites in the interslab layer. Komaba et al. (2010) compared the intercalation mechanism of $NaCrO_2$ vs $LiCrO_2$, and they showed that diffusion of Na^+ ions is faster than Li^+ ions in the oxide framework because of a longer alkali-oxygen bonding and larger interslab space of $NaCrO_2$ compared with those of $LiCrO_2$, as shown in Figure 4.7.

Moreover, the interstitial tetrahedral sites in $NaCrO_2$ are much larger and could be occupied with Cr^{6+} ions without their migration, and Cr^{4+} could be more thermodynamically stabilized in Na_xCrO_2. The low electrochemical activity of $LiCrO_2$ was attributed to the irreversible movement of Cr^{6+} induced

TABLE 4.3

Compositions, Phase Types, and the Specific Capacities Obtained for Na_xMnO_2

Compound	Phase Type	Synthesis Route	Temperature (°C)	Space Group	Specific Capacity (mAh.g⁻¹)at 1°C-rate	Voltage (V/Na/Na⁺)	References
$Na_{0.44}MnO_2$	Tunnel	Solid state	900	Pbam	140	2–3.8	Sauvage et al. (2007)
$Na_{0.6}MnO_2$	P2-layered /tunnel	Co-precipitation	450	P63/mmc/Pbam	193.6	1.5–4.3	Wu et al. (2017)
$Na_{0.7}MnO_2$	P2-layered	Co-precipitation	450	P63/mmc	166.6	1.5–4.3	Wu et al. (2017)
$Na_{0.7}MnO_2$	P2-layered	Hydrothermal	220	cmca	163.0	2.4.5	Su et al. (2013)
$Na_{0.44}MnO_2$	Tunnel	Reverse micro emulsion	850	Pbam	84.8	2–4.0	Liu et al. (2017)
$Na_{0.44}MnO_2$	Tunnel	Reverse micro emulsion	900	Pbam	50.0	2–4.0	Liu et al. (2017)
$Na_{2/3}MnO_2$	P2-layered	Solid state	600	C2/m	41.0	2–3.8	Ma et al. (2017)
$Na_{2/3}MnO_2$	P2-layered	Combustion	500	P63/mmc	80.0	2–4.3	Konarova et al. (2018)
$Na_{0.6}MnO_2$	P2-layered	Sol gel	800	P63/mmc	140 @0.1mAcm⁻²	2–3.8	Caballero et al. (2002)
$NaMnO_2$	O'3 layered	Solid state	800	C2/m	185 @C/10	2–3.8	Ma, Chen, and Ceder (2011)
$NaMnO_2$	O'3 layered	Solid state	700	C2/m	169	2–3.8	Ma et al. (2018)

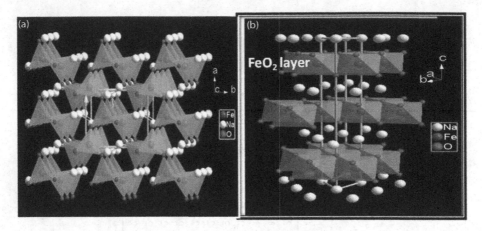

FIGURE 4.6 Crystal structures of the $NaFeO_2$: (a) β-$NaFeO_2$ and (b) α-$NaFeO_2$. Cell edges are illustrated by the turquoise lines. In the β-$NaFeO_2$ phase, tetrahedra of FeO_4 share corners in an ordered wurtzite arrangement, whereas the sodium ions occupy tetrahedral sites in the framework cavities (ICSD 27117).

TABLE 4.4

Summary of Some of the Phases of $NaFeO_2$

Compound	Phase Type	Synthesis Route	Temperature (°C)	Space Group	Specific Capacity (mAh.g⁻¹) at 1° C-Rate	Voltage (V/Na/Na⁺)	References
α-$NaFeO_2$	O3	Hydrothermal	220	$R\bar{3}m$	70@10 mA g⁻¹	1.5–4.0	Kataoka et al. (2015)
α-$NaFeO_2$	O3	Solid state	630	$R\bar{3}m$	70@10 mA g⁻¹	1.5–4.0	Kataoka et al. (2015)
α-$NaFeO_2$	O3	Solid state	650	$R\bar{3}m$	70@10 mA g⁻¹	1.5–4.0	Kataoka et al. (2015)
α-$NaFeO_2$	O3	Solid state	650	$R\bar{3}m$	85	1.5–3.6	Zhao et al. (2013)
α-$NaFeO_2$	O3	Solid state	700	$R\bar{3}m$	120@0.1mA cm⁻²	2.5–4.5	Takeda et al. (1994)
α-$NaFeO_2$	O3	Solid state	700	$R\bar{3}m$			
α-$NaFeO_2$	O3	Solid state	650	$R\bar{3}m$	78.4 15@ mA g⁻¹	2.0–3.6	Lee et al. (2015)

by Li^+ de-intercalation because of their similar size to Li^+. Phase transitions and structural change occur in $Na_{1-x}CrO_2$ due to the slippage of CrO_2 slabs without the breaking of Cr-O bonds depending on the Na compositions in the following sequence: rhombohedral O3 → monoclinic O′3 → monoclinic P′3, as shown by Chen et al. (2013), in Figure 4.8.

The excellent thermal stability of $NaCrO_2$ and the de-intercalated Na_xCrO_2 ($Na_{0.5}CrO_2$) in non-aqueous solvent between room temperature and 350°C have been shown by various authors using accelerating rate calorimetry (Xia and Dahn 2012; Chen et al. 2013). This occurs because the materials do not lose mass below 400°C, unlike its lithium counterpart $Li_{0.5}CoO_2$, that decomposes to $LiCoO_2$ and Co_3O_4 with associated mass loss (Dahn et al. 1994). In comparison to Li_xFePO_4, which is amongst the most thermodynamically stable electrode material in the LIB system, $Na_{0.5}CrO_2$ is more thermally stable (Komaba et al. 2010). Chromium oxides, such as Cr_2O_5, have also been used as cathodes in rechargeable SIBs (Feng et al. 2016) and delivered specific capacity of up to 310 mAh g⁻¹ at a current density of C/16 (or 20 mA g⁻¹). Table 4.5 summarizes some of the capacities obtained and the operational potential windows.

Other single layered oxides, such as Na_xNiO_2, Na_xTiO_2, and Na_xVO_2 have been successfully used in SIBs as positive electrodes. $NaNiO_2$ crystallizes into a layered monoclinic structure with C2/m symmetry. The Ni-O layer is obtained by edge-sharing of NiO_6 octahedron with Ni-O bonds having the Na ions between these layers, as shown in Figure 4.9a. Braconnier et al. (1982) first studied the Na^+ insertion/de-insertion into $NaNiO_2$ in the potential range of 1.8–3.4 V and found that only about 0.2 Na^+ can be moved in and out of the material, and that a multiple phase transformation (O′3-P′3-P″3-O″3) occurs.

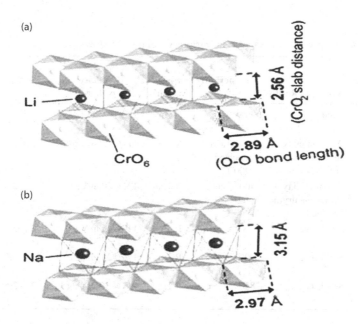

FIGURE 4.7 Crystal structure of (a) LiCrO$_2$ and (b) NaCrO$_2$. (Reprinted from *Electrochem. Commun.*, 12, Komaba, S. et al., Electrochemical Intercalation Activity of Layered NaCrO$_2$ vs. LiCrO$_2$, 355–358. Copyright 2010 with permission from Elsevier.)

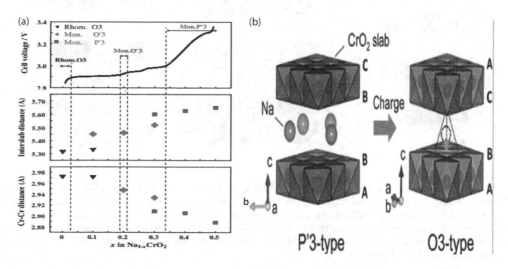

FIGURE 4.8 (a) Phase evolutions of the Na$_{1-x}$CrO$_2$ as a function of Na contents x-value; mean Cr–Cr interatomic distances (bottom), interslab distances (middle), and the galvanostatic charge curve of the NaCrO$_2$/NaFSA-KFSA/Na cell measured at 5 mA g^{-1} at 363 K (top). (Reprinted from *Materials and Research Bulletin*, 17, J. J. Braconnier, C. Delmas and P. Hagenmuller, 993. Copyright 2013 with permission from Elsevier.) (b) Crystal structures of P′3- and O3-type NaCrO$_2$. (Reprinted with permission from Kubota, K. et al., *Electrochemistry*, 82, 909–911. Copyright 2014 American Chemical Society.)

Wang et al. (2017) obtained charge and discharge capacities of up to 191 and 130 mA g^{-1} corresponding, respectively, to 0.81 Na$^+$ extraction, and 0.56 Na$^+$ intercalation has been obtained, while Vassilaras et al. (2013) achieved charge and discharge capacities of 199/147 mA g^{-1} and 121/119 mAh g^{-1} at 2.0–4.5 and 1.25–3.75, respectively. Wang et al. (2017) established that irreversible capacity due to phase transformation and structural distortion is obtained from a NaNiO$_2$ electrode when the material is charged beyond 4.0 V.

TABLE 4.5

Summary of Obtained Capacities for $NaCrO_2$ in Various Potential Windows

Compound	Phase Type	Synthesis Route	Temperature (°C)	Space Group	Specific Capacity (mAh.g^{-1})	Voltage (V/Na/Na$^+$)	References
$NaCrO_2$	O3	Solid state	900	R$\bar{3}$m	110 @C/20	2.5–3.6	Kubota et al. (2015)
$NaCrO_2$	O3	Water-in-oil type emulsion-drying method	400°C and calcined at 900°C	R$\bar{3}$m	111@ 20 mA g^{-1}	2.0–3.6	Yu et al. (2015)
$C-NaCrO_2$	O3	Water-in-oil type emulsion-drying method	400°C and calcined at 900°C	R$\bar{3}$m	121@20 mA g^{-1}	2.0–3.6	Yu et al. (2015)
$NaCrO_2$	O3	Solid state	900	R$\bar{3}$m	110@25 mA g^{-1}	2.0–3.6	Xia and Dahn (2012)
$NaCrO_2$	O3	Solid state	900	R$\bar{3}$m	120@25 mA g^{-1}	2.0–3.6	Komaba et al. (2010)
$NaCrO_2$	O3	Solid state	850	R$\bar{3}$m	113@ 125 mA g^{-1}	2.5–3.5	Chen et al. (2013)

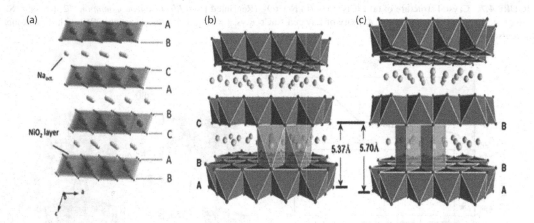

FIGURE 4.9 (a) Schematic structure of the O3-type NaNiO$_2$ crystal structure. (Reprinted from *Nano Energy*, 34, Wang, L., Wang, J., Zhang, X., Ren, Y., Zuo, P., Yin, G., & J. Wang, Unravelling the Origin of Irreversible Capacity Loss in NaNiO$_2$ for High Voltage Sodium Ion Batteries, 215–23. Copyright 2017 with permission from Elsevier.); (b) NaVO$_2$ (O3); and (c) Na$_{0.7}$VO$_2$ (P2) with indication of anionic layers stacking sequences, Na ion and interlayer distance. (Reprinted from *Electrochem. Solid-State Lett.*, 14, Didier, C. et al., Electrochemical Na-Deintercalation from NaVO2, A75–78. Copyright 2011 with permission from Elsevier.)

O3-NaVO$_2$ and P2-Na$_{0.7}$VO$_2$ prepared mainly by solid state and by the chemical reduction of NaVO$_3$ (Didier et al. 2011; Hamani et al. 2011; Guignard et al. 2013) have been used as positive electrodes for SIBs. While the O3-phase NaVO$_2$ crystallizes into a distorted monoclinic structure, the P2-phase Na$_{0.7}$VO$_2$ forms crystals with P63/mmc space group symmetry, as shown in Figure 4.9b and c. Both phases have been shown to reversibly intercalate up to 0.5 Na$^+$ (Chamberland and Porter 1988; Didier et al. 2011; Hamani et al. 2011; Guignard et al. 2013).

4.3.1.2 Multiple Metal Layered Oxides

Binary and multiple metal oxide systems are investigated to reduce the number of phase transitions encountered in single sodium transition metal layered oxide cathodes. Conscientious combination of transition metals to form multiple metal oxides has produced SIB cathodes with high capacity, operating voltage, and capacity retention. Combinations of electrochemically active transition metals such as Ni, Fe, Mn, V, and Co have been actively studied, and we present some of their performances based on highest reversible capacity, voltage profile, and operating voltage range.

Some of the interesting multi-layered oxides are those that contain Ni because the electrochemical performances are mostly based on the $Ni^{2+/3+/4+}$ redox reaction that offers a comparatively high operating voltage. Nevertheless, the electrochemical cycling of the single oxide Ni compounds undergoes phase transitions shown by the multiple plateaus they exhibit. Ivanova et al. (2016) studied the Li^+ insertion into P3-$Na_{3/4}Co_{1/3}Ni_{1/3}Mn_{1/3}O_2$ between 4.4 V and 1.8 V, and the oxide delivered a specific capacity of about 66 mAh g^{-1}, which corresponds to an intercalation of about 1/5 mol of Li^+ during discharging. On charging, a higher capacity (165 mAh g^{-1}) was obtained, which is an indication that both Li and Na ions are extracted during the charge. Kim et al. (2012) tested a SIB using a layered $Na[Ni_{1/3}Fe_{1/3}Mn_{1/3}]O_2$ cathode with a carbon-based anode in an ester carbonate electrolyte. The cell gave an average voltage of about 2.75 V (1.5–4.0 V), a moderate capacity of 100 mAh g^{-1} for 150 cycles at a 0.5°C-rate, and a capacity of 94 mAh g^{-1} at a 1°C-rate. Other Ni containing layered oxides, such as O3-phase $NaNi_{0.5}Mn_{0.5}O_2$ (Komaba et al. 2012) and P2-phase $Na_{2/3}Ni_{1/3}Mn_{2/3}O_2$ (Wang et al. 2013), have been studied. While the O3-phase $NaNi_{0.5}Mn_{0.5}O_2$ gave multiple plateau behaviour due to phase transitions, the P2-phase $Na_{2/3}Ni_{1/3}Mn_{2/3}O_2$ showed a smooth charge-discharge curve and cycling performance that greatly depended on the voltage window, as shown in Figure 4.10.

A P2-$Na_{0.67}Mn_{0.65}Ni_{0.2}Co_{0.15}O_2$ with a high performance and wide voltage range of 1.5–4.2 V was studied by Li et al. (2015). Cycling the cell at 0.05°C and 0.5°C-rates gave a capacity retention of about 85% and 78%, respectively, after 100 cycles. High capacities of up to 117, 93, and 70 mAh g^{-1} were also obtained at high current densities of 480 (2°C), 1200 (5°C), and 1920 mA g^{-1} (8°C). Biphasic (P3/P2-type) layered $Na_{0.66}Co_{0.5}Mn_{0.5}O_2$ that integrated the P2 phase into the P3-layered phase was prepared by Chen et al. (2015) in order to enhance the stability of the cathode materials. The material was very stable after 100 cycles in a high-voltage range (1.5–4.3 V), giving a discharge capacity of up to 126.6 mAh g^{-1} at 5°C-rates.

Other binary and ternary combinations O3-type phases, such as $Na_xCo_{1/2}Fe_{1/2}O_2$ (Yoshida et al. 2013), $Na_xFe_{1/2}Mn_{1/2}O_2$ (Chen et al. 2015), $Na_xNi_{1/3}Co_{1/3}Fe_{1/3}O_2$ (Vassilaras et al. 2014), $Na[Ni_{0.4}Fe_{0.2}Mn_{0.4-x}Ti_x]O_2$ (Sun et al. 2014), and $Na(Mn_{0.25}Fe_{0.25}Co_{0.25}Ni_{0.25})O_2$ (Li et al. 2014) have been studied, and they show various levels of structural stability and electrochemical performance. While $Na_xCo_{1/2}Fe_{1/2}O_2$, $Na_xNi_{1/3}Co_{1/3}Fe_{1/3}O_2$, and $Na[Ni_{0.4}Fe_{0.2}Mn_{0.4-x}Ti_x]O_2$ showed very good performance in terms of discharge capacities and cycle stability, $Na_xFe_{1/2}Mn_{1/2}O_2$ exhibited large hysteresis and poor cycle stability. $Na(Mn_{0.25}Fe_{0.25}Co_{0.25}Ni_{0.25})O_2$ showed a reversible structure evolution of O3-P3-O3′-O3″ on Na^+ de-intercalation.

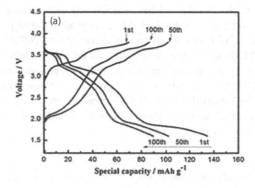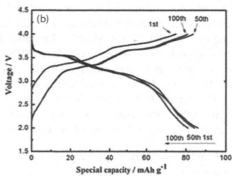

FIGURE 4.10 The plot of voltage profile vs obtained capacities of P2-$Na_{2/3}Ni_{1/3}Mn_{2/3}O_2$ cathode cycled at 0.1°C between (a) 1.6 V and 3.8 V and (b) 2.0 V and 4.0 V. (Reprinted from *Electrochem. Acta*, 113, Wang, H. et al., Electrochemical properties of P2-$Na_{2/3}$ [$Ni_{1/3}Mn_{2/3}]O_2$ cathode material for sodium ion batteries when cycled in different voltage ranges, 200–204. Copyright 2017 with permission from Elsevier.)

Talaie and his group (Talaie et al. 2015; Duffort et al. 2015) studied the phase transitions and the effect of Ni substitution on P2-$Na_{0.67}[Ni_yMn_{0.5+y}Fe_{0.5-2y}]O_2$ (y = 0, 0.10, 0.15). The results showed that P2-$Na_{0.67}Fe_{0.2}Mn_{0.65}Ni_{0.15}O_2$ is the optimal cathode material, delivering more capacity and energy. In addition, the Ni-substituted materials are less prone to charge/discharge polarization.

4.3.2 Phosphates Polyanions

Phosphates polyanion materials have gained increasing interest as electrodes for SIBs because of their very good electrochemical storage performance, as well as their structural stability (Masquelier and Croguennec 2013). The P-O bonds in the phosphates frameworks are very stable, reducing the likelihood of oxygen liberation, while ensuring the safety and long term stability of SIBs. Secondly, the interstices in the phosphate polyanion framework are spacious, hence, volumetric and phase transitions are less. The powerful inductive effect of the phosphate polyanion (PO_4^{3-}) affects the redox couple, giving rise to higher operating potential values vs Na/Na^+ (Fang et al. 2017). Phosphate framework materials used in SIBs can be classified into single phosphates, pyrophosphates, and mixed phosphates.

4.3.2.1 Single Phosphates

Single-phosphate materials, such as $LiFePO_4$, are the first to be investigated and commercialized, and this aroused interest in its sodium counterpart $NaFePO_4$ to be actively investigated for SIBs. Single phosphates polyanion materials can be further classified into two major groups: olivine-structured and NASICON-structured.

4.3.2.1.1 Olivine Phosphates AMPO$_4$ (M = Fe, Mn, Co, Ni)

The olivine structures with a *Pnma* space group have a vertex-sharing MO_6 octahedron and PO_4 tetrahedron that share one edge and all vertices with MO_6 octahedra (Ong et al. 2011). Olivine $NaFePO_4$ has gained a lot of interest as a potential cathode material for SIBs due to its high theoretical specific capacity (154 mAh g^{-1}), as well as good operational voltage ($\approx$2.8 V). However, solid state synthesis does not produce pure olivine-phased $NaFePO_4$, otherwise known as the triphylite phase, but a maricite phase is thermodynamically favoured. Studies have shown that the maricite phase has poor electrochemical activity (Avdeev et al. 2013). This is because it's one-dimensional, FeO_6 octahedra share edges, while the Na ions are located at isolated tetrahedral sites, hence, creating barriers for Na-ion migration, as shown in Figure 4.11. However, the triphylite $NaFePO_4$, which is mostly obtained by ion exchange reactions using $LiFePO_4$, has shown good electrochemical performance (Moreau et al. 2010; Oh et al. 2012; Fang et al. 2015).

Zhu et al. (2018) reported a sodium iron phosphate ($Na_{0.71}Fe_{1.07}PO_4$) as a good potential cathode material for SIBs. The phosphate exhibited a very small volume change (~ 1%) during de-sodiation and yielded a capacity of about 78 mA·h·g^{-1} when cycled at 50°C. It also showed high stability by maintaining its specific capacity even after 5000 cycles at 20°C. The surface of olivine $NaFePO_4$ was modified with polythiophene by Ali et al. (2016), while Li, Liu, and Guo (2014) developed an olivine composite $NaFePO_4$/graphene composite for positive electrodes in SIBs. All these modifications enhanced the electrochemical performance of the phosphate material for SIBs. Studies have shown that olivine $NaFePO_4$ gives an intermediate phase ($Na_{0.7}FePO_4$) around 2.95 V during charging in contrast to its lithium-based counterpart $LiFePO_4$ (Pan et al. 2013). $FePO_4$ (Walczak et al. 2018) obtained from de-lithiation of $LiFePO_4$ as well as amorphous $FePO_4$ (Shiratsuchi et al. 2006; Zhao et al. 2012) and its carbon composites (Chen et al. 2012; Moradi et al. 2015) have shown good electrochemical performance for SIBs.

4.3.2.1.2 NASICON Structured Phosphates

The NASICON (Natrium Super Ionic Conductor) family of compounds was originally proposed by Hong and Goodenough as Na^+ ion solid electrolytes (Goodenough et al. 1976) because of the high Na^+ ionic conductivity and has been extensively studied as both Li-ion and SIB cathodes (Delmas et al. 1988). They can be represented generally as $XM1M2(PO_4)_3$. The position of "X" in the crystal structure can be occupied by alkali ions such as Li^+, Na^+, K^+, Rb^+, and Cs^+ and alkaline earth ions such as Mg^{2+}, Ca^{2+},

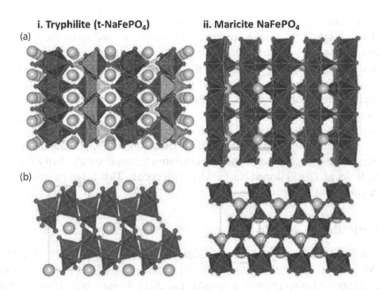

FIGURE 4.11 Schematic presentation of orthorhombic structured triphylite $NaFePO_4$ (left) and maricite $NaFePO_4$ (right) polymorphs (a and b). FeO_6 octahedra, PO_4 tetrahedra, and Na atoms are depicted. (Reprinted with permission from Avdeev, M. et al., *Inorg. Chem.*, 52, 8685–8693, 2013. Copyright 2013 American Chemical Society.)

Sr^{2+}, and Ba^{2+}. It can also be occupied by other ions such as H^+, H_3O^+, NH_4^+, Cu^+, Cu^{2+}, Ag^+, etc., and it can be vacant as well. The M1 and M2 site positions are taken by divalent ions such as Zn^{2+}, Cd^{2+}, Ni^{2+}, Mn^{2+}, and Co^{2+}, trivalent ions such as Fe^{3+}, Sc^{3+}, Ti^{3+}, V^{3+}, Cr^{3+}, Al^{3+}, In^{3+}, Ga^{3+}, Y^{3+}, and Lu^{3+}, tetravalent ions (Ti^{4+}, Zr^{4+}, Hf^{4+}, Sn^{4+}, Si^{4+}, Ge^{4+}), as well as pentavalent ions (V^{5+}, Nb^{5+}, Ta^{5+}, Sb^{5+}, As^{5+}) of the transition metals to balance the charge neutrality (Anantharamulu et al. 2011). All the compositions almost maintain a rhombohedral crystal structure with space group *R3-c* which consists of three-dimensional frameworks of $M1O_6$ and $M2O_6$ octahedra sharing corners with PO_4 tetrahedra, as shown in Figure 4.12. These frameworks form large channels and numerous voids, which can accommodate alkali cations. Aragón et al. (2017) showed these large channels in their vanadium superstoichiometric NASICON phosphate.

4.3.2.1.3 Other Phosphates

Other phosphates, such as pyrophosphate represented as $Na_2MP_2O_7$ (M = Fe, Mn, Co) (Niu et al. 2019) and sodium fluorophosphates, generally represented as $AMPO_4F$ (where A is an alkali metal and

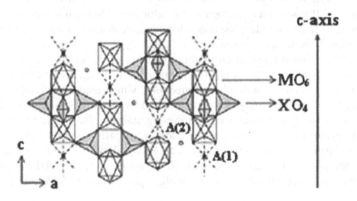

FIGURE 4.12 The structure of NASICON showing two types of sites: A1 and A2. (With kind permission from Springer Science + Business Media: *J. Mater. Sci.*, Wide-Ranging Review on Nasicon Type Materials, 46, 2011, 2821–2837, Anantharamulu, N.)

M is a 3d transition metal), are potential cathode materials for SIBs (Dacek et al. 2016). There are different polymorphs of $Na_2MP_2O_7$, such as triclinic, orthorhombic, and tetragonal structures. Amongst these, studies have shown that the triclinic phase is the most thermodynamically stable. The structure of the triclinic phase of $Na_2MP_2O_7$ consists of corner- and edge-sharing Fe_2O_{11} dimers, as well as bridged P_2O_7 groups creating a three-dimensional migration path for Na. Barpanda et al. showed that $Na_2FeP_2O_7$ has high thermal stability and is therefore a safe cathode yielding a discharge capacity of up to 90 mAh g^{-1} (Barpanda et al. 2013).

Some of the sodium fluorophosphates that have been studied include $Na_xV_2(PO_4)_2F_3$ (Dacek et al. 2016), Na_2FePO_4F (Sharma et al. 2018), Na_2MnPO_4F (Recham et al. 2009), and Na_2CoPO_4F (Kubota et al. 2014), amongst others. To overcome the low inherent electronic conductivity that exists in sodium fluorophosphates, Jin et al. (2019) doped Na_2FePO_4F with cobalt. This helped to enhance the conductivity for practical applications.

4.3.3 Other Cathodes Materials

Prussian blue analogue, generally represented chemically with $A_xM[Fe(CN)_6]_{1-y}B_y \cdot mH_2O$ (where A is alkaline metal; M is transition metal; B is $[Fe(CN)_6]$ vacancies; $0 \leq x \leq 2$, $y < 1$) (Qian et al. 2018; Wessells et al. 2012), as well as sulphates (Barpanda et al. 2014; Jungers et al. 2019) have been studied as cathodes in SIBs. Prussian blue is a type of metal organic framework structure that easily gets inserted/de-inserted by various ions, such as lithium, sodium, potassium, and ammonium, in an aqueous environment. Prussian blue analogues have a perovskite type of crystal structure with large interstitial sites, that can host and transport alkaline metal ions. They have been actively studied in recent times for SIBs (Wheeler et al. 2019; Xie et al. 2019), giving excellent electrochemical properties due to their unique open-framework crystal structure.

The alluaudite-type $Na_2Fe_2(SO_4)_3$ (*C2/c* space group) (Wessells et al. 2012) and Kröhnkite-type $Na_2Fe(SO_4)_2 \cdot 2H_2O$ (space group = P21/c, No. 14) (Jungers et al. 2019) of the sulphates have shown appreciable electrochemical properties for SIBs.

4.4 Conclusion and Perspective

SIBs have emerged as a key alternative to LIBs based on their inherent properties that are similar to a LIB system. It is, however, challenged by the heavier Na$^+$, which lead to slower migration of the ions in materials and consequent lower capacity. Most research efforts are focused on engineering cathode materials that can enhance easy Na$^+$ intercalation/de-intercalation in electrodes. In this chapter, we have reviewed some of the successful cathode materials synthesized and used in SIBs. These ranged from layered oxides (single and multi-layered), phosphate polyanions (olivine, NASICON, fluorophosphates, and pyrophosphates) to Prussian blue analogues and sulphates. Although a SIB consists of a cathode, anode, and the electrolyte, the cathode is the major driving force that determines the overall performance of the SIB and is sometimes the most expensive.

The layered oxides are used more often than not because of the easy accessibility of the ions in the layered structure resulting in good electrochemical properties. However, some of the layered oxides are challenged with irreversible phase transitions, which lead to structural degradation with long cycling. The phosphates polyanion-based compounds, especially sodium vanadium fluorophosphates exhibit high operating potential and high cycling stability. However, due to the low intrinsic conductivity, their performance depends on the addition of carbonaceous materials. Prussian blue analogues with the large open channel structure can easily allow Na$^+$ to intercalate/de-intercalate, and as such, are good potential cathodes for SIBs. Nonetheless, their low intrinsic gravimetric densities make them unsuitable as cathode materials for high specific energy SIBs for portable and mobile applications.

Therefore, none of the cathode material will meet all the requirements of a battery system, which ranges from high energy and power density, long cycling stability, safety to cost effectiveness. Hence, the performance of the cathode materials can be improved by doping to increase conductivity and ion migration speed, using nano-composites to improve the structure stability, as well as reduce the ion migration path. These will lead to enhanced electrochemical performance.

ACKNOWLEDGEMENTS

The authors are grateful for the financial support for this work by UNESCO-UNISA Africa Chair in Nanosciences and Nanotechnology, College of Graduate Studies, University of South Africa (UNISA) for the award of Post-Doctoral Fellowship ACN (90406558) and VRSP Fellowship FIE (90407830).

REFERENCES

Ali, G., J. H. Lee, D. Susanto, S. W. Cho, B. W. Cho, K. W. Nam, and K. Y. Chung. 2016. "Polythiophene-Wrapped Olivine $NaFePO_4$ as a Cathode for Na-Ion Batteries." *ACS Applied Materials & Interfaces* 8 (24): 15422–29.

Anantharamulu, N., K. K. Rao, G. Rambabu, B. V. Kumar, V. Radha, and M. Vithal. 2011. "A Wide-Ranging Review on Nasicon Type Materials." *Journal of Materials Science* 46: 2821–37.

Aragón, M. J., P. Lavela, G. F. Ortiz, R. Alcántara, and J. L. Tirado. 2017. "Insight into the Electrochemical Sodium Insertion of Vanadium Superstoichiometric NASICON Phosphate." *Inorganic Chemistry*, 56 (19): 11845–53.

Avdeev, M., Z. Mohamed, C. D. Ling, J. Lu, M. Tamaru, A. Yamada, and P. Barpanda. 2013. "Magnetic Structures of $NaFePO_4$ Maricite and Triphylite Polymorphs for Sodium-Ion Batteries." *Inorganic Chemistry* 52: 8685–93.

Barpanda, P., G. Liu, C. D. Ling, M. Tamaru, M. Avdeev, S.-C.Chung, Y. Yamada, and A. Yamada. 2013. "$Na_2FeP_2O_7$: A Safe Cathode for Rechargeable Sodium-Ion Batteries." *Chemistry of Materials*, 2517: 3480–87.

Barpanda, P., G. Oyama, C. D. Ling, and A. Yamada. 2014. "Kröhnkite-type $Na_2Fe(SO_4)_2 \cdot 2H_2O$ as a Novel 3.25 V Insertion Compound for Na-Ion Batteries." *Chemistry of Materials* 26: 1297–99.

Berthelot, R., D. Carlier, and C. Delmas. 2010. "Electrochemical Investigation of the P2–NaxCoO$_2$ Phase Diagram." *Nature Materials*, 10 (1): 74–80.

Braconnier, J. J., C. Delmas, and P. Hagenmuller. 1982. "Etude par desintercalation electrochimique des systemes Na_xCrO_2 et Na_xNiO_2 des systems Na_xCrO_2 et Na_xNiO_2." *Materials Research Bulletin*, 17: 993.

Brazil. R. Oct 2014. "A Salt Ion Battery." *Chemistry and Industry*, 78(10): 24–27.

Breeze, P. 2014. "Energy Storage." In *Power Generation Technologies*, 2nd ed., Vol. 78, 24–27. Wiley Online Library.

Caballero, A., L. Hernan, J. Morales, L. Sánchez, J. Santos Pena, and M. A. G. Aranda. 2002. "Synthesis and Characterization of High-Temperature Hexagonal P2-$Na_{0.6}MnO_2$ and Its Electrochemical Behaviour as Cathode in Sodium Cells." *Journal of Materials Chemistry*, 12 (4): 1142–47.

Cao, Y., L. Xiao, W. Wang, D. Choi, Z. Nie, J. Yu, L. V. Saraf, Z. Yang, and J. Liu. 2011. "Reversible Sodium Ion Insertion in Single Crystalline Manganese Oxide Nanowires with Long Cycle Life." *Advanced Materials*, 23(28): 3155–60.

Carlier, D., M. Blangero, M. Menetrier, M. Pollet, J.-P. Doumerc, and C. Delmas. 2009. "Sodium Ion Mobility in Na_xCoO_2 (0.6 < x < 0.75) Cobaltites Studied by ^{23}Na MAS NMR." *Inorganic Chemistry*, 48: 7018–25.

Chamberland, B. L., and S. K. Porter. 1988. "A Study on the Preparation and Physical Property Determination of $NaVO_2$." *Journal of Solid State Chemistry*, 73 (2): 398–404.

Chen, C.-Y., K. Matsumoto, T. Nohira, R. Hagiwara, A. Fukunaga, S. Sakai, K. Nitta, and S. Inazawa. 2013. "Electrochemical and Structural Investigation of $NaCrO_2$ as a Positive Electrode for Sodium Secondary Battery Using Inorganic Ionic Liquid NaFSA-KFSA." *Journal of Power Sources*, 237: 52–57.

Chen, H., G. Hautier, and G. Ceder. 2012. "Synthesis, Computed Stability, and Crystal Structure of a New Family of Inorganic Compounds: Carbonophosphates." *Journal of the American Chemical Society* 13448: 19619–27.

Chen, X., X. Zhou, M. Hu, J. Liang, D. Wu, J. Wei, and Z. Zhou. 2015. "Stable Layered P3/P2 $Na_{0.66}Co_{0.5}Mn_{0.5}O_2$ Cathode Materials for Sodium-Ion Batteries." *Journal of Materials Chemistry A*, 3: 20708–714.

Dacek, S. T., W. D. Richards, D. A. Kitchaev, and G. Ceder. 2016. "Structure and Dynamics of Fluorophosphate Na-Ion Battery Cathodes." *Chemistry of Materials*, 2815: 5450–60.

Dahn, J. R., E. W. Fuller, M. Obrovac, and U. von Sacken. 1994. "Thermal Stability of $LixCoO_2$, $LixNiO_2$ and λ-MnO_2 and Consequences for the Safety of Li-Ion Cells." *Solid State Ionics*, 69: 565.

Das, B. K., and R. Goplan. 2019. "Intercalation-Based Layered Materials for Rechargeable Sodium-ion Batteries." In *Layered Materials for Energy Storage and Conversion*, edited by D. Geng, Y. Cheng, and G. Zhang, 71–94. Royal Society of Chemistry.

Delmas, C., A. Nadiri, and J. L. Soubeyroux. 1988. "The Nasicon-Type Titanium Phosphates $Ati_2(PO_4)_3$ (A=Li, Na) as Electrode Materials." *Solid State Ionics*, 28–30 (1): 419–423.

Delmas, C., J. J. Braconnier, C. Fouassier, et al. 1981. "Electrochemical Intercalation of Sodium in Na_xCoO_2 Bronzes." *Solid State Ionics*, 3–4: 165–69.

Didier, C., M. Guignard, Z. C. Denage, O. Szajwaj, S. Ito, A. I. Saadoune, B. J. Darriet, and C. Delmas. 2011. "Electrochemical Na-Deintercalation from $NaVO_2$." *Electrochemical and Solid-State Letters*, 14 (5): A75–78.

Ding, J. J., Y. N. Zhou, Q. Sun, X. Q. Yu, X. Q. Yang, and Z. W. Fu. 2013. "Electrochemical Properties of P2-phase $Na_{0.74}CoO_2$ Compounds as Cathode Material for Rechargeable Sodium-Ion Batteries." *Electrochimica Acta*, 87 (1): 388–93.

Duffort, V., E. Talaie, R. Black, and L. F. Nazar. 2015. "Uptake of CO_2 in Layered P2-$Na_{0.67}$ $Mn_{0.5}Fe_{0.5}O_2$: Insertion of Carbonate Anions." *Chemistry of Materials* 27: 2515–24.

Ellis, B. L., and L. F. Nazar. 2012. "Sodium and Sodium-Ion Energy Storage Batteries." *Current Opinion in Solid State and Materials Science* 16 168–77.

Fang, Y., J. Zhang, L. Xiao, X. Ai, Y. Cao, and H. Yang. 2017. "Phosphate Framework Electrode Materials for Sodium Ion Batteries." *Advanced Science*: 1600392

Fang, Y., Q. Liu, L. Xiao, X. Ai, H. Yang, and Y. Cao. 2015. "High-Performance Olivine $NaFePO_4$ Microsphere Cathode Synthesized by Aqueous Electrochemical Displacement Method for Sodium Ion Batteries." *ACS Applied Materials & Interfaces*, 7: 17977.

Feng, X.-Y., P.-H. Chien, A. M. Rose, J. Zheng, I. Hung, Z. Gan, and Y.-Y. Hu. 2016. "Cr_2O_5 as New Cathode for Rechargeable Sodium Ion Batteries." *Journal of Solid State Chemistry*, 242: 96–101.

Ferrara, C., C. Tealdi, V. Dall'Asta, D. Buchholz, L. G. Chagas, E. Quartarone, V. Berbenni, and S. Passerini. 2018. "High-Performance Na0.44MnO2 Slabs for Sodium-Ion Batteries Obtained through Urea-Based Solution Combustion Synthesis." *Batteries* 4: 8; doi:10.3390/batteries4010008.

Fleischer, M. 1963. "Recent Estimates of the Abundances of the Elements in the Earth's Crust." United States Department of the Interior, *Geological Survey Circular* 285: 3.

Goodenough, J. B., H. Y-P. Hong, and J. A. Kafalas. 1976. "Fast Na^+-Ion Transport in Skeleton Structures." Materials Research Bulletin, 11 (2): 203–20.

Grey, I. E., R. J. H., and A. W. Hewat. 1990. "A Neutron Powder diffraction Study of the β to γ Phase Transformation in $NaFeO_2$." *Zeitschrift Für Kristallographie*, 193(1/2): 51–69.

Guignard, M., C. Didier, J. Darriet, P. Bordet, E. Elkaïm, and C. Delmas. 2013. "P2-Na_xVO_2 System as Electrodes for Batteries and Electron-Correlated Materials." *Nature Materials* 12: 74–80.

Guo, S., Y. Sun, J. Yi, K. Zhu, P. Liu, Y. Zhu, G. Zhu, M. Chen, M. Ishida, and H. Zho. 2016. "Understanding Sodium-ion Diffusion in Layered P2 and P3 Oxides via Experiments and First-Principles Calculations: A Bridge Between Crystal Structure and Electrochemical Performance." *NPG Asia Materials* 8(2016): 266.

Hamani, D., M. Ati, J.-M. Tarascon, and P. Rozier. 2011. "Na*x*VO2 as Possible Electrode for Na-Ion Batteries." *Electrochemistry Communications*, 13: 938–41.

Han, D. W., J. H. Ku, R. H. Kim, D. J. Yun, S. S. Lee, and S. G. Doo. 2014. "Aluminium Manganese Oxides with Mixed Crystal Structure: High-Energy-Density Cathodes for Rechargeable Sodium Batteries." *ChemSusChem*, 7 (7): 1870–75.

Hwang, S., Y. Lee, E. Jo, K. Y. Chung, W. Cho, S. M. Kim, and W. Chang. 2017. "Investigation of Thermal Stability of P2-Na_xCoO_2 Cathode Materials for Sodium Ion Batteries." *ACS Applied Material Interfaces*, 922: 18883–88.

Ivanova, S., E. Zhecheva, R. Kukeva, D. Nihtianova, L. Mihaylov, G. Atanasova, and R. Stoyanova. 2016. "Layered P3-$Na_{3/4}$ $Co_{1/3}Ni_{1/3}Mn_{1/3}O_2$ versus Spinel $Li_4Ti_5O_{12}$ as a Positive and a Negative Electrode in a Full Sodium–Lithium Cell." *ACS Applied Material Interfaces*, 8: 17321–33.

Jin, D., H. Qiu, F. Du, Y. Wei, and X. Meng. 2019. "Co-Doped Na_2FePO_4F Fluorophosphates as a Promising Cathode Material for Rechargeable Sodium-Ion Batteries." *Solid State Sciences*, 93: 62–69.

Jungers, T., A. Mahmoud, C. Malherbe, F. Boschinia, and B. Vertruyena. 2019. "Sodium Iron Sulfate Alluaudite Solid Solution for Na-Ion Batteries: Moving Towards Stoichiometric $Na_2Fe_2(SO_4)_3$." *Journal of Materials Chemistry A*, 7: 8226–33.

Kang, S. M., J.-H. Park, A. Jin, Y. H. Jung, J. Mun, and Y.-E. Sung. 2018. "Na+ /vacancy Disordered P2-$Na_{0.67}Co_{1-x}Ti_xO_2$: High Energy and High Power Cathode Materials for Sodium Ion Batteries." *ACS Applied Material Interfaces*, 10 (4): 3562–70.

Kataoka, R., K. Kuratani, M. Kittta, N. Takeichi, T. Kiyobayashi, and M. Tabuchi. 2015. "Influence of the Preparation Methods on the Electrochemical Properties and Structural Changes of Alpha-Sodium Iron Oxide as a Positive Electrode Material for Rechargeable Sodium Batteries." *Electrochimica Acta*, 182: 871–877.

Kikkawa, S., S. Miyazaki, and M. Koizumi. 1985. "Sodium Deintercalation from α-$NaFeO_2$." *Materials Research Bulletin*, 20: 373–77.

Kim, D., E. Lee, M. Slater, W. Lu, S. Rood, and C. S. Johnson. 2012. "Layered $Na[Ni_{1/3}Fe_{1/3}Mn_{1/3}]O_2$ Cathodes for Na-Ion Battery Application." *Electrochemistry Communications*, 18: 66–69.

Komaba, S., C. Takei, T. Nakayama, A. Ogata, and N. Yabuuchi. 2010. "Electrochemical Intercalation Activity of Layered $NaCrO_2$ vs. $LiCrO_2$." *Electrochemistry Communications* 12: 355–58.

Komaba, S., N. Yabuuchi, T. Nakayama, A. Ogata, T. Ishikawa, and I. Nakai. 2012. "Study on the Reversible Electrode Reaction of $Na1-xNi0.5Mn0.5O_2$ for a Rechargeable Sodium-Ion Battery." *Inorganic Chemistry*, 51: 6211–20.

Konarova, A., J. U. Choia, Z. Bakenov, and S.-T. Myung. 2018. "Revisit of Layered Sodium Manganese Oxide: Achievement of High Energy by Ni Incorporation." *Journal of Materials Chemistry A*, 6 (18): 8558–67.

Kubota, K., I. Ikeuchi, T. Nakayama, C. Takei, N. Yabuuchi, H. Shiiba, M. Nakayama, and S. Komaba. 2015. "New Insight into Structural Evolution in Layered $NaCrO_2$ during Electrochemical Sodium Extraction." *Journal of Physical Chemisty C*, 119 (1):166–75.

Kubota, K., K. Yokoh, N. Yabuuchi, and S. Komaba. 2014. "Na_2CoPO_4F as a High-voltage Electrode Material for Na-Ion Batteries." *Electrochemistry*, 82 (10): 909–11.

Kuratani, K., K. Tatsumi, and N. Kuriyama. 2007. "Manganese Oxide Nanorod with 2 × 4 Tunnel Structure: Synthesis and Electrochemical Properties." *Crystal Growth & Design*, 7 (8): 1375–1377.

Lee, E., D. E. Brown, E. E. Alp, Y. Ren, J. Lu, J.-J. Woo, and C. S. Johnson. 2015. "New Insights into the Performance Degradation of Fe-Based Layered Oxides in Sodium-Ion Batteries: Instability of Fe^{3+}/Fe^{4+} Redox in α-$NaFeO_2$." *Chemistry of Materials*, 27(19): 6755–64.

Lei, Y., X. Li, L. Liu, and G. Ceder. 2014. "Synthesis and Stoichiometry of Different Layered Sodium Cobalt Oxides." *Chemical Materials* 26: 5288–96.

Li, D., H. K. Liu, and Z. Guo. 2014. "Synthesis and Electrochemical Performance of Olivine $NaFePO_4$/ Grahene Composite as Cathode Materials for Sodium Ion Batteries." *17th ECS International Meeting on Lithium Batteries* MA2014-04 243

Li, X., D. Wu, Y.-N. Zhou, L. Liu, X.-Q. Yang, and G. Cedar. 2014. "O3-Type $Na(Mn_{0.25}Fe_{0.25}Co_{0.25}Ni_{0.25})O_2$: A Quaternary Layered Cathode Compound for Rechargeable Na Ion Batteries." *Electrochemistry Communications* 49: 51–54.

Li, Z.-Y., R. Gao, L. Sun, Z. Hu, and X. Liu. 2015. "Designing an Advanced P2-$Na_{0.67}Mn_{0.65}Ni_{0.2}Co_{0.15}O_2$ Layered Cathode Material for Na-Ion Batteries." *Journal of Materials Chemistry A*, 3: 16272–78.

Liang, F., X. Qiu, Q. Zhang, Y. Kang, A. Koo, K. Hayashi, K. Chen, D. Xue, K. N. Hui, H. Yadegari, and X. Sun. 2018. "A Liquid Anode for Rechargeable Sodium-Air Batteries with Low Voltage Gap and High Safety." *Nano Energy*, 49(2018): 574–79.

Liu, Q., Z. Hu, M. Chen, Q. Gu, Y. Dou, Z. Sun, S. Chou, S. Dou, and S. Xue. 2017. "Multiangular Rod-Shaped $Na_{0.44}MnO_2$ as Cathode Materials with High Rate and Long Life for Sodium-Ion Batteries." *ACS Applied Materials and Interfaces*, 9 (4): 3644–52.

Liu, S., S. Liu, and J. Luo. 2016. "Carbon-Based Cathodes for Sodium-Air Batteries." *New Carbon Materials*, 31(3): 264–70.

Ma, G., Y. Zhao, K. Huang, Z. Ju, C. Liu, Y. Hou, and Z. Xing. 2016. "Effects of the Starting Materials of $Na_{0.44}MnO_2$ Cathode Materials on Their Electrochemical Properties for Na-Ion Batteries." *Electrochimica Acta*, 222: 36–43.

Ma, T., G.-L. Xu, X. Zeng, Y. Li, Y. Ren, C. Sun, S. M. Heald, J. Jorne, K. Amine, and Z. Chen. 2017. "Solid State Synthesis of Layered Sodium Manganese Oxide for Sodium-Ion Battery by In-Situ High Energy X-Ray Diffraction and X-Ray Absorption Near Edge Spectroscopy." *Journal of Power Sources*, 341: 114–21.

Ma, T., G.-L. Xu, Y. Li, B. Song, X. Zeng, C.-C. Su, W. L. Mattis, F. Guo, Y. Ren, R. Kou, C. Sun, S. M. Heald, H.-H. Wang, R. Shahbazian-Yassar, J. Jorne, Z. Chen, and K. Amine. 2018. "Insights into the Performance Degradation of O-Type Manganese-Rich Layered Oxide Cathodes for High-Voltage Sodium-Ion Batteries." *ACS Applied Energy Materials*, 1 (10): 5735–45.

Ma, X., H. Chen, and G. Ceder. 2011. "Electrochemical Properties of Monoclinic NaMnO$_2$." *Journal of Electrochemical Society*, 158(12): A1307–12.

Masquelier, C., and L. Croguennec. 2013. "Polyanionic (Phosphates, Silicates, Sulfates) Frameworks as Electrode Materials for Rechargeable Li (or Na) Batteries." *Chemical Reviews* 1138: 6552–91.

Matsui, M., F. Mizukoshi, and N. Imanishi. 2015. "Improved Cycling Performance of P2-Type Layered Sodium Cobalt Oxide by Calcium Substitution." *Journal of Power Sources*, 280: 205–209.

Mendiboure, A., C. Delmas, and P. Hagenmuller. 1985. "Electrochemical Intercalation and Deintercalation of Na$_x$MnO$_z$ Bronzes." *Journal of Solid State Chemistry*, 57 (1985): 323–331.

Moradi, M., Z. Li, J. Qi, W. Xing, K. Xiang, Y. M. Chiang, and A. M. Belcher. 2015. "Improving the Capacity of Sodium Ion Battery Using a Virus-Templated Nanostructured Composite Cathode." *Nano Letters* 15 (5): 2917–21.

Moreau, P., D. Guyomard, J. Gaubicher, and F. Boucher. 2010. "Structure and Stability of Sodium Intercalated Phases in Olivine FePO$_4$." *Chemistry of Materials* 22: 4126–28.

Niu, Y., Y. Zhang, and M. Xu. 2019. "A Review on Pyrophosphate Framework Cathode Materials for Sodium-Ion Batteries." *Journal of Materials Chemistry A*, 7: 15006–25.

Oh, S.-M., S.-T. Myung, J. Hassoun, B. Scrosati, and Y.-K. Sun. 2012. "Reversible NaFePO$_4$ Electrode for Sodium Secondary Batteries." *Electrochemistry Communications* 22: 149–52.

Ong, S. P., V. L. Chevrier, G. Hautier, A. Jain, C. Moore, S. Kim, X. Ma, and G. Cede. 2011. "Voltage, Stability and Diffusion Barrier Differences Between Sodium-Ion and Lithium-Ion Intercalation." *Materials, Energy and Environmental Science*, 4: 3680

Ono, Y., R. Ishikawa, Y. Miyazaki, Y. Ishii, Y. Morii, and T. Kajitani. 2002. "Crystal Structure, Electric and Magnetic Properties of Layered Cobaltite β-Na$_x$CoO$_2$." *Journal of Solid State Chemistry*, 166: 177, 181.

Pan, H., Y.-S. Hu, and L. Chen. 2013. "Room-Temperature Stationary Sodium-Ion Batteries for Large-Scale Electric Energy Storage." *Energy & Environmental Science*, 6(8): 2338.

Parant, J.-P., R. Olazcuaga, M. Devalette, C. Fouassier, and P. Hagenmuller. 1971. "Sur quelques nouvelles phases de formule Na$_x$MnO$_2$ (x ⩽ 1)." *Journal of Solid State Chemistry*, 3(1):1–11.

Paulsen, J. M., R. A. Donaberger, and J. R. Dahn. 2000. "Layered T2-, O6-, O2-, and P2-Type A$_{2/3}$[M'$^{2+}_{1/3}$M$^{4+}_{2/3}$] O$_2$ Bronzes, A = Li, Na; M'= Ni, Mg; M = Mn, Ti." *Chemical Materials*, 12 (8): 2257–67.

Qian, J., C. Wu, Y. Cao, Z. Ma, Y. Huang, X. Ai, and H. Yang. 2018. "Prussian Blue Cathode Materials for Sodium-Ion Batteries and Other Ion Batteries." *Advanced Energy Materials*, 8 (17): 1702619.

Radin, M. D., J. Alvarado, Y. S. Meng, and A. Van der Ven. 2017. "Role of Crystal Symmetry in the Reversibility of Stacking-Sequence Changes in Layered Intercalation Electrodes." *Nano Letters*, 17(12), 7789–95.

Rai, A. K., L. T. Anh, J. Gim, V. Mathew, and J. Kim. 2014. "Electrochemical Properties of Na$_x$CoO$_2$(x~0.71) Cathode for Rechargeable Sodium-Ion Batteries." *Ceramics International Part B*, 40(1): 2411–2417.

Recham, N., J.-N. Chotard, L. Dupont, K. Djellab, M. Armand, and J.-M. Tarascon. 2009. "Materials Ionothermal Synthesis of Sodium-Based Fluorophosphate Cathode." *Journal of the Electrochemical Society*, 156 (12): A993–99.

Sauvage, F., L. Laffont, J.-M. Tarascon, and E. Baudrin. 2007. "Study of the Insertion/Deinsertion Mechanism of Sodium into Na$_{0.44}$MnO$_2$." *Inorganic Chemistry*, 46: 3289–94.

Sharma, L., K. Nakamoto, R. Sakamoto, S. Okada, and P. Barpanda. 2018. "Na$_2$FePO$_4$F Fluorophosphate as Positive Insertion Material for Aqueous Sodium-Ion Batteries." *ChemElectroChem* 5: 1–7.

Shibata, T., Y. Fukuzumi, W. Kobayashi, and Y. Moritomo. 2015. "Fast Discharge Process of Layered Cobalt Oxides Due to High Na+ Diffusion." *Science Reports* 5: 9006.

Shiratsuchi, T., S. Okada, J. Yamaki, and T. Nishida. 2006. "FePO$_4$ Cathode Properties for Li and Na Secondary Cells." *Journal of Power Sources*, 159, (1,13): 268–71.

Skyllas-Kazacos, M. 2010. "Electro-Chemical Energy Storage Technologies for Wind Energy Systems." In *Stand-Alone and Hybrid Wind Energy Systems*, 323–365. Woodhead Publishing Limited.

Smirnova, O. A., M. Avdeev, V. B. Nalbandyan, V. V. Kharton, and F. M. B. Marques. 2006. "First Observation of the Reversible O3↔P2 Phase Transition-Crystal Structure of the Quenched High-Temperature Phase $Na_{0.74}M_{0.58}Sb_{0.42}O_2$." *Materials Research Bulletin*, 41(6):1056–62.

Steinbock, L. and C.-H. Dustmann. 2001. "Investigation of the Inner Structures of ZEBRA Cells with a Microtomograph." *Journal of Electrochemical Society*, 148(2): A132–36.

Stoyanova, R., D. Carlier, M. Sendova-Vassileva, M. Yoncheva, E. Zhecheva, D. Nihtianova, and C. Delmas. 2010. "Stabilization of Over-Stoichiometric Mn^{4+} in Layered $Na_{2/3}MnO_2$." *Journal of Solid State Chemistry*, 183: 1372–79.

Su, D., C. Wang, H.-J. Ahn, and G. Wang. 2013. "Single Crystalline $Na_{0.7}MnO_2$ Nanoplates as Cathode Materials for Sodium-Ion Batteries with Enhanced Performance." *Chemistry—A European Journal*, 19(33): 10884–89.

Sun, X., Y. Jin, C.-Y. Zhang, J.-W. Wen, Y. Shao, Y. Zang, and C.-H. Chen. 2014. "$Na[Ni_{0.4}Fe_{0.2}Mn_{0.4-x}Ti_x]O_2$: A Cathode of High Capacity and Superior Cyclability for Na-Ion Batteries." *Journal of Materials Chemistry A*, 2: 17268–71.

Takahashi, Y., Y. Gotoh, and J. Akimoto. 2003. "Single-Crystal Growth, Crystal and Electronic Structure of $NaCoO_2$." *Journal of Solid State Chemistry*, 172: 22–26.

Takeda, Y., J. Akagi, A. Edagawa, M. Inagaki, and S. Naka. 1980. "A Preparation and Polymorphic Relations of Sodium Iron Oxide ($NaFeO_2$)." *Materials Research Bulletin*, 15: 1167–72.

Takeda, Y., K. Nakahara, M. Nishijima, N. Imanishi, O. Yamamoto, M. Takano, and R. Kanno. 1994. "Sodium De-Intercalation from Sodium Iron Oxide." *Materials Research Bulletin*, 29: 659.

Talaie, E., V. Duffort, H. L. Smith, B. Fultz, and L. F. Nazar. 2015. "Structure of the High Voltage Phase of Layered P2-$Na_{2/3-z}[Mn_{1/2}Fe_{1/2}]O_2$ and the Positive Effect of Ni Substitution on Its Stability." *Energy Environmental Science*, 8: 2512.

Tang, J., A. D. Dysart, and V. G. Pol. 2015. "Advancement in Sodium-Ion Rechargeable Batteries." *Current Opinion in Chemical Engineering* 9(2015): 34–41.

Tsuchiya, Y., and N. Yabuuchi. 2016. "P2- and P3-Type $Na_{ax}Cr_{rx}Ti_{i1-x}O_2$Layered Oxides as Bi-Functional Electrode Materials for Rechargeable Sodium Batteries." *Chemistry of Materials*, 28(19): 7006–16.

Vassilaras, P., A. J. Toumar, and G. Ceder. 2014. "Electrochemical Properties of $NaNi_{1/3}Co_{1/3}Fe_{1/3}O_2$ as a Cathode Material for Na-ion Batteries." *Electrochemistry Communications*, 25 (38): 79–81.

Vassilaras, P., X. Ma, X. Li, and G. Ceder. 2013. "Electrochemical Properties of Monoclinic $NaNiO_2$." *Journal of The Electrochemical Society*, 160 (2): A207–11.

Walczak, K., A. Kulka, B. Gędziorowski, M. Gajewska, and J. Molenda. 2018. "Surface Investigation of Chemically Delithiatied $FePO_4$ as a Cathode Material for Sodium Ion Batteries." *Solid State Ionics*, 319: 186–93.

Wang, H., B. Yang, X.-Z. Liao, J. Xu, D. Yang, D. Yang, Y.-S. He, and Z.-F. Ma. 2013. "Electrochemical Properties of P2-$Na_{2/3}[Ni_{1/3}Mn_{2/3}]O_2$ Cathode Material for Sodium Ion Batteries When Cycled in Different Voltage Ranges." *Electrochemical Acta*, 113: 200–204.

Wang, L., J. Wang, X. Zhang, Y. Ren, P. Zuo, G. Yin, and J. Wang. 2017. "Unravelling the Origin of Irreversible Capacity Loss in $NaNiO_2$ for High Voltage Sodium Ion Batteries." *Nano Energy*, 34: 215–23.

Wang, P.-F., Y. You, Y.-X. Yin, and Y.-G. Guo. 2018. "Layered Oxide Cathodes for Sodium-Ion Batteries: Phase Transition, Air Stability, and Performance." *Advanced Energy Materials*, 8(8): 1701912.

Wang, T., D. Su, D. Shanmukaraj, T. Rojo, M. Armand, and G. Wang. 2018. "Electrode Materials for Sodium-Ion Batteries: Considerations on Crystal Structures and Sodium Storage Mechanisms." *Electrochemical Energy Reviews* 1: 200–37.

Wessells, C. D., S. V. Peddada, M. T. McDowell, R. A. Huggins, and Y. Cui. 2012. "The Effect of Insertion Species on Nanostructured Open Framework Hexacyanoferrate Battery Electrodes." *Journal of Electrochemical Soceity*, 159: A98–103.

Wheeler, S., I. Capone, S. Day, C. Tang, and M. Pasta. 2019. "Low-Potential Prussian Blue Analogues for Sodium-Ion Batteries: Manganese Hexacyanochromate." *Chemistry of Materials* 317: 2619–2626.

Wu, Z.-G., J.-T. Li, Y.-J. Zhong, X.-D. Guo, L. Huang, B.-H. Zhong, D. A. Agyeman, J.-M. Lim, D. Kim, M. Cho, and Y.-M. Kang. 2017. "Mn-Based Cathode with Synergetic Layered-Tunnel Hybrid Structures and Their Enhanced Electrochemical Performance in Sodium Ion Batteries." *ACS Applied Material Interfaces*, 9(25): 21267–75.

Xia, X., and J. R. Dahn. 2012. "A Study of the Reactivity of De-Intercalated P2-Na$_x$CoO$_2$ with Non-Aqueous Solvent and Electrolyte by Accelerating Rate Calorimetry." *Journal of Electrochemical Society* 159 (5): A647–50.

Xia, X., and J. R. Dahn. 2012. "NaCrO$_2$ Is a Fundamentally Safe Positive Electrode Material for Sodium-Ion Batteries with Liquid Electrolytes." *Electrochemical and Solid-State Letters*, 15 (1): A1–A4.

Xiang, X., K. Zhang, and J. Chen. 2015. "Recent Advances and Prospects of Cathode Materials for Sodium-Ion Batteries." Advanced Materials 27: 5343–64.

Xie, B., P. Zuo, L. Wang, J. Wang, H. Huo, M. He, J. Shu, H. Li, S. Lou, and G. Yin. 2019. "Achieving Long-life Prussian Blue Analogue Cathode for Na-ion Batteries via Triple-Cation Lattice Substitution and Coordinated Water Capture." *Nano Energy*, 61: 201–10.

Xu, M., Y. Niu, Y. Li, S. Baoab, and C. M. Lia. 2014. "Synthesis of Sodium Manganese Oxides with Tailored Multi-Morphologies and Their Application in Lithium/Sodium ion Batteries." *RSC Advances*, 4: 30340–45.

Xu, X., D. Zhou, X. Qin, K. Lin, F. Kang, B. Li, D. Shanmukaraj, T. Rojo, M. Armand, and G. Wang. 2018. "A Room-Temperature Sodium–Sulfur Battery with High Capacity and Stable Cycling Performance." *Nature Communications*, 9, Article number: 3870.

Yabuuchi, N., H. Yoshida, and S. Komaba. 2012. "Crystal Structures and Electrode Performance of Alpha-NaFeO$_2$ for Rechargeable Sodium Batteries." *Electrochemistry*, 80(10): 716–19.

Yabuuchi, N., K. Kubota, M. Dahbi, and S. Komaba. 2014. "Research Development on Sodium-Ion Batteries." *Chemical Review*, 114(23): 11636–82.

Yabuuchi, N., M. Kajiyama, J. Iwatate, H. Nishikawa, S. Hitomi, R. Okuyama, R. Usui, Y. Yamada, and S. Komaba. 2012. "P2-type Na$_x$[Fe$_{1/2}$Mn$_{1/2}$]O$_2$ made from Earth-abundant Elements for Rechargeable Na Batteries." *Nature Materials*, 11: 512–17.

Yoshida, H., N. Yabuuchi, and S. Komaba. 2013. "NaFe$_{0.5}$Co$_{0.5}$O$_2$ as High Energy and Power Positive Electrode for Na-ion Batteries." *Electrochemistry Communications*, 34, 60–63.

Yu, C.-Y., J.-S. Park, H.-G. Jung, K.-Y. Chung, D. Aurbach, Y.-K. Sun, and S.-T. Myung. 2015. "NaCrO$_2$ Cathode for High-Rate Sodium-Ion Batteries." *Energy & Environmental Science*, 8: 2019–26.

Zhang, X., W. Yang, X. Chen, and Y. Ma. 2009. "Sodium Manganese Oxide Nanobelts with a 2 × 4 Tunnel Structure: One-Step Hydrothermal Synthesis and Electrocatalytic Properties." *Journal of Nanoscience and Nanotechnology*, 9: 5860–64.

Zhao, J. M., Z. L. Jian, J. Ma, F. C. Wang, Y. S. Hu, W. Chen, L. Q. Chen, H. Z. Liu, and S. Dai. 2012. "Monodisperse Iron Phosphate Nanospheres: Preparation and Application in Energy Storage." *ChemSusChem*, 5, 1495–1500.

Zhao, J., L. Zhao, N. Dimov, S. Okada, and T. Nishida. 2013. "Electrochemical and Thermal Properties of α-NaFeO$_2$ Cathode for Na-Ion Batteries." *Journal of The Electrochemical Society*, 160 (5): A3077–81.

Zhu, X., T. Mochiku, H. Fujii, K. Tang, Y. Hu, Z. Huang, B. Luo, K. Ozawa, and L. Wang. 2018. "A New Sodium Iron Phosphate as a Stable High-Rate Cathode Material for Sodium Ion Batteries." *Nano Research*, 11 (12): 6197–205.

5

Magnesium Battery

E. Sheha

CONTENTS

5.1 Introduction 81
5.2 Magnesium Electrolytes 83
5.3 Mg Batteries 85
    5.3.1 Mg/V$_2$O$_5$ Battery 85
    5.3.2 Magnesium–Sulphur Battery 87
    5.3.3 Magnesium/CuS Battery 90
    5.3.4 Magnesium/Organic Cathode Battery 92
    5.3.5 Magnesium/Alloy Battery 94
5.4 Summary and Outlook 97
References 99

5.1 Introduction

In the context of the global trend towards vehicle electrification and the removal of fossil fuels, for reasons of climate, sustainable earth, environment, and to reach a green planet, global energy demand is growing strongly towards renewable energy sources and, in particular, electrochemical energy storage. Currently, rechargeable lithium batteries are on the throne as a source of energy storage due to their ability to deliver high energy density and low self-discharge rate. However, there are some issues related to the safety factor, temperature tolerance, and cost that require the search for an alternative to the lithium battery, especially that the intercalation chemistry of the lithium battery has reached its energy density ceiling. In light of these variables, there is a glimmer of hope coming from the magnesium battery and the possibility of using it as a post-lithium battery, particularly for green energy storage and power applications. The magnesium battery has potential advantages compared to its lithium battery counterpart (see Table 5.1) (Zhang et al. 2018), these include: its bivalent nature, which leads to a high volumetric capacity of 3833 mAh cm^{-3} (~ twice of lithium 2061 mAh cm^{-3}); magnesium is abundant in the earth's crust; and, unlike lithium, magnesium does not form dendrites during reversible insertion/extraction, which results in safety concerns as an anode material. More importantly, magnesium metal possesses a low reduction voltage ~2.37 V (vs a normal hydrogen electrode). Despite all these positive signs of Mg, there are major challenges that need to be overcome before realizing a practical magnesium battery. The first challenge is the synthesis of electrolytes with a chemical nature compatible with a Mg anode. In general, as a result of the chemical activity of magnesium, most solvents of magnesium salts interact with it forming non-conducting passivation layers that do not allow the magnesium ions to shuttle, and this leads to the failure of the battery. The second challenge lies in the architecture of a new lattice that can compensate for the charge imbalance caused by the intercalation of the divalent magnesium ion. Aurbach et al. (2000) presented an unprecedented prototype system of a rechargeable magnesium battery that shows promise for applications. This is based on the utilization of Chevrel-phase molybdenum chalcogenides properties that were reported in 1971. The reason behind the ability of a Chevrel-phase structure to understand the

TABLE 5.1

Key Parameter Comparisons Between Mg and Li Metals for Rechargeable Batteries

Metal Anode	Melting Point (°C)	Market Price [k\$ ton⁻¹]	Volumetric Capacity [mAh cm⁻³]	Theoretical Specific Capacity [mAh g⁻¹]	Electrochemical Redox Potential versus SHE [V]	Stability in Air?	Ionic Radii (Å)
Mg	650	2.7	3833	2206	–2.37	Yes	0.72
Li	180.5	64.8	2061	3860	–3.01	Not stable	0.76

Source: Zhang et al., *Small Methods*, 1800020. https://doi.org/10.1002/smtd.201800020.

charge imbalance is due to the introduction of multiple cations that come from: (1) successively axial sulphurs and Mo_6 cluster redox centres, (2) there are 12 vacant sites for a diffusion path, (3) the electrons move more freely in this structure, and (4) ionicity of sulphides is less than their counterparts of oxides, which enhances the Mg^{2+} mobility due to decreasing the electrostatic forces between it and the host ions (Yoo et al. 2013; Figure 5.1a). Mo_6S_8 intercalates Mg ions reversibly as follows:

$$Mo_6S_8 +_2 Mg^{2+} + _4e^- \leftrightarrow Mg_2Mo_6S_8$$

Aurbach used $Mg(AlCl_2(butyl)(ethyl))_2$/tetrahydrofuran (THF) electrolyte solution, the product reaction between the $(butyl)_2Mg$ Lewis base and $(ethyl)AlCl_2$ Lewis acid. Figure 5.1b shows deposition-dissolution on the anode side and cycling performance of their developed prototype system. Although the Aurbach cell shows promising results, like a very wide electrochemical window and very high Coulombic efficiency in inert metals, however, the Aurbach battery is still far from commercial applications due to the high cost of molybdenum metal, its low voltages ~ 1.1 V, as well as low theoretical specific capacities ~ 130 mAh g⁻¹, which are undesirable for high load density energy storage. Therefore, it is crucial to look for alternatives.

This chapter summarizes and presents significant achievements in the last few years regarding the development of Mg battery materials for realizing a practical rechargeable magnesium battery. Brilliant strategies to improve electrochemical performance that cross electrolyte optimization, surface modification, and structure design will be considered. In addition, the conclusion and future outlook on an advanced Mg battery for a post-lithium era will also be proposed.

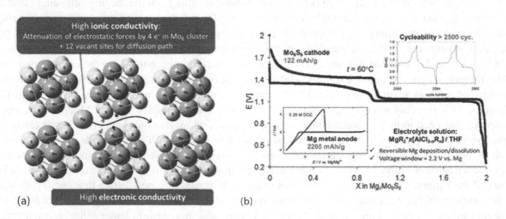

FIGURE 5.1   (a) Illustration of the movement of Mg^{2+} ions in a Mo_6S_8 structure. (b) Electrochemical performance of a Mg/Mo_6S_8 standard cell. (Yoo, H.D. et al., *Energy Environ. Sci.*, 6, 2265–2279. Copyright 2008, Reproduced by permission of Royal Society of Chemistry.)

5.2 Magnesium Electrolytes

The high reactivity of the Mg results in dual roles, the first is positive due to the high (negative) voltage, but the second is negative due to the chemical incompatibility of the conventional electrolyte solutions with the high reducing ability of the magnesium. Magnesium electrolytes are very sensitive to very small quantities of contaminants (~ ppm), its result is an insulating layer at the surface of the anode that inhibits the shuttle movement of magnesium ions. The low reduction potential stands behind why magnesium metal tends to form surface films. Generally, magnesium deposition and stripping is difficult in most non-aqueous electrolytes because of the electronically insulating layer (passive layer) that forms on Mg electrodes and freezes the Mg^{2+} movement to the Mg anode surface. So it is important to skip that obstacle by developing an ideal electrolyte for rechargeable Mg batteries that achieves reversible Mg deposition-dissolution, wide electrochemical windows, and high ionic conductivity. In this context, Aurbach et al. (2000) took advantage of the organometallic components as a strong reductive agent. Accordingly, the organometallic components act as a scavenger for reducible contaminants that help strongly to develop an electrolyte solution of $Mg(AlCl_2(butyl)(ethyl))_2/THF$ with a very high Coulombic efficiency. Although this electrolyte showed low ionic conductivity, approximately several millisiemens, 100% Coulombic efficiency on inert metals, and good anodic stability, the main limitations of an organohaloaluminate electrolyte are due to the relatively weak Al–C bond, which broke via C-H elimination, resulting in the electrochemical instability on non-inert metals in addition to their high cost (Muldoon et al. 2012). Their work extended to develop the all phenyl complex (APC) electrolyte solution with no β-located hydrogen (Shterenberg et al. 2014), composed of the reaction products between $PhxMgCl_2$–x and $PhyAlCl_3$–y with an anodic stability window ~3.3 V, a high specific conductivity (2 mS cm^{-1}) at room temperature, high Coulombic efficiency (~100%), and a low overpotential in Mg deposition processes (below 2 V). Figure 5.2a shows a typical cyclic voltammogram of standard APC solutions. But it is necessary to find alternatives to both the chloride and aluminium chloride, which corrode the components of the battery. Consequently, there is still a pressing need to develop an electrolyte that can overcome excess air sensitivity. In addition, this electrolyte should be free of chloride compounds and have low volatility to realize battery safety.

In this context, Mohtadi et al. (2012) reported a new class of halide-free electrolytes based on $Mg(BH_4)_2$ in both THF and dimethoxyethane (DME) solvents for magnesium-ion batteries. $Mg(BH_4)_2$ has a strong reductive nature that can be used as a scavenger for reducible molecules in electrolyte solutions in a similar way to Grignard reagents. $LiBH_4/Mg(BH_4)_2$ in DME showed unprecedented reversible Mg deposition and dissolution compared to those in THF. The addition of $LiBH_4$ enhanced the Coulombic efficiency and the current density. A prototype Mg battery was fabricated using a Chevrel-phase Mo_6S_8 cathode, $Mg/[LiBH_4/Mg(BH_4)_2/DME]/Mo_6S_8$. This prototype showed good electrochemical performance in the absence of chlorine and THF, Figure 5.2b, making this solution distinctive for developing a replacement category of ideal Mg electrolytes. Currently, the most analysis is targeted on improving the oxidative stability of $Mg(BH_4)_2$. Additionally, the precise nature of the electroactive species within the presence and therefore the absence of the additive is being studied to guide the look of $Mg(BH_4)_2$ primarily based electrolytes. This work provides a cornerstone for expanding the applications of $Mg(BH_4)_2$ and confirms the beauty and flexibility of the chemistry of borohydrides. Ha et al. (2014) reported a new family of electrolyte based on $Mg(TFSI)_2$ dissolved in dimethoxyethane-based solvents. Their study relied on the use of the unique characteristics of glyme-based solvents, such as a high dielectric constant and low viscosity. Glyme-based solvents have the power to create an applicable solvation sheath structure for Mg stripping/dissolution, high solvating power, the power to reduce and corrode the traditional collectors, low volatility, and a high anodic limit. A stainless steel working electrode (SS), a Mg reference electrode, and Mg counter electrode were designed in a three-electrode cell. Mg deposition/stripping was absented in the first scan, Figure 5.2c (inset), a reduction peak appeared at −0.4 V thanks to Mg deposition on the SS electrode, and an oxidation peak was observed at 0.5 V due to Mg stripping in the second negative scan, as shown in Figure 5.2c. A Mg battery was fabricated using a Chevrel-phase Mo_6S_8 cathode, a Mg metal anode, and with a glyme/diglyme/0.3 M $Mg(TFSI)_2$ electrolyte. A Mg/Mo_6S_8 cell delivered a reversible specific capacity of ~97 mAh g^{-1} at the second cycle, Figure 5.2d. Generally, this work provides another

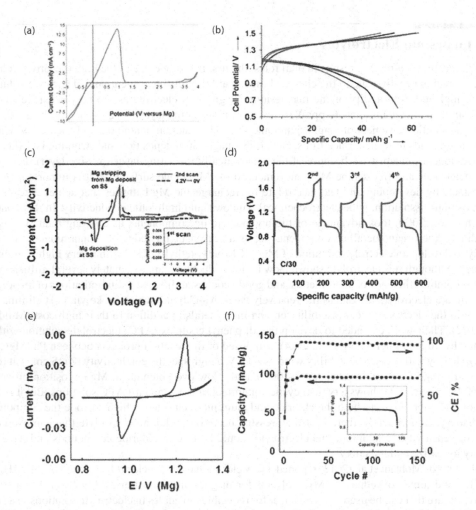

FIGURE 5.2 (a) CV curve of APC solutions. (Reprinted with permission from Shterenberg, I. et al., *Mrs Bulletin*, 39, 453–60, 2014. Copyright 2014, Cambridge University Press.) (b) Cycling performance of Mg/[LiBH₄/Mg(BH₄)₂/DME]/Mo₆S₈, cycle number 1 (56 mAh g⁻¹), 40 (42 mAh g⁻¹). (From Mohtadi, R. et al.: Magnesium Borohydride: From Hydrogen Storage to Magnesium Battery, *Angewandte Chemie International Edition*, 2012, 51, 9780–83. Copyright Wiley-VCH Verlag GmbH & Co. KGaA. Reprinted with permission.) (c) CV curves of a Mg/SS three-electrode cell in [0.3MMg(TFSI)₂]/[glyme/diglyme (1/1, v/v)] electrolyte (~0.2 mV/s). (d) Cycling performance of the Mg/0.3 M Mg(TFSI)₂ in glyme/diglyme/Mo₆S₈ cell (~C/30). (Reprinted with permission from Ha et al. 2014, 4063–73. Copyright 2014, American Chemical Society.) Electrochemical performance of a Mg(BH₄)₂/MgO/PEO electrolyte and (e) CV (0.05 mVs⁻¹) of a Mg insertion/extraction in Mo₆S₈. (f) Cycling performance of a solid-state Mg cell (inset discharge/charge curve). (Reprinted from *Nano Energy*, 12, Shao, Y. et al., Nanocomposite Polymer Electrolyte for Rechargeable Magnesium Batteries, 750–79. Copyright 2015, with permission from Elsevier.)

cornerstone for expanding the applications of Mg(TFSI)₂/glyme as an electrolyte in consideration for a rechargeable Mg batteries electrolyte. Most lithium-ion batteries have a special feature, which is the formation of a stable Li⁺ conductive solid electrolyte interface (SEI) that hinders electrolyte decomposition, in contrast to the Mg battery that suffers from the formation of an unstable SEI that entirely blocks the electrochemical activity of the Mg anode. So, from another point of view, replacing a routine liquid electrolyte with a solid polymer electrolyte (SPE) is highly desirable for many reasons: (1) a potentially high energy density due to its low volatility, (2) flexibility in dimensions/geometry, (3) high safety, (4) no passivation layer at a Mg/electrolyte interface (solvent free), and (5) easiness to form films. In this regard, Shao et al. (2015) developed an SPE using poly(ethylene oxide) (PEO), Mg(BH₄)₂, and MgO nanoparticles for secondary Mg batteries. The nanocomposite polymer electrolytes were fabricated by employing hot pressing or solution casting methods. A reversible Mg plating/stripping is achieved at 100°C, the high Coulombic efficiency ~98%, and the low overpotential ~0.2 V between the onset voltage of Mg plating and Mg stripping are key options for solid-state secondary Mg batteries and confirming high rechargeability and quick Mg^{2+} ion mobility, Figure 5.2e. A Mg battery was fabricated using a Chevrel-phase Mo_6S_8 cathode, Mg anode, along with the nanocomposite polymer electrolyte. A Mg/Mo_6S_8 cell pumped a reversible capacity of ~90 mAh g^{-1} for at least 150 cycles. The charge/discharge curves (inset Figure 5.2f) show very flat plateaus above 1 V. Their study attributed the reason behind the reversible Mg plating/stripping in $Mg(BH_4)_2$/MgO/PEO, to the formation of solvated $[MgBH_4]^+$ cation complexes. In summary, there is a great deal of work required in the next period to overcome the obstacles facing the development of an electrolyte ideal for a Mg battery. These obstacles include hazard properties, sensitivity to water and/or air contamination (detrimental results), initial capacity fading due to the unstable SEI film, non-corrosivity towards cheap current collectors, the passivation layer at a Mg electrolyte interface and volatile solvent hindering their practical application, cost, and constructing SPEs that effectively works at room temperature.

5.3 Mg Batteries

5.3.1 Mg/V_2O_5 Battery

V_2O_5 has an orthorhombic structure with layered framework made up of VO_5 pyramid units located at the edges and the interlayer bonds are weak. This unique structure gives V_2O_5 the flexibility to insert a range of cations, like Na$^+$ and Li$^+$, between the layers (up to 3 Li-ions/unit cell), and hence grants it a high specific capacity (~440 mAh g^{-1}). Recently, orthorhombic V_2O_5 structure was introduced as a host lattice for Mg ions (up to 0.5 Mg ions/unit cell) (Mukherjee et al. 2017). Sa, Kinnibrugh et al. (2016) studied the structural evolution during Mg^{2+} insertion/extraction into a bilayer structure of $V_2O_5\cdot nH_2O$ xerogel. Their study showed that there was shrinkage in the space between the layers of $V_2O_5\cdot nH_2O$ upon Mg insertion and expanded upon Mg extraction due to the robust electrostatic interaction between the bivalent ion and the host lattice. They tested the electrochemical performance using a Mg metal disk anode, a $V_2O_5\cdot nH_2O$ cathode, and ~0.07 mL $Mg(TFSI)_2$ in a diglyme electrolyte that was used per each coin cell, in addition to a current density of ~20 μA cm^{-2}. The cell displayed capacity fading after 10 cycles, Figure 5.3a. The cyclic voltammograms show the appearance of anodic and cathodic peaks around 3.0 V and 0.8 V, corresponding to the Mg insertion/extraction, respectively, as shown in Figure 5.3b. Their work attributed the overpotential to an increase in the internal resistance of the cell due to the formation of a layer on the surface between the magnesium and the electrolyte, which reduces the ion movement. This work extended to study the effect of the water level in the $(MgTFSI)_2$/(diglyme (G2)) at the plating/stripping Mg to identify the mysterious role of protons in improving cell efficiency (Sa, Wang et al. 2016). The simultaneous increase and reduction of ^1H nuclear magnetic resonance (NMR) spectra in the wet electrolyte during charge and discharge confirm the co-insertion/extraction of the proton in the lattice during cycling, Figure 5.3c and d. Although high capacity can be achieved due to a proton content electrolyte for the V_2O_5 cathode as displayed in the inset Figure 5.3c and d, this system is unable to include Mg metal as the anode because the presence of reducible molecules like water comes at the expense of Mg metal stability. In 2018, Son et al. (2018) presented an innovative approach to overcome that challenge via covering the surface of the magnesium anode with a thin Mg^{2+}-conducting polymeric film that works to separate the magnesium surface from the liquid electrolyte, which could successfully shield the Mg metal from being reduced. The artificial interphase as defined by the authors based thermal-cyclized polyacrylonitrile, Mg powder, carbon black, and Mg triflate in a weight ratio of 10:77:10:3. A Mg anode was thermally treated at 300°C for 1 hour under Ar to accelerate the cyclization of polymer units. Figure 5.4a illustrates the interaction between Mg powder and the artificial interphase and the structure scheme of the artificial interphase. Their work extended to use X-ray photoelectron spectroscopy, scanning electron microscopy (SEM), and energy dispersive spectroscopy (EDS)

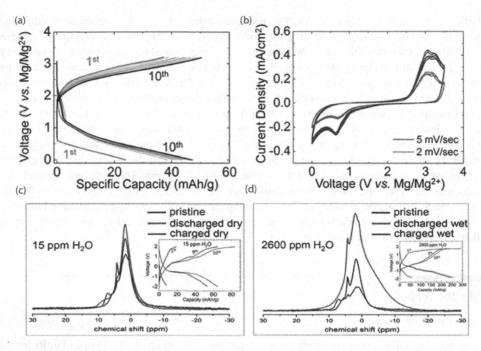

FIGURE 5.3 (a) Cycling performance of a Mg/1.0 M Mg(TFSI)₂ diglyme/V₂O₅•nH₂O cell. (b) Cyclic voltammetry of a V₂O₅•nH₂O versus Mg anode. (Reprinted with permission from Sa, Kinnibrugh et al. 2016, 2962–69. Copyright 2016, American Chemical Society.) ¹H NMR spectra of a V₂O₅ cathode at different states: (c) Mg(TFSI)₂/G2 electrolyte (15 ppm H₂O) (inset charge/discharge curves). (d) Mg(TFSI)₂/G2 electrolyte (2600 ppm H₂O) (inset charge/discharge curves). (Reprinted from *J. Power Sources*, 323, Sa, N. et al., Is Alpha-V₂O₅ a Cathode Material for Mg Insertion Batteries? 44–50. Copyright 2016, with permission from Elsevier.)

to characterize the covered and non-coated stainless-steel surface by the artificial interphase after electrochemical deposition. The results showed that the deposition process was achieved only in the case of a surface coating. A complete cell was constructed using a 0.5 M Mg(TFSI)₂/PC and 0.5 M Mg(TFSI)₂/[PC+H₂O] electrolyte, a magnesium anode, and a orthorhombic V₂O₅ cathode as a cathode in the presence and absence of the conductive artificial interphase at a rate of 29.4 mA g⁻¹. Figure 5.4b shows the electrochemical performance, it is notable that the efficiency of the cell, ability to recharge, and the cycling stability greatly improved in the presence of the conductive artificial interphase-shielded Mg anode, the cell still delivers 47 mAh g⁻¹ beyond 40 cycles. On the other hand, in the absence of the conductive artificial interphase, the cell displayed an increase in overpotential, and the capacity fades rapidly to zero after 40 cycles. Son et al. (2018) explained the discrepancy in electrochemical results before and after using the conductive artificial interphase for the formation of a passive layer, in the absence of the conductive artificial interphase, on the surface between the magnesium and the 0.5 M Mg(TFSI)₂/PC electrolyte, which hinders Mg²⁺ from the shuttle movement. Using X-ray diffraction (XRD) measurements, the authors demonstrated insertion/extraction of Mg²⁺ is via a V₂O₅ cathode through the shift in the position of (200) and (110) peaks to the lower Bragg angles values after magnesium insertion and restoring the original positions after magnesium extraction (Figure 5.4c). Despite that these look like impressive results, they have come under intense criticism, the most important of which is that the work presented consists of some experimental methods that blur the real electrochemical performance of the electrodes. For example, the total amount of magnesium retained, as well as the depth of discharge are not known, and, therefore, it is difficult to estimate the true cycling efficiency values. It is difficult to recognize the value of these results without conducting over 100 cycles with high reversibility. Therefore, this approach is still far from being realized (Attias et al. 2018). In summary, the results illustrated that the potential of V₂O₅ as a cathode

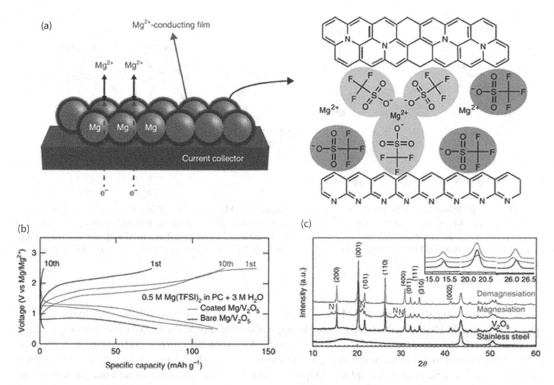

FIGURE 5.4 (a) Illustration of the interaction between Mg powder and the artificial interphase and the structure scheme of the artificial interphase. (b) Discharge/charge profile of Mg/V_2O_5 cells with and without the artificial interphase. (c) XRD analyses of a V_2O_5 cathode before and after magnesiation. Inset, the main peaks of a V_2O_5 cathode that shift to the left when magnesiated and restore original position after demagnesiation. (Reprinted with permission from Son, S.-B. et al., *Nat. Chem.*, 10, 532. Copyright 2018, Springer Nature.)

for a Mg battery should be reconsidered. The presence of water molecules is an important factor to demonstrate significant capacities for Mg intercalation, on the other side, reducible molecules like water come at the expense of Mg metal stability. Questions regarding the insertion/extraction of protons rather than Mg ions into V_2O_5 are still unanswered, and the subject needs deep research to assess the feasibility of using V_2O_5 as a cathode for a rechargeable Mg battery.

5.3.2 Magnesium–Sulphur Battery

Sulphur occupies a high order of abundance in the crust of the earth (ranks 7th of non-metallic elements and 17th of all elements). Sulphur has a range of positive features like high theoretical capacity ~1673 mAh g^{-1}, low cost, and non-toxicity, which make it to be an ideal cathode material for lithium-sulphur batteries. A lithium-sulphur battery provides high theoretical specific energy ~2600 Wh kg^{-1}, and while the practical specific energy is ~150–378 Wh kg^{-1}; however, the system suffers from self-discharge, shuttle effects, and bad cycle life (Zhu et al. 2018). On the other hand, the two-electron conversion reaction is behind the high volumetric energy density ~3833 mAh cm^{-3} of a Mg/S battery. The high theoretical capacity of a Mg/S battery comes from the ability of sulphur to react with two electrons via a conversion reaction as follows:

$$S_8 + 4e^- + 2Mg^{2+} \leftrightarrow 2MgS_4$$

$$MgS_4 + 2e^- + Mg^{2+} \leftrightarrow 2MgS_2$$

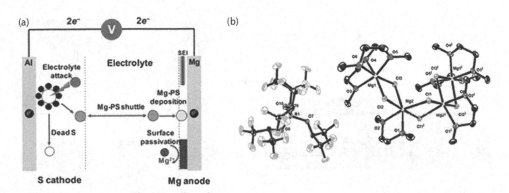

FIGURE 5.5 (a) Illustration of the obstacles facing the application of Mg/S batteries. (From Kong, L. et al.: A Review of Advanced Energy Materials for Magnesium–Sulfur Batteries, *Energ Env. Mater.*, 2018, 1, 100–112. Copyright Wiley-VCH Verlag GmbH & Co. KGaA. Reproduced with permission.) (b) Structure of $[Mg_4Cl_6(DME)_6][B(HFP)_4]_2$ using the Oak Ridge Thermal Ellipsoid Plot (ORTEP) program. (Du et al. 2017, 2616–2625. Reproduced by permission of The Royal Society of Chemistry.)

Moreover, the sluggish kinetics for Mg^{2+} intercalation may be overcome by the conversion reaction between Mg and S. However, the lack of suitable magnesium electrolytes with favourable electrochemical properties limits the achievement of a successful device. As shown in Figure 5.5a (Kong et al. 2018), coupling a magnesium metal anode with an organic electrolyte and cathodes of sulphur composite results in the formation of magnesium polysulphides (Mg-PSs) via a successive reduction in sulphur during cycling, which is one of the hurdles that hinder realizing practical Mg/S batteries. Kim et al. (2011) and Du et al. (2017) claimed that it is possible to realize a rechargeable Mg/S battery by developing non-nucleophilic hexamethyldisilazide magnesium chloride (HMDSMgCl). These non-nucleophilic electrolytes opened the door to address the inability to measure the electrochemical performance of Mg/S batteries and overcome the issue of incompatibility between the traditional electrolytes and electrophilic sulphur. In 2017, Du et al. (2017) tried to enhance the electrochemical performance of Mg/S cells by presenting a new type of non-nucleophilic organic electrolyte. This electrolyte is a product reaction of $B(HFP)_3$, $MgCl_2$, and Mg powder in DME yields $[Mg_4Cl_6(DME)_6][B(HFP)_4]_2$ (OMBB). Magnesium chloride plays a key role in improving the stability of the cation from its thermodynamic aspect and reduces the inactive species through ligand exchange between Cl^- ions and the DME. Their work extended to study the structure of the OMBB electrolyte using single-crystal XRD. The results showed a complex cation $[Mg_4Cl_6(DME)_6]^{2+}$ based on four Mg ions distributed octahedrally and linked to each other through two bridging chlorine atoms, Figure 5.5b. Their work extended to compare the electrochemical performance of Mg/S batteries using sulphur-based cathodes like in the 0.5 M OMBB electrolyte. Carbon nanotube (CNT), amorphous mesoporous carbon, and ordered mesoporous carbon as a different morphological form of carbon were mixed with sulphur to develop the cathodes. The excellent conductivity and wetting properties of CNTs were behind the highest discharge-charge specific capacity of up to 1247 mAh g^{-1} at a current density of 160 mA g^{-1} compared with the other sulphur-carbon composite cathodes and 10 times higher than that of the high-cost Mo_6S_8. The S-CNT cathode delivered a high capacity of 1019 mA h g^{-1} and capacity retention ~80.4% after the first 100 cycles, Figure 5.6a. The cycling performance of Mg/OMBB/S-CNT at a current density of 160 mA g^{-1} is shown in Figure 5.6b. The OMBB electrolyte gave outstanding results and opened new horizons for the development of a Mg/S battery compared with those reported. However, the corrosion problem of the traditional current collectors due to the presence of a chlorine ion in the structure of the electrolyte stands as a barrier to its practical application. In 2018, Zhao-Karger et al. (2018) presented a comprehensive insight into Mg-tetrakis(hexafluoroisopropyloxy)-borate $Mg[B(hfip)_4]_2(hfip = OC(H)(CF_3)_2)$ in DME as an adequate and practical electrolyte for a Mg/S battery. Their work extended to include the interfacial phenomena occurring between the $Mg[B(hfip)_4]_2$ electrolytes and a Mg anode, it also identified the reasons behind the hysteresis voltage in Mg–S batteries, optimized the ionic conductivity through the smoothness of the composition of the $Mg[B(hfip)_4]_2$ in DME,

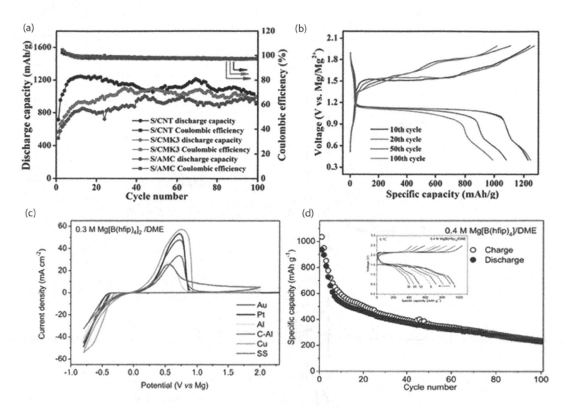

FIGURE 5.6 (a) Cycling performance of S-C cathodes using 0.5M(OMBB) electrolyte, (b) discharge/charge profiles of a Mg/OMBB/S-CNT cell. (Du et al. 2017, 2616–25. Reproduced by permission of The Royal Society of Chemistry.) (c) Mg deposition/dissolution in 0.3M(MgBhfip)/DME solution with changing the metal electrodes. (d) Cycling performance of Mg/[0.4M(MgBhfip)/DME]/sulphur-based cathode, inset (charge/discharge profiles). (Reprinted with permission from Zhao-Karger, Z. et al., *ACS Energy Lett.*, 3, 2005–13. Copyright 2018 American Chemical Society.)

assessed the practical application of the $Mg[B(hfip)_4]_2$ electrolytes, tested the polarization behaviour, and studied the long-term cycling stability. Ionic conductivity of σ_{ion} ~11 mS cm^{-1} was obtained at the concentration 0.3 $Mg[B(hfip)_4]_2$ in DME at 23°C. The authors observed in all solutions redox peaks which confirm Mg deposition/dissolution, the current densities gradually increase with the concentration of the $Mg[B(hfip)_4]_2$ in DME. The researchers observed that the electrochemical deposition value of magnesium changes by changing the type of the electrode, and there is a shift in the value of the onset of the deposition potential from –0.43 to –0.32 V by increasing the electrolyte concentration, Figure 5.6c. They interpreted this behaviour in light of the crystalline growth rate of magnesium varies with the sedimentation surface, as well as the dissociation energy of $[Mg(DME)_3]^{2+}$ ions.

Figure 5.6d displays the cycling performance of a Mg/S battery-based $Mg-[B(hfip)_4]_2$ electrolyte that still delivers over 200 mAh g^{-1} but the discharge/charge profiles suffer a voltage hysteresis ~0.63 V, which is reflected in reducing the efficiency of the cell, inset Figure 5.6d. Zhao-Karger et al. (2018) extended their study to use X-ray photoelectron spectroscopy (XPS) to investigate the type of reaction between the S cathode and MgBhfip electrolyte at the charged and discharged states. As shown in Figure 5.7, the characteristic 2p spectra of the parent sulphur split into two peaks at ~164.0 eV, and ~165.2 eV corresponds to $2p_{3/2}$ and $2p_{1/2}$ spectra, respectively. After discharging to 1.0 and 0.5 V, the intensity of the main peaks of the pristine sulphur decreased, and new peaks which correspond to MgS_x and MgS_2 were observed. After recharging to 2.5 V, the intensities of the pristine sulphur restored, while the signal intensities of MgS_x and MgS_2 reduced. This confirms that the transformation of S to MgS_x and MgS_2 in a $Mg-[B(hfip)_4]_2$ electrolyte is a reversible reaction.

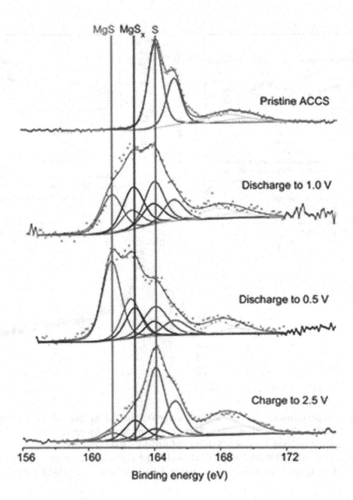

FIGURE 5.7   XPS spectra of the sulphur (2p) cathodes at different conversion states (discharge/charge). (Reprinted with permission from Zhao-Karger, Z. et al., *ACS Energy Lett.*, 3, 2005–13. Copyright 2018 American Chemical Society.)

In summary, a good step has been made in the development of an electrolyte suitable for a Mg/S battery and in understanding the nature of its interaction. However, short cycling stability, dissolution, and shuttling of active materials, remain major challenges to transfer a Mg/S battery from the lab to the market. New strategies should be proposed to overcome these obstacles, which can be inspired by their lithium battery counterparts, e.g., using highly concentrated electrolytes and engineering an artificial interphase at an anode to suppress the soluble magnesium polysulphide shuttle effects and achieve better cycling stability.

5.3.3 Magnesium/CuS Battery

CuS has recently attracted considerable attention in Mg battery applications as a conversion-type cathode. Theoretically, CuS can deliver a high specific capacity ~560 mAh g⁻¹ and one of the most important features of CuS is the low cost. The high theoretical capacity of CuS comes from the conversion reaction as follow:

$$2CuS + Mg^{2+} + 2e^- \leftrightarrow Cu_2S + MgS$$

$$Cu_2S + Mg^{2+} + 2e^- \leftrightarrow_2 Cu + MgS$$

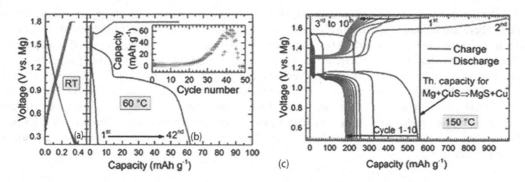

FIGURE 5.8 Discharge/charge profiles of a CuS cycled (a) 25°C; (b) 60°C; and (c) 150°C. (Duffort, V. et al., *Chem. Commun.*, 52, 12458–12461, 2016. Reproduced by permission of The Royal Society of Chemistry.)

Duffort et al. (2016) reported that a CuS cathode is capable of accommodating Mg^{2+} electrochemically and reversibly (over 200 mAh g^{-1}) with an APC (2PhMgCl-AlCl$_3$) in a tetraglyme electrolyte (boiling point ~275°C), at 150°C. A molybdenum current collector was used to circumvent chlorine corrosion. As shown in Figure 5.8a and b, CuS delivers only about 10% of the theoretical capacity, which was obtained at 60°C for 40 cycles. However, CuS delivers 98% of the theoretical capacity of ~550 mAh g^{-1} at 150°C, Figure 5.8c. Their study concluded that the temperature plays a dual and contradictory role, the first helps to overcome the low ionic mobility of Mg^{2+}, the second increases the kinetics of parasitic reactions. In 2018, Wu et al. (2018) studied the electrochemical performance of commercial CuS nanoparticles (Aladdin, 99.99%) at room temperature. Copper sulphide nanoparticles show a high reversible capacity of 175 mAh g^{-1} over 350 cycles at a current density of 50 mAg^{-1} at room temperature in magnesium bis(hexamethyldisilazide) (Mg(HMDS)$_2$)/AlCl$_3$ electrolyte, Figure 5.9a and b. As shown in Figure 5.9c

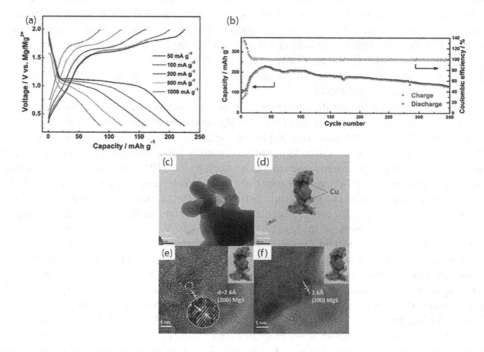

FIGURE 5.9 (a) Discharge/charge curves of a CuS-I electrode, (b) cycling performance of the CuS-I electrode (50 mA g^{-1}); TEM images of CuS-I nanoparticles (c); pristine (d); discharged state (e); and (f) magnified images of (d). (Wu, M. et al., *Nanoscale*, 2018. Reproduced by permission of The Royal Society of Chemistry.)

and d, the transmission electron microscopy (TEM) observation of pristine CuS displayed uniform distribution. There are no major noticeable changes after long-term cycling, where the particle size was kept lower than 100 nm, the particle surface became coarse and the surface morphology appeared homogenous.

This confirms that the CuS particles are stable in their positions and have strong links between each other that enable them to overcome mechanical changes during cycling. The Mg/S phase was identified by the lattice fringe that was indexed to the (200) planes (spacing ~2.6 Å), Figure 5.9e and f. Their conclusion revolved around their finding that the conversion reaction is kinetically affected by reducing the particle size of CuS to the small size. In summary, downsizing CuS particles might be an effective strategy to improve the kinetics of the conversion reaction and guarantee a large surface area.

5.3.4 Magnesium/Organic Cathode Battery

Organic materials have a range of positive features such as versatility, low cost and sustainability, environmental friendliness, ease of preparation, and modification to tune their electrochemical and physical properties. Organic materials are characterized by their ability to store the charges via an electrochemical conversion reaction compared to the inorganic hosts, which depend on the de/intercalation and diffusion of ions. These good features qualify organic materials to be promising electrode materials for rechargeable magnesium batteries with a high electrochemical performance, where it can overcome the challenge of the slow diffusion of Mg^{2+} inside inorganic hosts, resulting in hysteresis overpotential, capacity fading, and low current capabilities. However, early research of organic cathodes such as 2,5-dimethoxy-1,4-benzoquinone in magnesium batteries showed poor cyclability and showed an abrupt capacity decay. Aiming to develop high performance cathodes for rechargeable magnesium-ion batteries, Pan et al. (2016) succeeded to modify the structure of anthraquinone $C_{14}H_8O_2$ to develop 2,6-polyanthraquinone (26PAQ) and 1,4-polyanthraquinone (14PAQ) based. Figure 5.10a shows the structure of the anthraquinone $C_6H_8O_2$ before and after modification. As shown, the idea of storing the magnesium in these materials depends on the resonant reaction of the two double bonds in the inner ring of PAQsulphide (S) from a binary C=O to a single C–O bond before and after reaction with magnesium, respectively. Using HMDS:hexamethyldisilazide/THF as an electrolyte, the electrochemical performance between the developed polyanthraquinone series were compared. The authors reported the redox reactions of 14PAQ and 26PAQ using cyclic voltammetry, as shown in Figure 5.10b and c. 14PAQ displayed two reduction phases at ~1.57 and 1.48 V and a broad oxidation peak at ~2.0 V, while 26PAQ displayed the broad oxidation peak at 1.75 V, which shifted about 0.25 V compared with 14PAQ.

This shift in the value of the oxidation voltage can be attributed to the low binding energy between the magnesium ion and the oxygen atom of the 26PAQ compared to the 14PAQ. It is noteworthy that the magnesium cell-based 26PAQ cathode displayed a specific capacity more than 100 mAh g^{-1} in the first 100 cycles at a rate equal to 130 mAg^{-1}, but the capacity value continued to decrease by exceeding the 100 cycles. On the other hand, the magnesium cell-based 14PAQ cathode displayed stable specific capacity for more than 1000 cycles at relatively high current rates, Figure 5.10d and e. This high electrochemical performance of the magnesium cell-based 14PAQ cathode compared with the standard cathode Mo_6S_8 opens the gate for an organic cathode as potential candidate electrodes for Mg batteries. These results are also beneficial in that they are working to encourage researchers to devote more efforts to optimize the physical and the electrochemical properties of the organic cathode materials via designing and tuning new structures of these materials. Aiming to understand the electrochemical conversion reaction mechanism in organic cathodes, Vizintin et al. (2018) and Bančič et al. (2018) tried to follow the changes that occur in the active bonds of the PAQS and naphthalenehydrazine diimide polymer cathodes through the development of an in situ infrared spectrometer during discharge/charge processes of the Mg battery. As shown in Figure 5.11, the intensity of C=O and –C=C– stretching modes of PAQS during cell work decreases upon discharge and restores their values upon charge, additional peak indexed to –C–O– Mg^{2+} was observed at 1376 cm^{-1}. The cornerstone of their explanation is that the conversion reaction of a PAQS cathode during discharge/charge processes of the Mg battery is based on

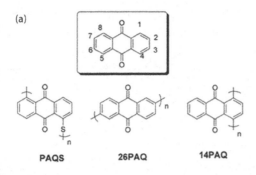

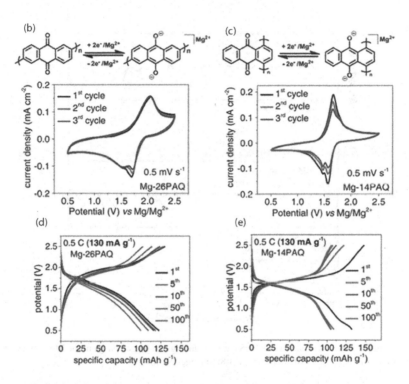

FIGURE 5.10 (a) Illustration of the chemical structure of anthraquinone, PAQS, 26PAQ, and 14PAQ. Cyclic voltammetry of Mg cells-based (0.5 mV s-1) (b) 26PAQ and (c) 14PAQ; and discharge/charge profiles of Mg cells-based (d) 26PAQ and (e) 14PAQ. (From Pan, B. et al.: Polyanthraquinone-Based Organic Cathode for High-Performance Rechargeable Magnesium-Ion Batteries, *Adv. Energy Mater.*, 2016, 6, 1600140. Copyright Wiley-VCH Verlag GmbH & Co. KGaA. Reproduced with permission.)

the reversible redox reaction of the carbonyl bond from C=O (double) to C–O– (single). They attributed the poor electrochemical performance of a Mg/PAQS cell compared to a Li/PAQ cell to the large barrier Mg^{2+} solvation shell that restricts the spread of Mg^{2+} ions deep in the polymer. Their work extended to a comparison between the theoretical vibrational modes, which were predicted by the theoretical density functional theory (DFT) calculations based on harmonic frequencies calculations, and the experimental vibrational modes of PAQS in Mg systems, the results showed a significant match between the two results.

In summary, benefitting from their low cost and environmental benignancy, organics can be promising cathodes for rechargeable magnesium batteries. Yet, some challenges remain: the development of

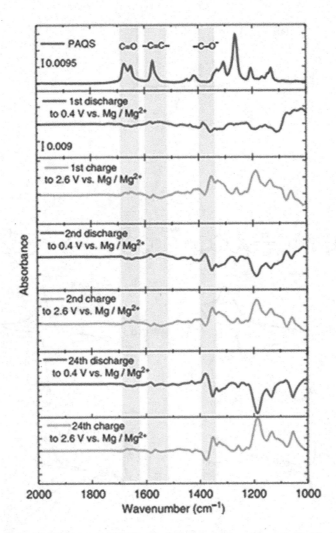

FIGURE 5.11   In situ Infrared (IR) spectra of PAQS cathode during discharge/charge in the Mg battery. (Reprinted with permission from Vizintin et al. 2018, 661. Copyright 2018, Springer Nature.)

electrolyte systems to suppress electrode dissolution, design and engineering new structures that are electrically starved for magnesium ions, the low specific capacity, understanding the nature of the conversion reaction, and solving the problem of electrode instability.

5.3.5 Magnesium/Alloy Battery

Alloying-type anode materials are among the established anode systems that recently attracted high attention as promising anode materials for magnesium-ion batteries, this is due to their ability to alloy with magnesium to form Mg-metal-alloy phases that can deliver high specific capacity. One of the key elements in this regard is Sn (high theoretical capacity ~903 mAh g^{-1}), a low diffusion barrier, and small volume expansions qualify the Sn to be one of the important alloying-type anode materials for a rechargeable magnesium battery. However, the sluggish kinetics and the low reversibility are some of the obstacles to the application of Sn. There was an attempt to improve the electrochemical properties of Sn by Niu et al. (2018) by synthesizing the nanoporous phase of Bi_6Sn_4 and Bi_4Sn_6 (NP-Bi-Sn). This nanoporous dual phase shows high porosity and high density of spaces between grains, which enabled it to absorb volume changes during cycling, and improve the kinetics of Mg^{2+} diffusion. The authors used 0.4 M APC as an electrolyte and ex-situ XRD, TEM, and selected area electron diffraction (SAED) techniques to characterize the NP-Bi_6Sn_4 before and after magnesiation processes, as shown in Figure 5.12. The XRD pattern of a pristine NP-Bi_6Sn_4 phase agreed well with the TEM and SAED results, Figure 5.12a–c, where the fingerprint of the phase coincided with 111 and 110 zone axes of Sn and Bi, respectively. As shown in Figure 5.12a, after the magnesiation process, the XRD pattern displayed new phases which matched well with cubic Mg_2Sn, and hexagonal Mg_3Bi_2 phases according to the standard files of joint committee on powder diffraction standards (JCPDS) no. 65-2997 and no. 65-8732, respectively. After magnesiation, TEM images and SAED patterns matched well with the XRD results and confirmed the existence of the two dual phases Mg_2Sn and Mg_3Bi_2 (Figure 5.12d–g). As shown in Figure 5.12h and k, the NP-Bi_6Sn_4 and NP-Bi_4Sn_6 dual phase electrodes delivered high specific capacity > 400 mAh g^{-1} and stable cycling in comparison to the single phases. The main reason behind the great difference between

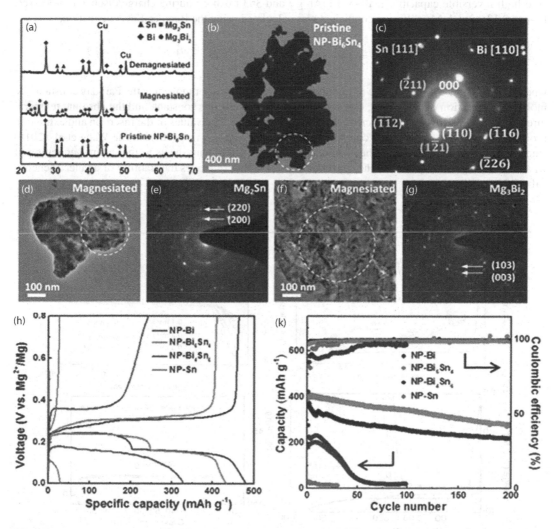

FIGURE 5.12 (a) XRD curves of NP-Bi_6Sn_4 before and after magnesiation. (b) TEM image and (c) SAED pattern of NP-Bi_6Sn_4 (d), (f). TEM images (e), (g). SAED patterns Mg/NP-Bi_6Sn_4. (h) Charge/discharge curves of NP-Bi, NP-Bi-Sn (50 mAg^{-1}), and NP-Sn (20 mA g^{-1}), (k) cycling performance of NP-Bi, NP-Bi-Sn (200 mAg^{-1}), and NP-Sn (20 mAg^{-1}). (Reprinted from *Energy Storage Mater.*, Niu, J. et al., Dual Phase Enhanced Superior Electrochemical Performance of Nanoporous Bismuth-tin Alloy Anodes for Magnesium-ion Batteries. Copyright 2018, with permission from Elsevier.)

the experimental specific capacity value of the NP-Bi$_6$Sn$_4$ compared to the theoretical value can be attributed to the non-participation of all Bi and Sn atoms in the reaction.

The reason behind the high electrochemical performance of the dual phases of Bi-Sn alloys compared to single phases of Sn and Bi can be explained in the light of the high porosity of the dual phases relative to the single phases. The high porosity generates fast kinetics of Mg^{2+} in between pores of the dual phase and can absorb volume changes upon cycling and hence prohibit the electrochemical milling. Wang et al. (2018) tested the Mg storage mechanism of bismuth oxyfluoride (BiOF) nanosheets. They synthesised BIOF via a solvothermal method. Their study demonstrated BiOF under a successive conversion reaction from BiOF to Bi and BiMg$_2$ during the Mg^{2+} ion insertion process, Figure 5.13a. They confirmed the reversibility of this conversion reaction using cyclic voltammetry at various scan rates. Cyclic voltammetry (CV) spectra (Figure 5.13b), of the first cathodic scan displayed two reduction peaks at ~0.85 and 0.02 V, represented the conversion reaction from BiOF to Bi and Mg$_3$Bi$_2$, respectively, while the CV spectra of the first anodic scan confirmed the de-alloying process from Bi and Mg$_3$Bi$_2$ to BiOF via the oxidation that was observed at ~0.45 V. After a few cycles of activation processes, this electrode delivered high reversible capacity equal ~335 mAh g^{-1} and 353 mAh g^{-1} during charge/discharge processes, Figure 5.13c and d. Mg diffusion coefficient was calculated using the Randles-Sevcik equation,

$$I_p = 0.4463n^{3/2}F^{3/2}CSR^{-1/2}T^{-1/2}D_{cv}^{1/2}v^{1/2}$$

where n, F, C, S, R, T, D_{cv}, and v are the number of active electrons per molecule, the Faraday constant, the molar concentration of Mg^{2+} ions, electrode's surface area, the gas constant, and the absolute temperature, the chemical diffusion coefficient, and the scanning rate. D_{Mg} values of the insertion and extraction processes are equal ~1.78×10^{-14} cm^2 s^{-1} and 1.28×10^{-14} cm^2 s^{-1}, respectively. The Wang et al. (2018) study attributed the high electrochemical performance of BiOF nanosheets to the in situ generation of Bi atoms during the sequence of the conversion reaction that improves the stability of an electrochemical reaction.

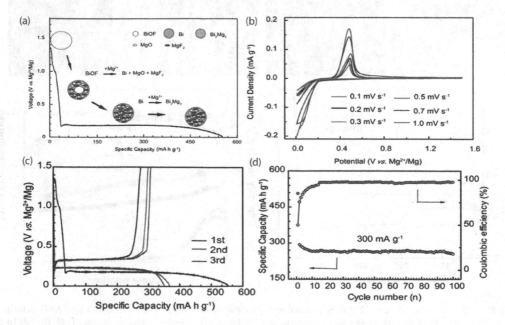

FIGURE 5.13 (a) The phase evolution of a BiOF electrode at different discharge states. (b) The CV curves of a Mg/BiOF at different scan rates, (c) charge/discharge profiles of a Mg/BiOF cell, and (d) cycling performance of a Mg/BiOF cell. (Wang, W. et al., *Chem. Commun.*, 1714–1717, 2018. Reproduced by permission of The Royal Society of Chemistry.)

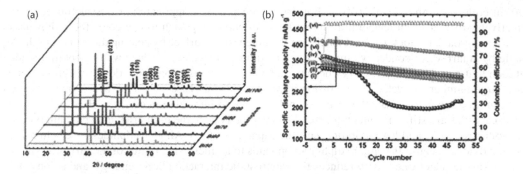

FIGURE 5.14 XRD patterns of the samples: (a) Bi/RGO. (b) Cycle life stability and Columbic efficiency of the samples are represented for 50 cycles: Bi100 (i), Bi95 (ii), Bi80 (iii), Bi70 (iv), Bi60 (v), and Bi50 (vi). (Penki, T.R. et al., *New J. Chem.*, 2018, 5996–6004. Reproduced by permission of The Royal Society of Chemistry.)

Penki et al. (2018) synthesized x wt.% bismuth (Bi)/$_{(1-x)}$wt.% reduced graphene oxide (RGO) nano-composites by a solvothermal method at 100°C under a N_2 atmosphere as an anode material for Mg-ion batteries. Their strategy is based upon the ability of RGO to increase the electrochemical performance via minimization of mechanical strain during magnesiation and demagnesiation. Figure 5.14a shows the XRD patterns of $_x$wt.%(Bi)/$_{(1-x)}$wt.%(RGO), and the main peaks of Bi are indexed to a rhombohedral phase (JCPDS No. 05-0519). The absence of the characteristic diffraction peak of the graphene sheets in the composites reflects less agglomeration of graphene sheets. The composite-based 60% Bi and 40% RGO displayed a high electrochemical performance compared to the other composite samples, Figure 5.14b. The high electronic conductivity of the Bi/RGO composites and their ability to absorb the mechanical changes gave it high accreditation. In summary, alloying-type anode materials possess good merits such as high theoretical capacity and low reduction potentials. We can overcome the low electrochemical stability issue after long cycling by reducing the Mg diffusion lengths to mitigate kinetic hurdles. This motivation can be verified via developing the innovative design of nanoporous, dual phase alloys that can promote the development of alloying-type anode materials for practical rechargeable Mg batteries.

5.4 Summary and Outlook

High cost and scarcity of lithium resources represent high barriers to the ability of the lithium battery to meet the demands of power applications and the smart grid. Magnesium has many features that qualify it to be a strong competitor to the lithium battery in the near future such as low cost, the abundance of Mg, the high volumetric capacity of Mg metal, and the safety benefit due to lack of dendrite formation. Although magnesium is gaining the high volumetric capacity of its bivalent nature, it causes a range of obstacles to its practical application, most notably the lack of a thermodynamically stable electrolyte at low potential and the development of a suitable functional cathode to accommodate the bivalent Mg^{2+} ions with reasonable capacity. Therefore, there must be substantial funding for researchers to overcome these obstacles. Many electrolytes have been developed in recent years, such as Mg organohaloaluminate in THF, APC electrolytes, $LiBH_4/Mg(BH_4)_2$ in DMEs, non-nucleophilic hexamethyldisilazide magnesium chloride, and non-nucleophilic magnesium tetrak is(hexafluoroisopropyloxy) borate. Although these electrolytes achieved good results, varying from reversible Mg deposition/stripping, wide electrochemical windows, high ionic conductivity, and high Coulombic efficiency, there are still many tough drawbacks such as large environmentally benign properties, sensitivity to water and/or air contamination (detrimental results), capacity fading as a result of the growth of the passive layer at a Mg electrolyte interface, corrosivity towards low cost current collectors, and volatile solvent hindering their practical application. Constructing SPEs that effectively work at room temperature might be an effective approach to overcome these drawbacks. Another approach is blocking the reduction of the magnesium surface by coating the surface of the Mg anode by smart Mg^{2+}-conductive interphase. An additional challenge is how we can obtain the ideal cathode that can deliver high energy density and accepts the insertion/extraction of magnesium ions for a large number of cycles. Mg diffusion in Mo_6S_8 Chevrel phase shows the best kinetics because metallic Mo_6 can compensate the charge imbalance due to intercalation of Mg^{2+}. However, Chevrel-phase cathodes are still far from commercial applications due to the high price molybdenum issue, its low voltages ~1.1 V, and low theoretical specific capacities ~130 mAh g^{-1}. Sulphides have better kinetics than oxides because the larger sulphur anion leads to an increasing radius of the diffusion channel and the smaller electronegativity reduces the electrostatic interaction between Mg^{2+} and anions of the host lattice. The nature of an electronic configuration of V_2O_5 gives it the ability to adopt multiple oxidation states and gives its structure the flexibility to accept insertion/extraction of a variety of monovalent and multivalent cations. As a result, a large number of researchers were attracted in the last period to study the possibility of using V_2O_5 as a cathode for the magnesium battery, most of these studies reported that high reversible capacity can be achieved only in the presence of water. Still, some issues remain to be addressed: does Mg^{2+} diffuse within V_2O_5 lattice or just at the surface? What is the ratio of intercalated protons and Mg^{2+}?

Without a doubt, access to the era of a magnesium battery beyond the era of a lithium battery requires shifting from insertion to conversion cathodes such as sulphur and CuS. Despite the significant improvements regarding the properties of Mg/S batteries that have been achieved, due to the introduction of non-nucleophilic electrolytes like boron-based electrolytes, there is a lot of research needed to overcome some drawbacks, such as the shuttle phenomenon of polysulphides, the slow conversion reaction kinetics, the incomplete reversibility of the conversion reaction, the electro-chemical milling of the cathode due to the volume changes, and elucidate the nature of the reaction between the cathode and the electrolyte to avoid the undesired reactions. Organic cathodes are inexpensive, environmentally friendly, and show faster kinetics compared to the inorganic compounds due to the weak interaction between Mg^{2+} and organic molecules. Flexible structure and tenability of the organic cathode materials qualify these cathodic materials to deliver high capacity, high rate, and high diffusion of Mg^{2+}. Although operando infrared spectroscopy gave us valuable information about the nature of the conversion reaction of the organic cathode, there are still some challenges remaining: the electrode dissolution, understanding the nature of the conversion reaction, and solving the problem of electrode instability. Alloying-type anode materials possess good merits, such as high theoretical capacity and low reduction potentials. The recent research in this area focused on overcoming some challenges that hinder an applicable alloying-type anode, such as the low electro-chemical stability issue after long cycling. Suggested solutions such as reducing the Mg diffusion lengths, raise the value of porosity, engineering dual phase alloys showed a marked electrochemical improvement. In this chapter, we have tracked the most important results published in the last five years regarding the development of cathodes and electrolytes of a magnesium battery and discussed the progress and challenges. It is therefore clear that in order to realize a commercialized magnesium battery, future work should focus on addressing the problems of the anode-electrolyte passive layer, as well as engineering a new cathode structure that can reconcile with the strong electrostatic interaction due to the introduction of the bivalent cation. The main requirements for realizing a practical magnesium battery are not far off, more work is required to: (1) develop ideas out of the box to develop chloride-free electrolyte solutions that can possess high electrochemical stability and Mg^{2+} that can behave reversibly. (2) Engineering a Mg^{2+}-conductive solid polymer interface at the surface of a Mg anode in order to block the reducible molecules from reaching the magnesium anode and (3) engineer high electrochemical performance cathode materials. All these frameworks guarantee the access to a Mg battery era. We hope this chapter will shed light on the fundamental science of developing Mg batteries.

REFERENCES

Attias, Ran, Michael Salama, Baruch Hirsch, Yosef Goffer, and Doron Aurbach. 2018. "Anode-Electrolyte Interfaces in Secondary Magnesium Batteries." *Joule.* https://doi.org/10.1016/j.joule.2018.10.028.

Aurbach, D., Z. Lu, A. Schechter, Y. Gofer, H. Gizbar, R. Turgeman, Y. Cohen, M. Moshkovich, and El Levi. 2000. "Prototype Systems for Rechargeable Magnesium Batteries." *Nature* 407 (6805): 724. doi:10.1038/35037553.

Bančič, Tanja, Jan Bitenc, Klemen Pirnat, Anja Kopač Lautar, Jože Grdadolnik, Anna Randon Vitanova, and Robert Dominko. 2018. "Electrochemical Performance and Redox Mechanism of Naphthalene-Hydrazine Diimide Polymer as a Cathode in Magnesium Battery." *Journal of Power Sources* 395: 25–30. https://doi.org/10.1016/j.jpowsour.2018.05.051.

Du, Aobing, Zhonghua Zhang, Hongtao Qu, Zili Cui, Lixin Qiao, Longlong Wang, Jingchao Chai, Tao Lu, Shanmu Dong, and Tiantian Dong. 2017. "An Efficient Organic Magnesium Borate-Based Electrolyte with Non-nucleophilic Characteristics for Magnesium–Sulfur Battery." *Energy & Environmental Science* 10 (12): 2616–25. https://doi.org/10.1039/C7EE02304A.

Duffort, Victor, Xiaoqi Sun, and Linda F. Nazar. 2016. "Screening for Positive Electrodes for Magnesium Batteries: A Protocol for Studies at Elevated Temperatures." *Chemical Communications* 52 (84): 12458–61. https://doi.org/10.1039/C6CC05363G.

Ha, Se-Young, Yong-Won Lee, Sang Won Woo, Bonjae Koo, Jeom-Soo Kim, Jaephil Cho, Kyu Tae Lee, and Nam-Soon Choi. 2014. "Magnesium (II) Bis (trifluoromethane sulfonyl) Imide-Based Electrolytes with Wide Electrochemical Windows for Rechargeable Magnesium Batteries." *ACS Applied Materials & Interfaces* 6 (6): 4063–73. https://doi.org/10.1021/am405619v.

Kim, Hee Soo, Timothy S. Arthur, Gary D. Allred, Jaroslav Zajicek, John G. Newman, Alexander E. Rodnyansky, Allen G. Oliver, William C. Boggess, and John Muldoon. 2011. "Structure and Compatibility of a Magnesium Electrolyte with a Sulphur Cathode." *Nature Communications* 2: 427. 10.1038/ncomms1435.

Kong, Long, Chong Yan, Jia-Qi Huang, Meng-Qiang Zhao, Maria-Magdalena Titirici, Rong Xiang, and Qiang Zhang. 2018. "A Review of Advanced Energy Materials for Magnesium–Sulfur Batteries." *Energy & Environmental Materials* 1 (3): 100–12. https://doi.org/10.1002/eem2.12012.

Mohtadi, Rana, Masaki Matsui, Timothy S. Arthur, and Son-Jong Hwang. 2012. "Magnesium Borohydride: From Hydrogen Storage to Magnesium Battery." *Angewandte Chemie International Edition* 51 (39): 9780–83. https://doi.org/10.1002/anie.201204913.

Mukherjee, Arijita, Niya Sa, Patrick J. Phillips, Anthony Burrell, John Vaughey, and Robert F. Klie. 2017. "Direct Investigation of Mg Intercalation into the Orthorhombic V2O5 Cathode Using Atomic-Resolution Transmission Electron Microscopy." *Chemistry of Materials* 29 (5): 2218–26. https://doi.org/10.1021/acs.chemmater.6b05089.

Muldoon, John, Claudiu B. Bucur, Allen G. Oliver, Tsuyoshi Sugimoto, Masaki Matsui, Hee Soo Kim, Gary D. Allred, Jaroslav Zajicek, and Yukinari Kotani. 2012. "Electrolyte Roadblocks to a Magnesium Rechargeable Battery." *Energy & Environmental Science* 5 (3): 5941–50. doi:10.1039/C2EE03029B.

Niu, Jiazheng, Hui Gao, Wensheng Ma, Fakui Luo, Kuibo Yin, Zhangquan Peng, and Zhonghua Zhang. 2018. "Dual Phase Enhanced Superior Electrochemical Performance of Nanoporous Bismuth-Tin Alloy Anodes for Magnesium-Ion Batteries." *Energy Storage Materials.* https://doi.org/10.1016/j.ensm.2018.05.023.

Pan, Baofei, Jinhua Huang, Zhenxing Feng, Li Zeng, Meinan He, Lu Zhang, John T. Vaughey, Michael J. Bedzyk, Paul Fenter, and Zhengcheng Zhang. 2016. "Polyanthraquinone-Based Organic Cathode for High-Performance Rechargeable Magnesium-Ion Batteries." *Advanced Energy Materials* 6 (14): 1600140. https://doi.org/10.1002/aenm.201600140.

Penki, Tirupathi Rao, Geetha Valurouthu, S. Shivakumara, Vijay Anand Sethuraman, and N. Munichandraiah. 2018. "In Situ Synthesis of Bismuth (Bi)/Reduced Graphene Oxide (RGO) Nanocomposites as High-Capacity Anode Materials for a Mg-Ion Battery." *New Journal of Chemistry* 42 (8): 5996–6004. https://doi.org/10.1039/C7NJ04930G.

Sa, Niya, Tiffany L. Kinnibrugh, Hao Wang, Gopalakrishnan Sai Gautam, Karena W. Chapman, John T. Vaughey, Baris Key, Timothy T. Fister, John W. Freeland, and Danielle L. Proffit. 2016. "Structural Evolution of Reversible Mg Insertion into a Bilayer Structure of $V_2O_5 \cdot nH_2O$ Xerogel Material." *Chemistry of Materials* 28 (9): 2962–69. https://doi.org/10.1021/acs.chemmater.6b00026.

Sa, Niya, Hao Wang, Danielle L. Proffit, Albert L. Lipson, Baris Key, Miao Liu, Zhenxing Feng, Timothy T. Fister, Yang Ren, and Cheng-Jun Sun. 2016. "Is Alpha-V_2O_5 a Cathode Material for Mg Insertion Batteries?" *Journal of Power Sources* 323: 44–50. https://doi.org/10.1016/j.jpowsour.2016.05.028.

Shao, Yuyan, Nav Nidhi Rajput, Jianzhi Hu, Mary Hu, Tianbiao Liu, Zhehao Wei, Meng Gu, Xuchu Deng, Suochang Xu, and Kee Sung Han. 2015. "Nanocomposite Polymer Electrolyte for Rechargeable Magnesium Batteries." *Nano Energy* 12: 750–79. https://doi.org/10.1016/j.nanoen.2014.12.028.

Shterenberg, Ivgeni, Michael Salama, Yossi Gofer, Elena Levi, and Doron Aurbach. 2014. "The Challenge of Developing Rechargeable Magnesium Batteries." *Mrs Bulletin* 39 (5): 453–60. https://doi.org/10.1557/mrs.2014.61.

Son, Seoung-Bum, Tao Gao, Steve P. Harvey, K. Xerxes Steirer, Adam Stokes, Andrew Norman, Chunsheng Wang, Arthur Cresce, Kang Xu, and Chunmei Ban. 2018. "An Artificial Interphase Enables Reversible Magnesium Chemistry in Carbonate Electrolytes." *Nature Chemistry* 10 (5): 532. 10.1038/s41557-018-0019-6.

Vizintin, Alen, Jan Bitenc, Anja Kopač Lautar, Klemen Pirnat, Jože Grdadolnik, Jernej Stare, Anna Randon-Vitanova, and Robert Dominko. 2018. "Probing Electrochemical Reactions in Organic Cathode Materials via in Operando Infrared Spectroscopy." *Nature Communications* 9 (1): 661. https://doi.org/10.1038/ncomms1435.

Wang, Wei, Lin Liu, Peng-Fei Wang, Tong-Tong Zuo, Ya-Xia Yin, Na Wu, Jin-Ming Zhou, Yu Wei, and Yu-Guo Guo. 2018. "A Novel Bismuth-Based Anode Material with a Stable Alloying Process by the Space Confinement of an In Situ Conversion Reaction for a Rechargeable Magnesium Ion Battery." *Chemical Communications* 54 (14): 1714–17. https://doi.org/10.1039/C7CC08206A.

Wu, Mengyi, Yujie Zhang, Ting Li, Zhongxue Chen, Shunan Cao, and Fei Xu. 2018. "Copper Sulfide Nanoparticles as High-Performance Cathode Material for Magnesium Secondary Batteries." *Nanoscale*. https://doi.org/10.1039/C8NR03375G.

Yoo, Hyun Deog, Ivgeni Shterenberg, Yosef Gofer, Gregory Gershinsky, Nir Pour, and Doron Aurbach. 2013. "Mg Rechargeable Batteries: An On-going Challenge." *Energy & Environmental Science* 6 (8): 2265–79. DOI: 10.1039/C3EE40871J.

Zhang, Zhonghua, Shamu Dong, Zili Cui, Aobing Du, Guicun Li, and Guanglei Cui. 2018. "Rechargeable Magnesium Batteries Using Conversion-Type Cathodes: A Perspective and Minireview." *Small Methods*: 1800020. https://doi.org/10.1002/smtd.201800020.

Zhao-Karger, Zhirong, Runyu Liu, Wenxu Dai, Zhenyou Li, Thomas Diemant, B. P. Vinayan, Christian Bonatto Minella, Xingwen Yu, Arumugam Manthiram, and R. Jürgen Behm. 2018. "Toward Highly Reversible Magnesium–Sulfur Batteries with Efficient and Practical Mg [B (Hfip) 4] 2 Electrolyte." *ACS Energy Letters* 3 (8): 2005–13. https://doi.org/10.1021/acsenergylett.8b01061.

Zhu, Jian, Jianli Zou, Hua Cheng, Yingying Gu, and Zhouguang Lu. 2018. "High Energy Batteries Based on Sulfur Cathode." *Green Energy & Environment*. https://doi.org/10.1016/j.gee.2018.07.001.

6

Graphene-Based Electrode Materials for Supercapacitor Applications

Moshawe J. Madito, Katlego Makgopa, Christopher B. Mtshali, and Abdulhakeem Bello

CONTENTS

6.1 Introduction 101
6.2 Graphene Synthesis Methods 104
    6.2.1 Chemical Modification of GO for Graphene-Based Supercapacitors 104
    6.2.2 CVD Graphene Growth for Graphene-Based Supercapacitor Electrodes 105
6.3 Characterization of Graphene 106
    6.3.1 Raman Spectroscopy 106
    6.3.2 Transmission Electron Microscopy and Scanning Electron Microscopy 107
    6.3.3 X-Ray Photoelectron Spectroscopy 108
6.4 Principle and Performance of Supercapacitors 109
6.5 Applications of Graphene-Based Electrodes in Supercapacitors 111
    6.5.1 0D Graphene Dots or Powders 112
    6.5.2 1D Carbon Nanotubes 112
    6.5.3 2D Graphene Sheets/Films 118
    6.5.4 3D Graphene-Based Foams and Aerogels 119
6.6 Conclusion and Outlook 125
References 125

6.1 Introduction

Supercapacitors (or electrochemical capacitors) have attracted considerable recent attention due to their high-power density, fast dynamics of charge propagation, long cycle life performance (>10,000 cycles), low maintenance, and safe and reliable performance for energy storage applications (Lemine et al. 2018). Also, supercapacitors have the potential to complement or eventually replace the rechargeable batteries for some energy storage applications. But supercapacitors are plagued with a lower energy density (i.e., ~5–8 Wh kg^{-1}) as compared to rechargeable lithium-ion batteries (120–200 Wh kg^{-1}) (Simon and Gogotsi 2008; Naoi and Simon 2008). Therefore, there are great research efforts towards enhancing the performance of supercapacitor devices to have considerable energy densities comparable to that of lithium-ion batteries. According to their charge storage mechanism, there are two types of supercapacitors, i.e., electrochemical double-layer capacitor (EDLC) and pseudocapacitor (Figure 6.1). The operation mechanism of the former involves the non-Faradaic separation of charges at the 'double-layer' (i.e., electrode/electrolyte interface), while the latter involves fast Faradaic charge transfer between the electrolyte and the electrode material at the interface, involving reversible redox reactions, electroadsorption, or intercalation processes (Figure 6.1) (Elzbieta Frackowiak 2001). Carbon-based materials have been mostly used as the electrode materials for EDLCs due to their high specific surface area, good electrical conductivity, and

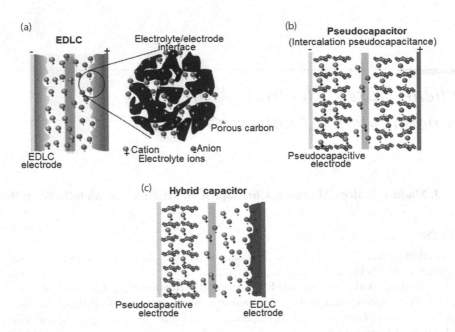

FIGURE 6.1   (a) Schematic of (a) an EDLC and enlargement of porous carbon electrode interface, (b) pseudocapacitor, and (c) hybrid capacitor. (Reprinted from *Mater. Today*, 21, Lin, Z. et al., Materials for supercapacitors: When li-ion battery power is not enough, 419–436, Copyright 2018, with permission from Elsevier.)

large pore size distribution (micro, meso, and macropores) (Y. Wang et al. 2018; Liu et al. 2018), while pseudocapacitors are from redox-active materials, such as transition metal oxides or hydroxides and conductive polymers. Hybrid materials or devices incorporating EDLCs and pseudocapacitor materials (Figure 6.1) are thought to give the next-generation high-performance supercapacitor devices (Chidembo et al. 2010; Toupin, Brousse, and Be 2004).

Graphene-related electrode materials (or hybrid materials incorporating graphene and pseudocapacitive materials) based on the structural complexity of graphene, zero-dimension (0D), one-dimension (1D), two-dimension (2D), and three-dimension (3D), have been extensively investigated as a support of the redox reactions of the nanoparticles of transitional metal oxides/hydroxides and conducting polymers for superior electrochemical performance, as a result of the synergistic effect between graphene and the composite materials.

Briefly, graphene, a 2D one-atom-thick sheet of sp²-bonded carbon atoms in a honeycomb crystal lattice with atoms arranged in a hexagonal pattern, since its discovery, has emerged as one of the most exciting materials for research (Huang et al. 2012). Graphene can be wrapped up into 0D buckyballs, rolled into a 1D nanotube, or stacked into 3D graphite, hence, it is a building block for other carbon allotropes (see Figure 6.2) (Wan, Huang, and Chen 2012).

Moreover, graphene has unique electronic properties: it is a zero-gap semiconductor that exhibits linear dispersion at a high symmetry point in the reciprocal space, resulting in effective dynamics of electrons similar to that of massless relativistic Dirac fermions (Novoselov et al. 2005). Also, graphene has excellent optical, thermal, mechanical, and electrochemical properties that are superior to other allotropes of carbon, such as graphite, carbon nanotubes (CNTs), and fullerene as shown in Table 6.1 (Rao et al. 2009; Schedin et al. 2007; Pisana et al. 2007; Nair et al. 2008; Novoselov et al. 2005). The recognition of graphene's properties promulgated several graphene-based materials for potential application in electronics, energy storage, catalysis, gas sorption, storage, separation, and sensing. This chapter will comprehensively summarize the recent research efforts towards graphene-based electrodes for enhancing the performance of the supercapacitor devices.

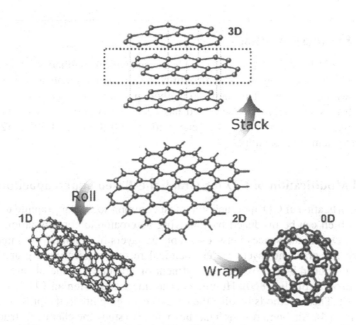

FIGURE 6.2 Schematic illustration: Graphene is the building block for other carbon allotropes. (Reprinted with permission from Wan, X. et al., *Acc. Chem. Res.*, 45, 598–607, 2012. Copyright 2012 American Chemical Society.)

TABLE 6.1

A Comparison of Intrinsic Properties of Graphene with Those of Various Carbon Allotropes

Allotropes of Carbon	Graphite	Diamond	Fullerene	Carbon Nanotubes	Graphene
Dimensionality	Three (3D)	Three (3D)	Zero (0D)	One (1D)	Two (2D)
Hybridization	sp^2	sp^3	Mainly sp^2	Mainly sp^2	sp^2
Crystal system	Hexagonal	Octahedral	Tetragonal	Icosahedral	Hexagonal
Experimental specific surface area ($m^2\ g^{-1}$)	~10–20	~20–160	~80–90	~1300	~2675
Density ($g\ cm^{-3}$)	2.09–2.23	3.5–3.53	1.72	>1	>1
Electrical conductivity ($S\ cm^{-1}$)	Anisotropic 2,3 ×10^4a, ×10^6b	–	10^{-10}	Depends on the particular structure	2000
Electronic properties	Conductor	Insulator, semiconductor	Insulator	Metallic or semiconducting	Semimetal, zero gap semiconductor
Thermal conductivity ($W\ m^{-1}K^{-1}$)	1500–2000a 5–10c	900–2320	0.4	3500	4848–5300
Hardness tenacity	High	Ultrahigh	Highly elastic	High flexible elastic	Highest flexible elastic (single layer)
Optical properties	Uniaxial	Isotropic	Non-linear optical response	Structural dependent	97.7% optical transmittance

Key: a, b, c = directions relative to the plane.

Source: Rao, C.N. et al., *Angewandte Chemie*, 48, 7752–7777, 2009; Schedin, F. et al., *Nat. Mater.*, 6, 652–655, 2007; Pisana, S. et al., *Nat. Mater.*, 6, 198–201, 2007; Nair, R.R. et al., *Science*, 320, 1308, 2008; Wu, Z.-S., et al., *Nano Energy*, 1, 107–131, 2012; Novoselov, K.S.A. et al., *Nature*, 438, 197–200, 2005.

6.2 Graphene Synthesis Methods

Graphene is synthesized by several processes, such as mechanical exfoliation of graphite, chemical vapour deposition (CVD), unzipping of CNTs, also through reduction of graphene oxide, etc., for various applications, as shown in Figure 6.3. The chemical method via reduction of graphite oxide is considered a scalable approach to synthesizing graphene and has been widely utilized to synthesize chemically derived graphene also known as reduced graphene oxide (rGO) (C. Li and Shi 2012), especially for application in electrochemical capacitors.

6.2.1 Chemical Modification of GO for Graphene-Based Supercapacitors

In the chemical modification of GO approach, the most common source of graphite used for oxidation is flake graphite, which can be produced by removing heteroatomic contaminations from naturally occurring graphite. Due to the spaces between graphene layers in graphite, the intercalating agents can reside between the graphene layers under chemical reactions, forming a graphite intercalation compound. GO is produced by the oxidative treatment of graphite via one of three principal methods developed by Brodie (Brodie 1860), Hummers (Hummers and Offeman 1958), and Staudenmaier (Staudenmaier 1898). These methods involve the oxidation of graphite in the presence of strong acids and oxidants (Figure 6.4). Although research has been done to study the chemical structure of graphite oxide, several models are still being proposed (Park and Ruoff 2009). Graphite oxide thus consists of a layered structure of 'graphene oxide' sheets that are strongly hydrophilic, such that intercalation of water molecules between the layers readily occurs (Buchsteiner, Anton Lerf, and Pieper 2006). The interlayer distance between the graphene oxide sheets increases reversibly from 0.6 to 1.2 nm with increasing relative humidity (Buchsteiner, Anton Lerf, and Pieper 2006). Notably, graphite oxide can be completely exfoliated to produce aqueous colloidal suspensions of graphene oxide sheets by simple sonication (Stankovich et al. 2006) and by stirring the water/graphite oxide mixture for a long enough

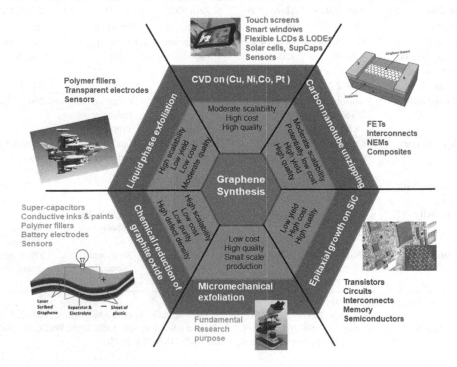

FIGURE 6.3 A process flowchart of graphene synthesis methods and the range of its potential applications.

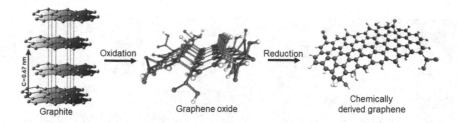

FIGURE 6.4 Schematic illustration of graphite oxidation into graphene oxide followed by its reduction into reduced graphene oxide. (Reprinted with permission from Tung, V.C. et al., *Nat. Nanotechnol.*, 4, 25–29, 2009, Copyright 2008, Springer Nature.)

time (Jung et al. 2007). GO consists of graphene sheets decorated mostly with epoxide and hydroxyl groups. Rapid heating of GO results in its expansion and delamination caused by the rapid evaporation of the intercalated water and the evolution of gases from pyrolysis of the oxygen-containing groups. Such thermal treatment has recently been suggested to be capable of producing individual functionalized graphene sheets (Jung et al. 2007). In addition to thermal treatment, chemical treatment has also been shown to reduce graphene oxide sheets into graphene using various reducing agents like sodium borohydride and hydrazine (Figure 6.4) (Tung et al. 2009). GO is electrically insulating and thermally unstable. It has been demonstrated that the electrical conductivity of GO can be restored by chemical reduction.

Furthermore, irradiation of the thin film of graphite oxide dispersed in water and drop-cast onto a flexible substrate can produce graphene-based electrochemical capacitors (El-kady et al. 2012). In this approach, irradiation of the film with an infrared laser inside a LightScribe CD/DVD optical drive reduces the GO to a laser-scribed graphene. An advantage of a laser-scribed graphene is that it can be directly used as an electrode in electrochemical capacitors without any additional binders or conductive additives.

6.2.2 CVD Graphene Growth for Graphene-Based Supercapacitor Electrodes

CVD has proved to be the most popular method for graphene growth, as it yields high-quality and large-area graphene films with controlled thickness, and it is reproducible technique. CVD graphene growth on Ni foam has received attention as an efficient process to synthesize 3D graphene foams, which are highly conductive with fewer defects for the fabrication of 3D nanocomposite-type electrodes. The high conductivity and a high specific surface area of a 3D graphene foam network can promote the rapid access of electrolyte ions to the electrode surface and facilitate the fast electron transport between the active materials and current collectors in supercapacitor electrodes. Figure 6.5A shows typical field emission scanning electron microscopy (FESEM) images of vertical graphene nanosheets (VGNSs) grown on the Ni foam by a plasma-assisted chemical vapour deposition system. In Figure 6.5Aa–Ac, it can be seen that the dense and uniform structure of the VGNS is covering the entire Ni foam. From the magnified image (Figure 6.5Ad–Af), it can be observed that the VGNSs exhibit strong binding to the Ni foam with a vertical orientation and form a honeycomb structure enclosed by the neighbouring VGNS walls (Su et al. 2016). Figure 6.5B shows typical FESEM images at different magnifications of horizontal graphene grown on the Ni foam by a conventional chemical vapour deposition method. The high magnification images show typical wrinkles and ripples of graphene sheets, which are attributed to differences in thermal expansion coefficients between graphene and the Ni substrate. Recently, the inherent 3D open network structure of VGNSs showed great promise as electrode materials because of their excellent electrical conductivity and large surface area in contrast to the usual horizontal graphene (Miller et al. 2010). Also, the plasma-assisted chemical vapour deposition process can synthesise graphene in a lower temperature environment (400°C) compared to a conventional chemical vapour deposition method (~1000°C).

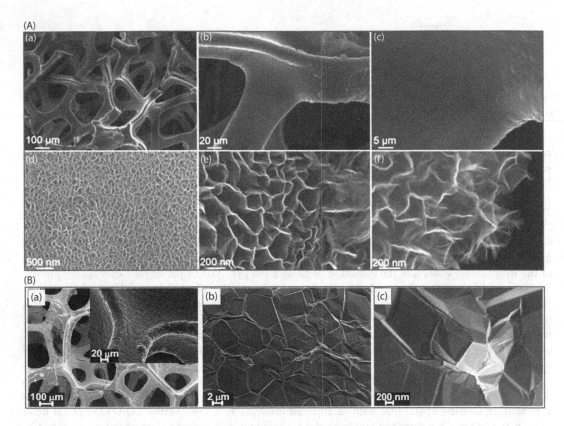

FIGURE 6.5 (A) VGNSs grown on the Ni foam by a plasma-assisted chemical vapour deposition system: (a–c) Low magnification FESEM images of VGNSs on Ni foam. (d–f) High magnification FESEM images of VGNS on Ni foam. (Reprinted with permission from Ref. Su et al. 2016, copyright Creative Commons Attribution 4.0 International License.) (B) Horizontal graphene sheet grown on the Ni foam by a conventional chemical vapour deposition method: (a–c) Low and high magnification FESEM images of a graphene sheet on Ni foam. (Reprinted from *J. Colloid Interface Sci.*, 484, Masikhwa, T.M. et al., High electrochemical performance of hybrid cobalt oxyhydroxide/nickel foam graphene, Copyright 2016, with permission from Elsevier.)

6.3 Characterization of Graphene

6.3.1 Raman Spectroscopy

Raman spectroscopy in graphene studies is a well-known, powerful technique to determine, among other things, the quality of graphene, the number of layers, the stacking order, and the interlayer interactions in a few layers of a graphene sample (Ferrari et al. 2006; Malard et al. 2009). In this spectroscopy technique, there is a scattering of photons (from a high-intensity laser light source) by phonons upon the interaction of photons with the chemical bonds within a material/sample (Figure 6.6a). The interaction of photons with the chemical bonds within a sample creates a time-dependent perturbation of the Hamiltonian (Ferrari and Basko 2013). The perturbation introduced by an incident photon of energy $\hbar\omega_L$ results in: (1) Rayleigh scattering (elastic scattering), (2) Stokes scattering with a photon energy $\hbar\omega_{Sc} = \hbar\omega_L - \hbar\Omega$ (where $\hbar\Omega$ is the scattered photon energy), and (3) anti-Stokes scattering with a photon energy $\hbar\omega_{Sc} = \hbar\omega_L + \hbar\Omega$ (Figure 6.6a) (Casiraghi et al. 2007; Ferrari and Basko 2013). Between Stokes and anti-Stokes scattering, Stokes is the most probable (Ferrari and Basko 2013), therefore, Raman spectra are Stokes measurements plots of the intensity of the scattered photon as a function of the difference between incident and scattered photon energy (known as Raman shift).

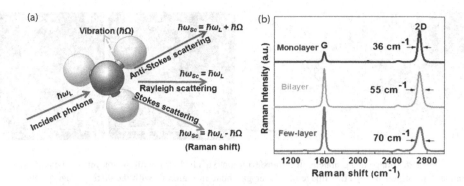

FIGURE 6.6 (a) Schematic diagram of light scattering in a vibrating molecule. (b) Raman spectra of as-grown CVD graphene on Ni foam. (Reprinted from *J. Colloid Interface Sci.*, 532, Ndiaye, N.M. et al., Three dimensional vanadium pentoxide/graphene foam composite as positive electrode for high performance asymmetric electrochemical supercapacitor, 395–406, Copyright 2018, with permission from Elsevier.)

In graphene and other graphitic materials, the appearance of the Raman modes is the result of electron-phonon coupling (Mafra et al. 2007; Saito et al. 2002). The main features that are observable in the Raman spectra of high-quality graphene are the G and 2D bands, as shown by Raman spectra of graphene on Ni foam in Figure 6.6b. However, in the case of disordered graphene or at the edge of a graphene sheet, a disorder-induced D-band appears in the Raman spectra at ~1350 cm^{-1}. Briefly, the origins of the Raman characteristic features of graphene (G, D, and 2D bands) are described as follows (Malard et al. 2009):

1. The G mode at ~1580 cm^{-1} involves the in-plane bond-stretching motion of pairs of carbon sp^2 atoms, and it originates from a first-order Raman scattering process (only one-phonon scattering is involved).

2. The D mode at ~1350 cm^{-1} (also known as a disorder-induced band) involves phonons near the K zone boundary and is only active in the presence of disorder in graphene. The band originates from a second-order Raman scattering process (double-resonance), involving one in-plane transverse optical mode phonon and one defect. The absence of the D-band in Raman spectra demonstrates high-quality graphene.

3. The 2D mode at ~2700 cm^{-1} originates from a second-order Raman scattering process (double-resonance) and involves two in-plane transverse optical mode phonons near the K zone boundary. The 2D mode also originates from triple resonance, and in contrast to the double resonance process, in the triple resonance process, all steps are resonant, and this could lead to a more intense 2D-band (relative to the G-band) as seen in monolayer graphene. Hence, this could be one of the reasons why the 2D-band is more intense than the G-band in monolayer graphene (see Figure 6.6b). To determine the number of layers in high-quality graphene, the 2D-to-G peak intensity ratio, 2D band line-shape, and peak width are analysed.

6.3.2 Transmission Electron Microscopy and Scanning Electron Microscopy

Transmission electron microscopy (TEM) is a technique that involves a beam of focused high-energy electrons (under ultra-high vacuum conditions), which is transmitted through a very thin sample and interacts with the sample as it passes through. The transmitted electrons are then used to generate an image of a sample. Different types of images can be obtained in TEM from different interactions of the electron beam with a sample. For instance, an image of electron diffraction patterns is obtained from elastically scattered electrons (diffracted beam), a bright field image from unscattered

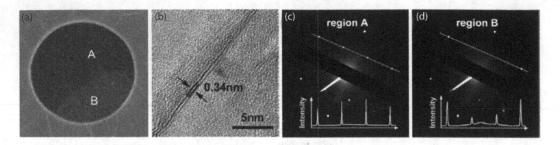

FIGURE 6.7   (a) SEM image of a graphene film transferred onto a TEM grid, with monolayer graphene in region A and bilayer graphene in region B. (b) TEM image of bilayer graphene in region B, with the distinct fingerprints showing an interlayer spacing of about 0.34 nm. (c and d) The selected area electron diffraction patterns and the corresponding intensities along the gray lines measured in regions A and B, respectively. (Reprinted with permission from Wu, Y. et al., *ACS Nano*, 6, 7731–7738, 2012. Copyright 2012, American Chemical Society.)

electrons (transmitted beam), and a dark field image from elastically scattered electrons (diffracted beam). Figure 6.7b shows a typical high-resolution TEM image of bilayer graphene with the distinct fingerprints showing an interlayer spacing of about 0.34 nm. Figure 6.7c and d shows selected area electron diffraction patterns (from graphene areas shown in Figure 6.7a), which show two sets of hexagonal diffraction spots (diffraction rings from crystalline graphene layers). The diffraction rings intensity profile which is indexed using the Miller-Bravais indices (*hkl*) for graphite shows peaks at $d = 0.123$ nm and peak $d = 0.213$ nm, which correspond to indices (1–210) for the outer ring (layer 1) and (1–110) for the inner ring (layer 2), respectively (Dato et al. 2008). In Bernal stacked bilayer graphene, the relative intensities of the spots in the outer ring are twice the intensities of the spots in the inner ring. On the contrary, in monolayer graphene, the relative intensities of the spots in the outer ring are equal to the intensities of the spots in the inner ring. Therefore, TEM gives direct information about the number of layers in graphene sheets.

Furthermore, SEM is a technique which involves a beam of focused electrons under ultra-high vacuum conditions and is used to analyse materials on micro to the nanometre scale. The analyses can yield information about topography, morphology, composition, and crystallography of materials (using other analysers intergraded in the SEM system, e.g., energy-dispersive X-ray spectroscopy). When a focused beam of high-energy primary electrons impinges the surface of a sample, among other things, it generates low energy secondary electrons. An image of the sample surface (about topography and morphology) is therefore constructed by measuring secondary electron intensity as a function of the position of the scanning primary electron beam. An SEM image of graphene sheet displays the image colour contrast between monolayer and multilayer graphene films (see Figure 6.7a), hence, it is also used to distinguish between monolayer and multilayer graphene.

6.3.3 X-Ray Photoelectron Spectroscopy

X-ray photoelectron spectroscopy (XPS) involves irradiation of the solid surface of a sample with a beam of X-rays (Al-Ka or Mg-Ka), while measuring the number and the kinetic energy of elastically scattered electrons (photoelectrons) that are emitted from the topmost surface atomic layers (1–10 nm) of the analysed sample. In XPS analysis, the measured binding energies (from the kinetic energies) of the detected electrons usually range between 0 and 1000 eV. For this energy range, the inelastic mean free path of the photoelectrons is in the order of a few nm, which corresponds to the topmost surface atomic layers of the analysed sample. The energies and intensities of the photoelectron peaks enable identification and quantification of the surface chemistry of a material. XPS of the carbonaceous materials typically shows a predominant carbon peak (C 1s) at a binding energy of about 285 eV (see Figure 6.8a and b for graphene oxide and reduced graphene oxide, respectively). Deconvolution of the C 1s peak gives information about the sp^2 C=C bond and oxide functional groups.

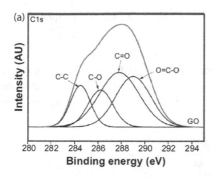 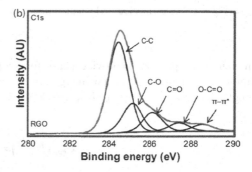

FIGURE 6.8 XPS spectra of C1s: (a) graphene oxide and (b) reduced graphene oxide. (Reprinted with permission from Manickam, S. et al., *Int. J. Nanomed.*, 10, 1505, 2015. Copyright 2015, Dove Medical Press Limited.)

6.4 Principle and Performance of Supercapacitors

Supercapacitors are typically two-terminal systems and can be either symmetric or asymmetric (hybrid) depending on the arrangement of the electrodes. The former involves two similar electrodes (i.e., material type, thickness, mass, etc.) sandwiched together, whereas the latter is made of two electrodes with different electrode materials. Under these conditions, an overall evaluation of the *two-electrode* system (without the utilization of a reference electrode) is obtained giving vital electrochemical information, such as energy densities and power densities that are not easily obtainable using data collected at individual electrodes in *three-electrode* cells (reference electrode included). Nevertheless, many researchers still opt for the *three-electrode* system since its measurements allow one to have fundamental information on the behaviour of the single electrode of a device (Conway 1999).

In electrochemical capacitors (ECs), the two porous electrodes are equivalent to two capacitors connected in a series, therefore, the resulting capacitance (C) obtained after polarization of the electrodes by applying potential difference can be expressed as follows:

$$\frac{1}{C_{cell}} = \frac{1}{C_+} + \frac{1}{C_-},$$ (6.1)

where C_{cell}, C_+, and C_- are the capacitance (in Farad = Coulomb/volt) of the resulting device or cell, positive and negative electrodes, respectively.

Equation 6.1 is typically used to obtain the capacitance from the cyclic voltammetry (CV) curves. From the slope of the discharge curve of the galvanostatic charge-discharge (GCD) profile, the following equation is adopted:

$$C_{cell}(F) = \frac{i}{\Delta V / \Delta t},$$ (6.2)

where i (A) is the applied current and $\Delta V / \Delta t$ (V s^{-1}) is the slope of the discharge curve after the initial iR drop.

In a symmetrical system where the two electrodes (positive and negative electrode) are similar with similar morphological and electronic properties ($C_+ = C_-$), the cell capacitance (C_{cell}), from Equation 6.1 will, therefore, be defined according to the following equation:

$$C_{cell} = \frac{C_e}{2},$$ (6.3)

where $C_e = C_+ = C_-$, and the electrode capacitance is calculated using the following equation:

$$C_e = 2C_{cell}. \qquad (6.4)$$

In electrochemical capacitors, to provide a basis for comparison between different electrode materials, it is a common practice to provide a specific (gravimetric) capacitance, which is related to the capacitance of one single electrode, $C_{e,sp}$ (F g^{-1}). Hence, dividing Equation 6.4 by the mass of the single electrode to obtain:

$$C_{e,sp}\left(F\ g^{-1}\right) = \frac{2C_{cell}}{m_e}, \qquad (6.5)$$

where $C_{e,sp}$ is the measured specific capacitance of each electrode and m_e (g) is the mass of the single electrode. It is worth noting that the capacitance of a single electrode derived from *three-electrode* (half-cell) measurements will be higher than the actual cell capacitance obtained from two-electrode (full-cell) measurements.

For a symmetric system, the specific (gravimetric) capacitance of the two electrodes (C_{sp}) is given by:

$$C_{sp}\left(F\ g^{-1}\right) = \frac{4C_{cell}}{M}, \qquad (6.6)$$

where M is the total mass of the active materials of the two electrodes (i.e., $M = 2m_e$ since the weight of each electrode is the same). The multiplier of four (4) only adjusts the capacitance of the cell and the combined weight of two electrodes to the capacitance and mass of a single electrode (Stoller and Ruoff 2010).

The other two relevant parameters of a capacitor apart from the C_{sp} are its energy and power density (Zhi et al. 2013). The energy (E) stored in a capacitor is related to the charge (Q) at each interface and the potential difference between the two plates. Therefore, energy is directly proportional to the capacitance as shown by the following equation:

$$E_{sp}\left(Wh\ kg^{-1}\right) = \frac{C_{cell}V^2}{2M}, \qquad (6.7)$$

where M (kg) is the total mass of the active materials of the two electrodes and V (V) is the maximum applied potential of electrochemical stability. The maximum power of the device is calculated using the following equation:

$$P_{max}\left(W\ kg^{-1}\right) = \frac{V^2}{4R_S M}, \qquad (6.8)$$

where R_S is the internal resistance calculated from the voltage drop at the beginning of a discharge curve and is given by the following equation:

$$R_S\left(\Omega\right) = \frac{\Delta V_{IR}}{2t}, \qquad (6.9)$$

and ΔV_{IR} is the voltage drop between the first two points from the start of the discharge curve.

For an asymmetric system, the above equations are also applied. However, before they are implemented to the asymmetric system, it is very critical first to perform '*mass-balancing*' from the *three-electrode* experiment for each of the electrodes. In the symmetric supercapacitor, the applied voltage is split equally between the two electrodes due to the use of the same material having the same mass in each electrode. In the asymmetric supercapacitors, however, the voltage split is dependent on the capacitance of the active material in each electrode. The capacitance is usually related to the mass and the specific capacitance of the active material (Cottineau et al. 2006; Chen et al. 2010). Thus, to split voltage equally, the mass balance between the two electrodes must be optimised using the following relationship: $q_+ = q_-$, where q_+ is the charge stored at the positive electrode and q_- is the charge stored at the negative electrode. Therefore, the following equation is used to express the stored charge:

$$q = C_{sp} \times m \times \Delta V \tag{6.10}$$

or:

$$\frac{m_+}{m_-} = \frac{C_{sp-}}{C_{sp+}} \times \frac{\Delta V_-}{\Delta V_+}, \tag{6.11}$$

where m, C_{sp}, and ΔV are the mass, specific capacitance, and potential range obtained from the charging/discharging process of three-electrode configuration of the individual positive and negative electrode, respectively.

Moreover, the electrolyte (an ionic conductor) serves as the medium for charge transfer between the positive and negative electrode. The choice of electrolyte solutions plays a crucial role when it comes to the performance of the ECs (G. Wang, Zhang, and Zhang 2012; Stoller and Ruoff 2010). The three types of electrolytes mainly used in ECs are aqueous, organic, or liquid salts (frequently known as ionic liquids). There are two main criteria involved in the selection of an electrolyte: (1) the electrochemical stability window and (2) the ionic conductivity.

Aqueous electrolytes (e.g., H_2SO_4, KOH, and Na_2SO_4) have a higher conductivity (up to ~1 S/cm), however, they suffer from the narrow electrochemical stability window (1.23 V) due to water electrolysis, leading to a relatively small (~1 V) operating potential window and, consequently, limiting the energy stored in the device (Kim et al. 2010). It is worth mentioning that the overpotential for electrolyte decomposition varies, depending on the carbon used, with the temperature playing an important role in the degradation mechanism (Béguin et al. 2014). On the other hand, the organic electrolytes composed of salt that dissolved in an organic solvent (i.e., propylene carbonate and acetonitrile) provide a wider electrochemical stability window (ranging from 2.7 to 2.8 V). However, they suffer relatively low ionic conductivity and high viscosity that result in lower specific capacitance ($100–150$ F g^{-1}) compared to aqueous electrolytes (Luo et al. 2011; Jiang and Wu 2014). However, the wide electrochemical stability window in organic electrolytes is advantageous to the ECs for high specific energy density. Furthermore, for ionic liquids (ILs), also called room-temperature ionic liquids (organic salts that are likely liquids at room temperature; Wilkes 2002), their desirable properties make them promising candidates for ECs electrolytes. These electrolytes have a very low vapour pressure because no solvent is required, hence, limiting environmental exposure and preventing the risk of explosion. ILs possess interesting properties, such as high thermal stability at elevated temperatures (beyond the ~80°C limit of organic electrolytes), low flammability, and a broad electrochemical stability window (ranging from 2 to 6 V, typically about 4.5 V). These properties are considerably higher than those of organic and aqueous electrolytes (Balducci et al. 2007; G. Wang, Zhang, and Zhang 2012). However, the ionic conductivity of ILs, specifically at room temperature, is lower than that of organic electrolytes, therefore reducing the power performance of IL-based ECs (Arbizzani et al. 2008; Weingarth et al. 2013).

6.5 Applications of Graphene-Based Electrodes in Supercapacitors

The carbon-based electrode materials, namely, 0D graphene dots or powders, 1D CNTs, 2D graphene sheets/films, 3D graphene-based foams and aerogels, activated carbon and carbon fibres, and porous carbon are mainly used in supercapacitor electrodes for designing devices with excellent cycling stability and high-power density due to their high surface area and good electrical conductivity. However, these electrode materials have a fast charge transfer, which leads to the energy density and overall device performance that are generally unsatisfactory. Therefore, one of the most intensive approaches to improve the energy density and overall device performance is the development of graphene-based composite electrodes (or hybrid electrodes) by integrating the metal oxides/hydroxides or conductive polymers with a graphene network. In this approach, graphene improves the electrode conductivity by providing a conductive channel for charge transfer, while the pseudocapacitance derived from the transition metal oxides (e.g., MnO_2, RuO_2, NiO, Co_3O_4, and Fe_3O_4) or hydroxides [$Ni(OH)_2$ and $Co(OH)_2$], as well as electrically conductive polymers (i.e., polyaniline (PANI), polypyrroles, and polythiophenes), improve the specific capacitance/energy density of the device (Ke and Wang 2016). Graphene is frequently reported in carbon-based supercapacitors based on their structural complexity (0D, 1D, 2D, and 3D) and nanocomposites, since they can be assembled into various carbon structures, and the electrochemical performances are mainly controlled by the structural and electrochemical properties of electrode materials (Zhai et al. 2011).

6.5.1 0D Graphene Dots or Powders

The 0D graphene dots or powders can be made through chemical exfoliation of graphite into GO, followed by the reduction of GO (Stoller et al. 2008). Ke et al. prepared Fe_3O_4-rGO nanocomposite powder via the electrostatic interaction between Fe_3O_4 nanoparticles and GO and the subsequent reduction of GO into rGO by a hydrothermal method (Ke et al. 2014). The resultant composites with oxide/hydroxide nanocrystals being intercalated between graphene sheets are preferable since the intercalation would provide spacers between the graphene layers (retaining active surface areas for charge storage), which tend to be restacked due to the high surface area in addition to the pseudocapacitance arising from the nanocrystals. The Fe_3O_4-rGO nanocomposite prepared by Ke et al. exhibited the largest capacitance of 169 F g^{-1} at the current density of 1 A g^{-1} compared to Fe_3O_4-GO (121 F g^{-1}), Fe_3O_4 (68 F g^{-1}), and rGO (101 F g^{-1}) (Ke et al. 2014). On the other hand, the Fe_3O_4 electrode exhibited the largest potential drop suggesting the largest internal resistance, while the IR drop was greatly lowered in the Fe_3O_4-rGO, which confirmed that the presence of rGO has greatly improved the overall conductivity of the electrode. Furthermore, the Fe_3O_4-rGO electrode exhibited good cycle stability, retaining over 88% of the initial capacity after 1000 charge-discharge cycles at a current density of 1 A g^{-1}. The observed improvement in the electrochemical performance of the Fe_3O_4-rGO nanocomposite electrode confirmed the synergistic effect between Fe_3O_4 clusters and rGO.

H.-W. Wang et al. demonstrated the electrochemical performance of Co_3O_4-rGO nanocomposites derived by a chemical precipitation approach from α-$Co(OH)_2$ supported by graphene oxide, and followed by thermal treatment (H.-W. Wang et al. 2011). Figure 6.9A displays the binary structure of the rGO/Co_3O_4 composite with mass ratio (rGO:Co_3O_4) of 0.22:1, which shows Co_3O_4 nanoparticles on rGO sheets (see Figure 6.9Aa). Figure 6.9Ba shows the CV curves of the rGO/Co_3O_4 composite at different scan rates in the range of 5–20 mV s^{-1} in 6 M KOH solution, which displays redox peaks attributed to Co_3O_4. In agreement with the CV curves, the galvanostatic discharge curves of the composite at various specific currents (Figure 6.9Bb) show non-linear slopes. This composite achieved a specific capacitance as high as 291 F g^{-1} at a current density of 1 A g^{-1}, and at a high current density of 8 A g^{-1}, 80% of the capacitance at 1 A g^{-1} was retained, demonstrating an excellent rate capability (Figure 6.9Bc). In addition, the rGO/Co_3O_4 composite maintained 90% of the initial specific capacitance over 1000 charge-discharge cycles at 1 A g^{-1} (Figure 6.9Bd), demonstrating excellent electrochemical stability.

6.5.2 1D Carbon Nanotubes

Carbon nanotubes have received significant attention because of their widespread applications, such as catalysts, sensors, supercapacitors, and so on in many fields (Zhai et al. 2011; Wu et al. 2017). There are two types of CNTs available, i.e., single-walled (SWNTs) and multi-walled (MWNTs), and in both nanotubes, the electronic transport occurs over long tube ranges without electronic scattering, hence, they have remarkable electrical conductivity. Because of their enhanced electronic conductivity, CNTs and CNT nanocomposites have been tested as electrode materials for supercapacitors (Zhai et al. 2011). Wu et al. have demonstrated a high-performance supercapacitor device based on polyaniline/vertical-aligned carbon nanotubes (PANI/VA-CNTs) nanocomposite electrodes, where the vertical-aligned structure was formed by the electrochemical induction (Wu et al. 2017). Figure 6.10A displays the morphology structure of PANI/VA-CNTs with vertical-aligned structure, which provides a straight and

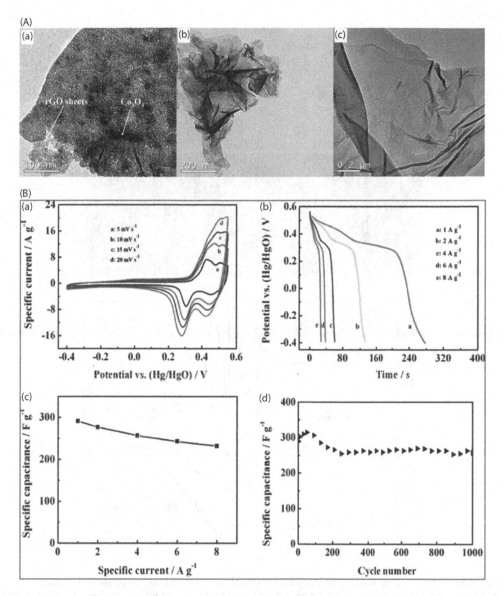

FIGURE 6.9 (A): (a and b) TEM images of rGO/Co_3O_4 composites, and (c) the residual sediment after HCl leaching of the rGO/Co_3O_4 composite. (B): (a) CV curves of the rGO/Co_3O_4 composite at various scan rates, (b) galvanostatic discharge curves, (c) specific capacitance of the rGO/Co_3O_4 composite at various specific currents, and (d) stability test measured at 1 A g^{-1}. (Reprinted from *Mater. Chem. Phys.*, 130, Wang, H.-W. et al., Preparation of reduced graphene oxide/cobalt oxide composites and their enhanced capacitive behaviors by homogeneous incorporation of reduced graphene oxide sheets in cobalt oxide matrix, 672–679, Copyright 2011, with permission from Elsevier.)

fast ion transport path channel, which is the key to high-performance supercapacitor devices, including high pseudocapacitance of polyaniline and high electric conductivity. In Figure 6.10Aa–Ad, an increase in the electrochemical-induced time from 0 to 60 min led to the gradual formation of the vertical-aligned nanostructure of disordered CNTs (D-CNTs), and at 60 min, the orderly and regularly vertical-aligned PANI/CNTs were formed. The inset TEM image in Figure 6.10Ad confirmed that the PANI was grown on CNTs. Moreover, two symmetric supercapacitors based on the PANI/VA-CNTs electrode and the D-CNTs electrode were fabricated using 1 M 1-eutyl-3-methylimidazolium tetrafluoroborate/ propylene carbonate (EMIBF4/propylene carbonate) organic electrolyte. Figure 6.10Ba displays the CV

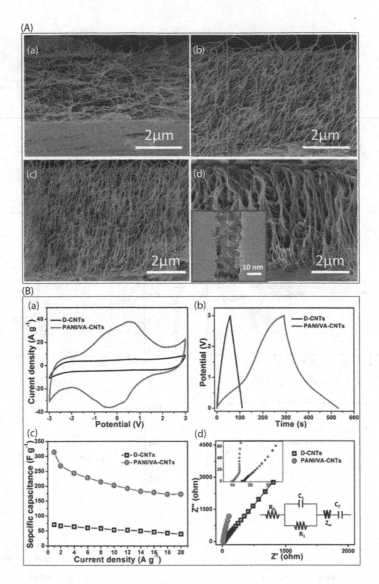

FIGURE 6.10 (A) Cross-sectional SEM images of PANI/VA-CNTs electrodes prepared within various electrochemical polymerization: (a) 0 min (D-CNTs), (b) 20 min, (c) 40 min, and (d) 60 min (the inset is the TEM image). (B) Electrochemical performance of the electrodes: (a) and (b) CV and GCD curves of PANI/VA-CNTs and D-CNTs at a scan rate of 100 mV s⁻¹ and current density of 1 A g⁻¹, respectively. (c) The calculated specific capacitance as a function of current density. (d) Electrochemical impedance spectroscopy (EIS) analysis of supercapacitors. The inset figures are the depressed semicircle of Nyquist plots and the equivalent circuit model. (Reprinted from Wu, G. et al., *Scientific Rep.*, 7, 43676, 2017. With permission.)

curves of the cells at a scan rate of $100\,\text{mV s}^{-1}$, which shows a high current response for the PANI/VA-CNTs electrode compared to the D-CNTs, demonstrating the effect of the structural modification (i.e., from disorder to vertical-aligned structure) and the effect of PANI coating. Also, the GCD curves of the PANI/VA-CNTs cell exhibited non-linear charge and discharge curves (Figure 6.10Bb) owing to the pseudocapacitive behaviour of PANI. The specific capacitance was calculated for PANI/VA-CNTs and D-CNTs symmetric cells from the discharge curves at different current densities, as shown in Figure 6.10Bc. From this figure, the specific capacitance of the PANI/VA-CNTs cell was 314.6 F g^{-1} at a current density of $1\,\text{A g}^{-1}$, which is much higher than that of the D-CNTs cell (70.6 F g⁻¹). At a high current density of $10\,A\,g^{-1}$, only 64.4% of capacitance ($202.7\,Fg^{-1}$) at low current density ($1\,A\,g^{-1}$) could be retained by the PANI/VA-CNTs cell, demonstrating a good rate capability.

Furthermore, Figure 6.10Be shows the Nyquist plots for the PANI/VA-CNTs and D-CNTs symmetric cells. From the fitting parameters of the equivalent circuit model (inset to Figure 6.10Be), the PANI/A-CNTs cell exhibited a smaller internal resistance, $R_0 = 9.6\,\Omega$ than the D-CNTs cell ($15.3\,\Omega$), which resulted from the good electric conductivity of the electrodes predominantly due to polyaniline. The Warburg diffusion impedance (Z_w), representing the ion diffusion ability, showed a lower resistance for PANI/VA-CNTs ($0.43\,\Omega$) compared to D-CNTs ($2.15\,\Omega$), which confirmed that the vertical-aligned structure of the PANI/VA-CNTs provides open pathways for ion faster diffusion. In agreement to the Z_w, the intercalation capacitance (C_2) of PANI/VA-CNTs ($18.3\,mF$) is much larger than that of D-CNTs ($2.74\,mF$), further confirming open pathways for ion faster diffusion.

A PANI/VA-CNTs cell showed good cyclic stability with a capacitance retention of 90.2% over 3000 cycles at a current density of $4\,A\,g^{-1}$ (Figure 6.11a). However, this is less than 98.3% of the D-CNTs cell, and this could be due to the redox reaction and volume change of PANI in the composite electrode. From the Ragone plot (Figure 6.11b), the PANI/VA-CNTs symmetric cell displayed a high energy density of $98.1\,Wh\,kg^{-1}$, which is much larger than that of the D-CNTs ($22.1\,Wh\,kg^{-1}$). It is evident that a supercapacitor based on PANI/VA-CNTs nanocomposite displays a high electrochemical performance developed through combining the structure and electrochemical activity of the electrode materials.

K. Wang et al. fabricated a flexible supercapacitor based on a composite electrode of cloth-supported SWCNTs and conducting PANI nanowire arrays (K. Wang et al. 2011). A non-woven wiper cloth was used as a flexible substrate to produce a conductive SWCNT/cloth composite by dip-coating it in the SWCNT ink, thereafter, PANI nanowire arrays were then deposited onto the surface of the SWCNT/cloth composite through dilute polymerization to obtain the PANI/SWCNT/cloth composite electrode (Figure 6.12A), which was used to assemble the supercapacitor device. In this case, 10 dip-coating cycles of SWCNT/cloth in the SWCNT ink were chosen to be sufficient for preparation of the composite electrode. Figure 6.12Ba and Bb displays SEM images of the SWCNT/cloth fibres after 10 dipping cycles at low and high magnification, respectively. In addition, Figure 6.12Bc and Bd presents the morphologies of the PANI/SWCNT/cloth composite electrode, which display vertically aligned PANI nanowires with ~50 nm average diameter and ~600 nm length (K. Wang et al. 2011).

Furthermore, Figure 6.13 presents the electrochemical performance of SWCNT/cloth, PANI/cloth, and PANI/SWCNT/cloth flexible supercapacitors with a $1\,M\,H_2SO_4$ aqueous electrolyte. From the CV and GCD curves (Figure 6.13a and b), SWCNT/cloth showed a typical EDLC behaviour (rectangular CV and triangular GCD curve with linear slopes), while the CV curves of PANI/cloth and PANI/SWCNT/cloth electrodes demonstrated redox peaks and non-linear GCD curves attributed to the pseudocapacitive behaviour of PANI. From the CV and GCD curves, the PANI/SWCNT/cloth displays a high current response, longer charge/discharge time, and high specific capacitance of $410\,F\,g^{-1}$ at

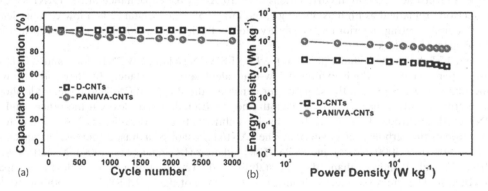

FIGURE 6.11 (a) Capacitance retention against cycle number at $4\,A\,g^{-1}$ and (b) Ragone plot of PANI/VA-CNTs and D-CNTs electrodes. (Reprinted from Wu, G. et al., *Scientific Rep.*, 7, 43676, 2017. With permission.)

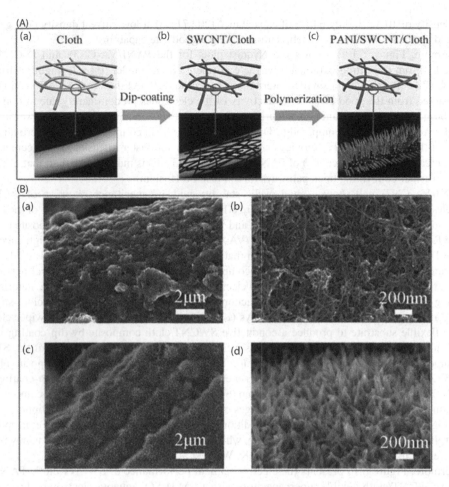

FIGURE 6.12 (A): Schematic illustration of the preparation of the PANI/SWCNT/cloth electrode: (a) wiper cloth; (b) SWCNT/cloth prepared by dip coating in SWCNT ink; (c) PANI/SWCNT/cloth prepared by chemical polymerization of PANI nanowire arrays. (B) SEM images: The microstructure of SWCNTs on a single viscose fiber of the wiper cloth at (a) low and (b) high magnification. The PANI nanowire arrays on SWCNT/cloth at (c) low and (d) high magnification. (Wang, K. et al., *J. Mater. Chem.*, 21, 16373. Reproduced by permission of The Royal Society of Chemistry.)

a current density of 0.5 A g⁻¹ (Figure 6.13c) compared to SWCNT/cloth (60 F g⁻¹) or PANI/cloth (290 F g⁻¹) electrodes. At a high current density of 10 A g⁻¹, the capacitance of the PANI/SWCNT/ cloth electrode remained as high as 390 F g⁻¹, retaining 95% of capacitance at a low current density (1 A g⁻¹), demonstrating superior rate capability. Hence, flexible supercapacitors can be considered as high-rate supercapacitors.

Figure 6.13d shows the Nyquist plots for the PANI/SWCNT/cloth, SWCNT/cloth, and PANI/cloth flexible supercapacitors, which exhibited low-equivalent series resistance. Furthermore, from the Ragone plot (Figure 6.13e), at the same power densities, the PANI/SWCNT/cloth electrode showed larger energy densities than the SWCNT/cloth and PANI/cloth electrodes. On the other hand, the SWCNT/cloth electrode showed excellent cyclic stability after 3000 cycles at 3 A g⁻¹ due to the stable nature of the carbonaceous materials. The PANI/cloth showed a drastic decrease in capacitance (39% retention over 3000 cycles) due to PANI degradation (Bélanger et al. 2000). Nonetheless, the PANI/SWCNT/cloth shows enhanced cyclic stability (90% retention over 3000 cycles) as compared to PANI/cloth. The improved electrochemical performance of the PANI/SWCNT/cloth electrode is evidently due to both the pseudocapacitance of the PANI and the well-defined nanostructure of the electrode.

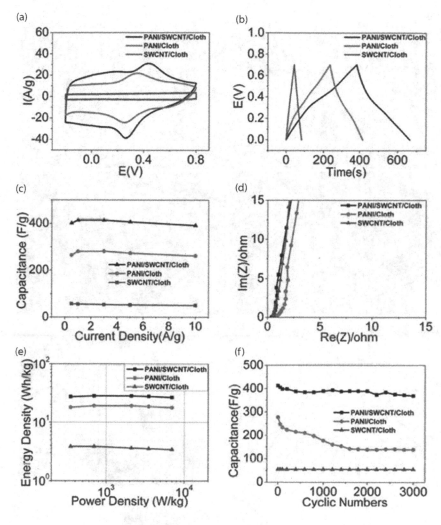

FIGURE 6.13 (a) CV curves at a scan rate of 100 mV s⁻¹. (b) GCD curves at 0.5 A g⁻¹. (c) Specific capacitance plots at different current densities. (d) Nyquist plots, (e) Ragone plot, and (f) cycle life at a current density of 3 A g⁻¹ for PANI/SWCNT/cloth, SWCNT/cloth, PANI/cloth flexible SCs. (Wang, K. et al., *J. Mater. Chem.*, 21, 16373. Reproduced by permission of The Royal Society of Chemistry.)

Ochai-Ejeh et al. decorated multi-walled carbon nanotubes with MnO_2 nanoparticles via a facile hydrothermal reflux technique to produce a MnO_2-CNT nanocomposite (Ochai-Ejeh et al. 2018). Figure 6.14Aa and Ab presents the SEM images of the multi-walled carbon nanotubes at low and high magnifications, respectively, and Figure 6.14Ac and Ad presents the SEM images of the nanocomposite at low and high magnifications, respectively. In agreement with the SEM images, the TEM images (Figure 6.14B) show the tube-like nanostructure of the multi-walled carbon nanotubes uniformly decorated with MnO_2 nanoparticles. Moreover, a hybrid device fabricated using a MnO_2-CNT nanocomposite as the positive electrode and 3D activated carbon as the negative electrode exhibited good electrochemical performance that was ascribed to the synergistic effects between the MnO_2-CNT nanocomposite electrode and the 3D microporous activated carbon electrode. The device displayed an energy density of ~25 Wh Kg⁻¹ and corresponding power density of 500 W Kg⁻¹ at 0.5 A g⁻¹ in 1 M Li_2SO_4 aqueous electrolyte. The device further exhibited an excellent stability of ~100% Coulombic efficiency after 10,000 charge-discharge cycles and excellent capacitance retention of >70% at 5 A g⁻¹.

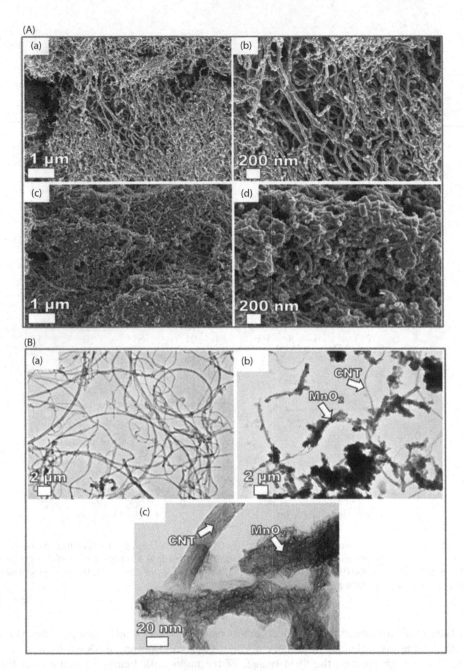

FIGURE 6.14   (A) SEM images: Multi-walled CNTs at (a) low and (b) high magnification; MnO$_2$-CNT nanocomposite at (c) low and (d) high magnification. (B) TEM images: (a) CNT and MnO$_2$-CNT at (b) low and (c) high resolution. (Reprinted from *Electrochim. Acta*, 289, Ochai-Ejeh, F. et al., Electrochemical performance of hybrid supercapacitor device based on birnessite-type manganese oxide decorated on uncapped carbon nanotubes and porous activated carbon nanostructures, 363–375, Copyright 2018, with permission from Elsevier.)

6.5.3 2D Graphene Sheets/Films

The processing of graphene sheets/films is typically hindered by agglomeration and restacking of graphene sheets, which can greatly reduce the surface areas and limit the diffusion of electrolyte ions between graphene layers (Ke and Wang 2016). Therefore, there have been numerous attempts to add spacers to improve the stacking of graphene sheets (Ke and Wang 2016). The most widely investigated spacers are carbon-based materials, metals or metal oxides/hydroxides, and conducting polymers (Ke and Wang 2016). Paek et al. modified graphene nanosheets (GNSs) with tin oxide (SnO_2) particles inserted in between the sheets of graphene layers (Paek et al. 2009). The GNSs were prepared via the chemical reduction of exfoliated graphite oxide, while SnO_2 nanoparticles were obtained by the hydrolysis of $SnCl_4$ with NaOH. After that, the reduced GNSs were dispersed in the ethylene glycol and reassembled in the presence of SnO_2 nanoparticles to prepare a SnO_2/GNS nanocomposite (Figure 6.15A). In the nanostructured SnO_2/GNS electrode, the nanopores between SnO_2 and GNS would be used for ion insertion/extraction during charge/discharge, resulting in the superior cyclic performances and eventually higher reversible capacities (Paek et al. 2009). With the restacking of graphene sheets being addressed, the capacitance of the electrode is greatly improved leading to enhanced energy storage ability. Figure 6.15Ba presents the SEM image of GNSs, and the inset to the figure shows the edge-side of GNSs in which the thickness of an individual stack is estimated to be <10 nm, suggesting that the self-restacked GNSs consist of several layers. In Figure 6.15Bb, the SEM image of the SnO_2/GNS displays a similar morphology to that of the GNS. Furthermore, Figure 6.15Bc and Bd presents TEM images of GNSs at different magnification, and from these images, the *d*-spacing between two graphene nanosheets is ~0.39 nm, which is larger than 0.335 nm of graphite, indicating that graphene interlayer distances are increased by the reassembling reaction. On the other hand, the TEM images (Figure 6.15Be and Bf) of the GNSs dispersed in the ethylene glycol solution in the presence of SnO_2 nanoparticles show the spherical shapes of SnO_2 nanoparticles (average particle size is ~5.4 nm) with GNSs distributed between the nanoparticles to produce a nanoporous composite with a large amount of void spaces. Although Paek et al. presented a nanostructured SnO_2/GNS electrode as a potential anode material for lithium-ion batteries which exhibited a reversible capacity of 810 mAh g^{-1}, it can be applied in supercapacitors for enhanced energy storage.

Yan et al. reported a synthesis of GNSs doped with PANI particles using *in situ* polymerization to produce a GNS/PANI nanocomposite (Yan et al. 2009). From the electrochemical analysis, it was found that the specific capacitance of GNS/PANI (1046 F g^{-1} at 1 mV s^{-1}) is much higher than that of GNSs (183 F g^{-1} at 1 mV s^{-1}) and pure PANI (115 F g^{-1} at 1 mV s^{-1}). The greatly enhanced specific capacitance of GNS/PANI is due to the synergistic effect between GNSs and PANI.

6.5.4 3D Graphene-Based Foams and Aerogels

To further improve the electrochemical performance of the graphene-based supercapacitors, considerable efforts were made to develop graphene-based macrostructures with 3D networks, i.e., graphene foams and aerogels (Ke and Wang 2016). These 3D graphene-based materials have great potential as high energy and power density supercapacitor electrodes due to their macrostructure consisting of micro-, meso-, and macro-interconnected pores, high surface areas, and fast ion/electron transport channels (Ke and Wang 2016).

Dong et al. prepared Co_3O_4 nanowires on graphene foam that were assembled into a 3D graphene foam structure, as shown in Figure 6.16 (Dong et al. 2012). Figure 6.16Aa presents a SEM image of a 3D porous structure of graphene foam with a smooth surface. In a graphene/Co_3O_4 nanocomposite (SEM images in Figure 6.16Ab and Ac), the graphene foam surface is uniformly covered by Co_3O_4 nanowires, which are 200–300 nm in diameter and few micrometres in length. Figure 6.16Ba presents the CV curves of the 3D graphene and graphene/Co_3O_4 nanocomposite electrodes measured at a scan rate of 50 mV s^{-1}. From this figure, the nanocomposite displays a high current response compared to the graphene foam, which indicates a large specific capacitance enhanced by the pseudocapacitance of the Co_3O_4 nanowires. In Figure 6.16Bb, the CV curves of the nanocomposite at an increasing scan rate revealed that the redox reactions of Co_3O_4 are rapid. From the charge-discharge curves of the graphene/Co_3O_4 nanocomposite electrode at different current densities (Figure 6.16Bc), the specific capacitance was calculated to be 768, 618, 552, and 456 F g^{-1} at the current densities of 10, 15, 20, and 30 A g^{-1}, respectively. The graphene/Co_3O_4 nanocomposite electrode further showed excellent electrochemical stability (Figure 6.16Bd). From the cyclic stability, the specific capacitance of the composite electrode can be further enhanced, and such an activation process may result from the more complete intercalation and deintercalation of ions after a few initial charge-discharge cycles. The electrochemical performance of the graphene/Co_3O_4 nanocomposite electrode was dramatically enhanced by a synergistic effect between the Co_3O_4 nanowires and 3D graphene foam, which was

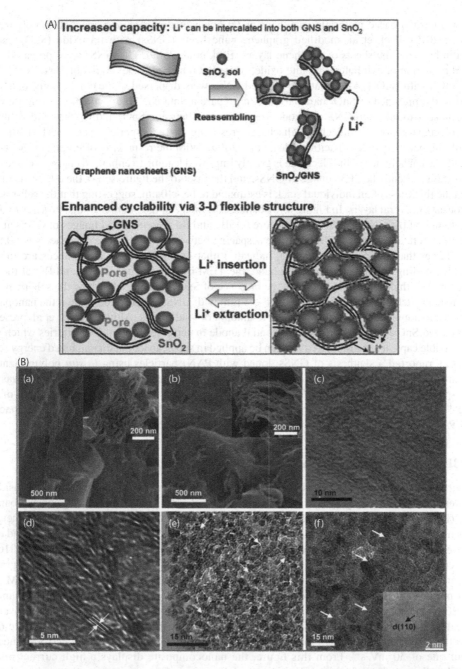

FIGURE 6.15 (A) Schematic illustration for the synthesis and structure of SnO_2/GNS. (B) SEM images: (a) GNS and (b) SnO_2/GNS. TEM images: (c) GNS, (d) GNS (high magnification), (e) as-prepared SnO_2/GNS, and (f) heat-treated SnO_2/GNS. (Reprinted with permission from Paek, S.-M. et al., *Nano Lett.*, 9, 2009. Copyright 2009, American Chemical Society.)

greatly influenced by the 3D structure of graphene foam, the nanoporous mesh of Co_3O_4 nanowires, and the rough surface of individual nanowires.

Figure 6.17 presents the SEM images and the electrochemical performance of the mesoporous nanosheets of cobalt oxyhydroxide (CoOOH) on 3D nickel foam graphene (Ni-FG) prepared via the in-situ hydrothermal method, while the nickel foam graphene was obtained by CVD (Masikhwa et al. 2016). The SEM images of graphene foam at high magnification (Figure 6.17Ab–Ac) show a smooth surface with the wrinkles and ripples of the sheets. The SEM images of CoOOH on the 3D Ni-FG (CoOOH/

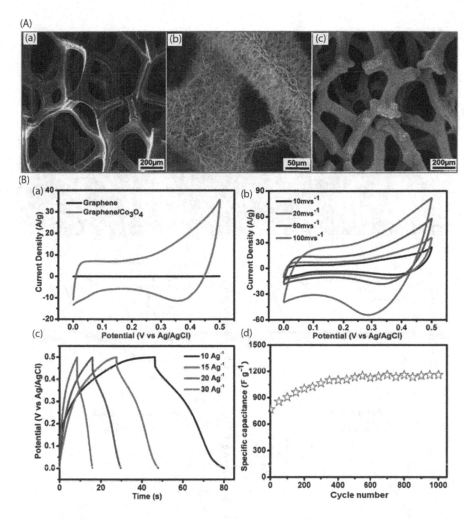

FIGURE 6.16 (A) SEM images: 3D graphene foam, (b and c) 3D graphene/Co_3O_4 nanowire composite. (B): (a) CV curves of the 3D graphene and graphene/Co_3O_4 nanocomposite electrodes at a scan rate of 50 mV s^{-1}. (b) CV curves of the nanocomposite at different scan rates in the range of 10–100 mV s^{-1}. (c) GCD curves of the nanocomposite at different specific currents in the range of 10–30 A g^{-1}. (d) Capacitance retention against cycle number at 10 A g^{-1}. (Reprinted with permission from Dong, X.C. et al., *ACS Nano*, 6, 3206–3213, 2012. Copyright 2012, American Chemical Society.)

Ni-FG) at low and high magnifications (Figure 6.17Ad–Af) show a homogenous coating of well-ordered intersected nanosheets of CoOOH. Figure 6.17Ba shows CV curves of Ni-F, Ni-FG, and CoOOH/Ni-FG electrodes at a scan rate of 50 mV s^{-1}. The CV curve of the CoOOH/Ni-FG electrode was dramatically enhanced by CoOOH nanosheets on 3D nickel foam graphene. The CV and GCD curves of the CoOOH/Ni-FG electrode at different scan rates and current densities, respectively (Figure 6.17Bb and 6.17Bc), revealed the redox reactions of CoOOH. From the charge-discharge curves of the CoOOH/Ni-FG electrode, the specific capacity was obtained, as shown in Figure 6.17Bd. At 0.5 A g^{-1}, the specific capacity of the CoOOH/Ni-FG electrode was calculated as 199 mA h g^{-1}. The cycling stability of the CoOOH/Ni-FG electrode showed excellent capacity retention of 98% after 1000 cycles at 10 A g^{-1}.

Moreover, in addition to 3D graphene foams, there is a unique class of porous and ultralight carbon materials known as 3D graphene aerogels, which shows high ratios of surface-area-to-volume and strength-to-weight (Ke and Wang 2016; Lemine et al. 2018). The graphene aerogels are typically used to fabricate high-performance and binder-free supercapacitor electrodes, flexible electrodes in particular. Xu et al. demonstrated flexible thin-film electrodes, based on functionalized graphene hydrogels, pressed on the gold-coated polyimide substrates (Xu et al. 2013). In these electrodes, the 3D continuous porous

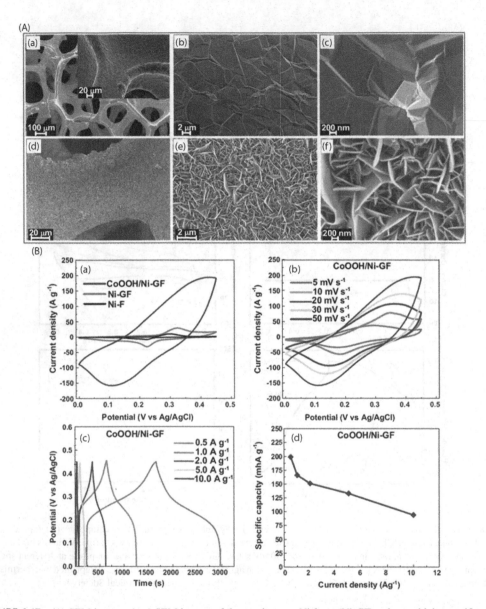

FIGURE 6.17 (A) SEM images: (a–c) SEM images of the graphene on Ni foam (Ni-GF) at low to high magnifications. (d–f) CoOOH on Ni foam graphene (CoOOH/Ni-FG) at low to high magnifications with CoOOH showing intersected nanosheets. (B) Electrochemical performance of the electrodes: (a) CV curves of Ni-F, Ni-FG, and CoOOH/Ni-FG at a scan rate of 50 mV s^{-1}. (b) CV curves of CoOOH/Ni-FG at scan rates ranging from 5 to 30 mV s^{-1}. (c) GCD curves of CoOOH/Ni-FG at current densities of 0.5–10 A g^{-1}. (d) Specific capacity of CoOOH/Ni-FG at different current densities. (Reprinted from *J. Colloid Interface Sci.*, 484, Masikhwa, T.M. et al., High electrochemical performance of hybrid cobalt oxyhydroxide/nickel foam graphene, Copyright 2016, with permission from Elsevier.)

network of graphene was maintained. The electrodes were assembled into an all-solid-state flexible supercapacitor using a H$_2$SO$_4$-polyvinyl alcohol gel electrolyte. The specific capacitance of the solid-state supercapacitor was calculated from the GCD curves as 412 F g^{-1} at 1 A g^{-1} and was found to be higher than that of previously reported solid-state devices based on carbon nanotubes and their composites, graphene films, and conducting polymers (Xu et al. 2013). The cycling stability of the device showed excellent capacitance retention of 87% after 10,000 cycles at 10 A g^{-1} under 150° bending angle.

Table 6.2 presents the electrochemical performance of some supercapacitor's electrodes based on 0D, 1D, 2D, and 3D graphene-based nanocomposites.

TABLE 6.2

Summary of Electrochemical Performances of Some Supercapacitor's Electrodes Based on 0D, 1D, 2D, and 3D Graphene-Based Nanocomposites

	Electrode Material	Cell Configuration	Electrolyte	Potential Window (V)	Specific Current/ Scan Rate	Specific Capacitance (F g^{-1})	Capacitance Retention	References
0D graphene dots or powders (nanocomposites)	Fe_3O_4/graphene//3D-graphene	2-electrode, asymmetric	$LiPF_6$-containing organic electrolyte	1.0–4.0	0.3 A g^{-1}	–	70% (1000 cycles) @ 2 A g^{-1}	F. Zhang et al. (2013)
	Fe_3O_4-rGO	3-electrode	1 M KOH	-1.0–0.0	1.0 A g^{-1}	169	88% (1000 cycles) @ 1 A g^{-1}	Ke et al. (2014)
	Co_3O_4-rGO	3-electrode	6 M KOH	-0.4–0.55	1.0 A g^{-1}	291	90% (1000 cycles) @ 1 A g^{-1}	H.-W. Wang et al. (2011)
	HRG/MnO_2	3-electrode	1 M Na_2SO_4	0.0–1.0	2 mV s^{-1}	211.5	75% (1000 cycles) @ 50 mV s^{-1}	Z. Li et al. (2011)
	Mesoporous graphene nanoball	3-electrode	1 M H_2SO_4	-0.5–0.3	5 mV s^{-1}	206	96% (10,000 cycles) @ 20 A g^{-1}	J.-S. Lee et al. (2013)
1D carbon nanotubes	D-CNTs	2-electrode, symmetric	$EMIBF_4$ organic electrolyte	0.0–3.0	1.0 A g^{-1}	70.6	90.2% (3000 cycles) @ 4 A g^{-1}	G. Wu et al. (2017)
	PANI/VA-CNTs	2-electrode, symmetric	$EMIBF_4$ organic electrolyte	0.0–3.0	1.0 A g^{-1}	314.6	98.3% (3000 cycles) @ 4 A g^{-1}	G. Wu et al. (2017)
	PANI/SWCNT/cloth	2-electrode, symmetric	1 M H_2SO_4	0.0–0.7	0.5 A g^{-1}	410	90% (3000 cycles) @ 3 A g^{-1}	K. Wang et al. (2011)
	MnO_2-CNT//AC (Multi-walled carbon nanotubes)	2-electrode, asymmetric	1 M Li_2SO_4	0.0–2.0	0.5 A g^{-1}	–	>70% (10,000 cycles) @ 5 A g^{-1}	Ochai-Ejeh et al. (2018)
	Aligned MWCNT/PANI	2-electrode, symmetric	H_3PO_4-PVA gel electrolyte	0.0–0.8	1.0 A g^{-1}	233	–	H. Lin et al. (2013)
	PEDOT/MWNT yarn	2-electrode, symmetric	1 M PVA-H_2SO_4 gel	0.0–0.8	–	~179 F cm^{-3}	–	J. A. Lee et al. (2013)
2D graphene sheets/films	GNS	3-electrode	6 M KOH	-0.7–0.3	1 mV s^{-1}	183	–	Yan et al. (2010)
	GNS/PANI nanocomposite	3-electrode	6 M KOH	-0.7 to 0.3	1 mV s^{-1}	1046	–	Yan et al. (2010)
	Graphene/PANI composite paper	2-electrode, symmetric	1 M H_2SO_4	-0.2–0.8	2 mV s^{-1}	233	–	D.-W. Wang et al. (2009)
	Graphene-hollow polypyrrole	3-electrode	2 M H_2SO_4	-0.2–0.8	5.0 A g^{-1}	504.0	>90% (10,000 cycles) @ 5 A g^{-1}	J. Zhang et al. (2013)
	Liquid electrolyte-mediated CCG	2-electrode, symmetric	$EMIBF_4$ organic electrolyte	0.0–3.5	0.1 A g^{-1}	261.3 F cm^{-3}	–	Yang et al. (2013)

(Continued)

TABLE 6.2 *(Continued)*

Summary of Electrochemical Performances of Some Supercapacitor's Electrodes Based on 0D, 1D, 2D, and 3D Graphene-Based Nanocomposites

	Electrode Material	Cell Configuration	Electrolyte	Potential Window (V)	Specific Current/ Scan Rate	Specific Capacitance ($F\,g^{-1}$)	Capacitance Retention	References
3D graphene-based foams and aerogels	Graphene/ Co_3O_4 nanocomposite	3-electrode	2 M KOH	0.0–0.5	$10\,A\,g^{-1}$	~1100	>90% (1000 cycles) @ $10\,A\,g^{-1}$	Dong et al. (2012)
	$Ni(OH)_2$/graphene foam	3-electrode	6 M KOH	0.0–0.5	$1.0\,A\,g^{-1}$	2420	~93% (1000 cycles) @ $10\,A\,g^{-1}$	Khaleed et al. (2016)
	3D graphene/PANI	3-electrode	1 M H_2SO_4	0.0–0.8	$0.5\,A\,g^{-1}$	740	87% (1000 cycles) @ $10\,A\,g^{-1}$	S. Wang et al. (2015)
	Functionalized graphene hydrogels	2-electrode, symmetric	PVA-H_2SO_4 gel	0.0–1.0	$1.0\,A\,g^{-1}$	412	87% (10,000 cycles) @ $10\,A\,g^{-1}$	Xu et al. (2013)
	Ni graphene foam/NiO	3-electrode	6 M KOH	0.0–0.5	$2\,mV\,s^{-1}$	783	84% (10,000 cycles) @ 10 mA	Bello et al. (2013)
	MoS_2/graphene foam	3-electrode	6 M KOH	0.0–0.5	$2\,A\,g^{-1}$	~225	–	Masikhwa et al. (2017)
	$Mn_3(PO_4)_2$/graphene foam	3-electrode	6 M KOH	0.0–0.4	$0.5\,A\,g^{-1}$	270	~90% (1000 cycles) @ $5.0\,A\,g^{-1}$	Mirghni et al. (2017)

Hydrothermally reduced graphene (HRG), chemically converted graphene (CCG).

6.6 Conclusion and Outlook

Significant progress has been made in applications of graphene-based electrodes in supercapacitors to improve their capacitance, energy, and power densities towards more efficient energy storage devices. Efforts have been made on the structural design, development of graphene-based composites electrodes (or hybrid electrodes), and electrodes performance evaluation. As demonstrated in this chapter, the development of graphene-based electrodes focused on graphene composites based on various forms of graphene ranging from 0D, 1D, 2D, and 3D. A 3D graphene/hydrogel framework advantageously serves as a 3D support of large capacity to uniformly anchor the transitional metal oxides/hydroxides, as well as electrically conductive polymers with well-defined size, shapes, and crystallinity for the enhanced electrochemical performance of the electrodes. In addition, in 3D graphene foam/hydrogels, the structures with improved rate stability and power density are achievable due to the porous structure and varying degrees of graphene sheet aggregation, which can affect the effective diffusion of electrolyte ions. The graphene composites need more research efforts to improve the interactions between graphene and their composite materials to optimize the pseudocapacitance and further improve the performance of supercapacitors. Additionally, an effort is needed in developing a cost-effective and environmentally friendly processing method that can synthesize high-quality graphene of a large quantity. The chemical method via reduction of graphite oxide is considered a scalable approach to synthesizing graphene and has been widely utilized to synthesize reduced graphene oxide, especially for application in supercapacitor electrodes. However, this method involves the oxidation of graphite in the presence of strong acids and oxidants. Therefore, there are health risks associated with such production of graphene and its derivatives due to the handling and possibilities of inhaling toxic reducing chemicals. Besides, before the application of this processing method in large scale production of graphene-based electrode materials, the stabilization of graphene sheets in various solvents to preserve their intrinsic properties and to break the bottleneck of re-stacking of graphene sheets must be addressed.

Moreover, to further develop graphene-based electrode materials for high-performance supercapacitor applications, the following prospects should be considered:

1. Compared to 0D, 1D, and 2D graphene forms, more focus should be on 3D graphene/hydrogels with tunable 3D networks based on an interconnected porous structure, which can be manipulated for a large surface area, more ion/charge pathways, and break the bottleneck of re-stacking of graphene sheets.
2. More attention should be paid to the structures of the graphene-based nanocomposite electrodes and the control of the interfacial interaction between graphene and pseudocapacitive materials to improve the overall processes across the interface.
3. Focus on the improvement of mechanical flexibility of graphene-based nanocomposite electrodes for the future development of flexible supercapacitors and the integration of graphene-based supercapacitors with other devices to develop self-powered hybrid systems.

REFERENCES

Arbizzani, Catia, Maurizio Biso, Dario Cericola, Mariachiara Lazzari, Francesca Soavi, and Marina Mastragostino. 2008. "Safe, High-Energy Supercapacitors Based on Solvent-Free Ionic Liquid Electrolytes." *Journal of Power Sources* 185 (2): 1575–79. doi:10.1016/J.JPOWSOUR.2008.09.016.

Balducci, Andrea, Romain Dugas, Pierre Louis Taberna, Patrice Simon, Dominique Plée, Marina Mastragostino, and Stefano Passerini. 2007. "High Temperature Carbon-Carbon Supercapacitor Using Ionic Liquid as Electrolyte." *Journal of Power Sources* 165: 922–27. doi:10.1016/j.jpowsour.2006.12.048.

Béguin, François, Volker Presser, Andrea Balducci, and Elzbieta Frackowiak. 2014. "Carbons and Electrolytes for Advanced Supercapacitors." *Advanced Materials (Deerfield Beach, Fla.)* 26 (14): 2219–51, 2283. doi:10.1002/adma.201304137.

Bélanger, Daniel, Xiaoming Ren, John Davey, Francisco Uribe, and Shimshon Gottesfeld. 2000. "Characterization and Long-Term Performance of Polyaniline-Based Electrochemical Capacitors." *Journal of the Electrochemical Society* 147 (8): 2923. doi:10.1149/1.1393626.

Bello, Abdulhakeem, Katlego Makgopa, Mopeli Fabiane, David Dodoo-Ahrin, Kenneth Ikechukwu Ozoemena, and Ncholu Manyala. 2013. "Chemical Adsorption of NiO Nanostructures on Nickel Foam-Graphene for Supercapacitor Applications." *Journal of Materials Science* 48 (19): 6707–12. doi:10.1007/s10853-013-7471-x.

Brodie, Benjamin Collins. 1860. "Sur Le Poids Atomique Du Graphite." *Annales de Chimie et de Physique* 59: 466.

Buchsteiner, Alexandra, Anton Lerf, and Jörg Pieper. 2006. "Water Dynamics in Graphite Oxide Investigated with Neutron Scattering." *The Journal of Physical Chemistry B* 110 (45): 22328–38. doi:10.1021/JP0641132.

Casiraghi, Cinzia, Achim Hartschuh, Elefterios Lidorikis, Hong Qian, Hayk Harutyunyan, Tobias Gokus, Kostya Sergeevich Novoselov, and Andrea Ferrari. 2007. "Rayleigh Imaging of Graphene and Graphene Layers." *Nano Letters* 7 (9): 2711–17. doi:10.1021/nl071168m.

Chen, Po-Chiang, Guozhen Shen, Yi Shi, Haitian Chen, and Chongwu Zhou. 2010. "Preparation and Characterization of Flexible Asymmetric Supercapacitors Based on Transition-Metal-Oxide Nanowire/Single-Walled Carbon Nanotube Hybrid Thin-Film Electrodes." *ACS Nano* 4 (8): 4403–11. doi:10.1021/nn100856y.

Chidembo, Alfred Tawirirana, Kenneth Ikechukwu Ozoemena, Bolade Oyeyinka Agboola, Vinay Gupta, Gregory Wildgoose, and Richard Guy Compton. 2010. "Nickel(Ii) Tetra-Aminophthalocyanine Modified MWCNTs as Potential Nanocomposite Materials for the Development of Supercapacitors." *Energy & Environmental Science* 3 (2): 228. doi:10.1039/b915920g.

Conway, Brian Evans. 1999. *Electrochemical Supercapacitors Scientific Fundamentals and Technological Applications*. New York: Kluwer Academic/Plenum.

Cottineau, Thomas, Mathieu Toupin, Timur Delahaye, Thierry Brousse, and Daniel Bélanger. 2006. "Nanostructured Transition Metal Oxides for Aqueous Hybrid Electrochemical Supercapacitors." *Applied Physics A* 82 (4): 599–606. doi:10.1007/s00339-005-3401-3.

Dato, Albert, Velimir Radmilovic, Zonghoon Lee, Jonathan Phillips, and Michael Frenklach. 2008. "Substrate-Free Gas-Phase Synthesis of Graphene Sheets 2008." *Nano Letters* 8 (7): 2012–16.

Dong, Xiao Chen, Hang Xu, Xue Wan Wang, Yin Xi Huang, Mary Chan-Park, Hua Zhang, Lian Hui Wang, Wei Huang, and Peng Chen. 2012. "3D Graphene-Cobalt Oxide Electrode for High-Performance Supercapacitor and Enzymeless Glucose Detection." *ACS Nano* 6 (4): 3206–13. doi:10.1021/nn300097q.

El-kady, Maher, Veronica Strong, Sergey Dubin, and Richard Kaner. 2012. "Laser Scribing of High-Performanceand Flexible Graphene-Based Electrochemical Capacitors." *Science* 335: 1326–30.

Elzbieta Frackowiak, and François Béguin. 2001. "Carbon Materials for the Electrochemical Storage of Energy in Capacitors." *Carbon* 39 (6): 937–50. doi:10.1016/S0008-6223(00)00183-4.

Ferrari, Andrea, and Denis Basko. 2013. "Raman Spectroscopy as a Versatile Tool for Studying the Properties of Graphene." *Nature Nanotechnology* 8 (4): 235–46. doi:10.1038/nnano.2013.46.

Ferrari, Andrea, Jannik Meyer, Vittorio Scardaci, Cinzia Casiraghi, Michele Lazzeri, Francesco Mauri, Stefano Piscanec, et al. 2006. "Raman Spectrum of Graphene and Graphene Layers." *Physical Review Letters* 97 (18): 187401.

Huang, Yi, Jiajie Liang, and Yongsheng Chen. 2012. "An Overview of the Applications of Graphene-Based Materials in Supercapacitors." *Small (Weinheim an Der Bergstrasse, Germany)* 8 (12): 1805–34. doi:10.1002/smll.201102635.

Jiang, De-en, and Jianzhong Wu. 2014. "Unusual Effects of Solvent Polarity on Capacitance for Organic Electrolytes in a Nanoporous Electrode." *Nanoscale* 6: 5545–50. doi:10.1039/c4nr00046c.

Jung, Inhwa, Matthew Pelton, Richard Piner, Dmitriy Dikin, Sasha Stankovich, Supinda Watcharotone, Martina Hausner, and Rodney Ruoff. 2007. "Simple Approach for High-Contrast Optical Imaging and Characterization of Graphene-Based Sheets." *Nano Letters* 7 (12): 3569–75. doi:10.1021/nl0714177.

Ke, Qingqing, Chunhua Tang, Yanqiong Liu, Huajun Liu, and John Wang. 2014. "Intercalating Graphene with Clusters of Fe_3O_4 Nanocrystals for Electrochemical Supercapacitors." *Materials Research Express* 1 (2). doi:10.1088/2053-1591/1/2/025015.

Ke, Qingqing, and John Wang. 2016. "Graphene-Based Materials for Supercapacitor Electrodes – A Review." *Journal of Materiomics* 2 (1): 37–54. doi:10.1016/J.JMAT.2016.01.001.

Khaleed, Abubakar Abubakar, Abdulhakeem Bello, Julien Dangbegnon, Faith Ugbo, Farshad Barzegar, Damilola Yusuf Momodu, Moshawe Jack Madito, Tshifhiwa Moureen Masikhwa, Okikiola Olaniyan, and Ncholu Manyala. 2016. "A Facile Hydrothermal Reflux Synthesis of $Ni(OH)_2GF$ Electrode for Supercapacitor Application." *Journal of Materials Science* 51 (12). doi:10.1007/s10853-016-9910-y.

Kim, Tae Young, Hyun Wook Lee, Meryl Stoller, Daniel Dreyer, Christopher Bielawski, Rodney Ruoff, and Kwang Suh. 2010. "High-Performance Supercapacitors Based on Poly(Ionic Liquid)-Modified Graphene Electrodes." doi:10.1021/NN101968P.

Lee, Jae Ah, Min Kyoon Shin, Shi Hyeong Kim, Hyun Cho, Geoffrey Spinks, Gordon Wallace, Márcio Lima, et al. 2013. "Ultrafast Charge and Discharge Biscrolled Yarn Supercapacitors for Textiles and Microdevices." *Nature Communications* 4 (1): 1970. doi:10.1038/ncomms2970.

Lee, Jung-Soo, Sun-I Kim, Jong-Chul Yoon, and Ji-Hyun Jang. 2013. "Chemical Vapor Deposition of Mesoporous Graphene Nanoballs for Supercapacitor." *ACS Nano* 7 (7): 6047–55. doi:10.1021/nn401850z.

Lemine, Aicha, Moustafa Zagho, Talal Altahtamouni, and Nasr Bensalah. 2018. "Graphene a Promising Electrode Material for Supercapacitors—A Review." *International Journal of Energy Research* 42 (14): 4284–4300. doi:10.1002/er.4170.

Li, Chun, and Gaoquan Shi. 2012. "Three-Dimensional Graphene Architectures." *Nanoscale* 4 (18): 5549–63. doi:10.1039/c2nr31467c.

Li, Zhangpeng, Jinqing Wang, Sheng Liu, Xiaohong Liu, and Shengrong Yang. 2011. "Synthesis of Hydrothermally Reduced $Graphene/MnO_2$ Composites and Their Electrochemical Properties as Supercapacitors." *Journal of Power Sources* 196 (19): 8160–65. doi:10.1016/J.JPOWSOUR.2011.05.036.

Lin, Huijuan, Li Li, Jing Ren, Zhenbo Cai, Longbin Qiu, Zhibin Yang, and Huisheng Peng. 2013. "Conducting Polymer Composite Film Incorporated with Aligned Carbon Nanotubes for Transparent, Flexible and Efficient Supercapacitor." *Scientific Reports* 3 (1): 1353. doi:10.1038/srep01353.

Lin, Zifeng, Eider Goikolea, Andrea Balducci, Katsuhiko Naoi, Pierre-Louis Taberna, Mathieu Salanne, Gleb Yushin, and Patrice Simon. 2018. "Materials for Supercapacitors: When Li-Ion Battery Power Is Not Enough." *Materials Today* 21 (4): 419–36. doi:10.1016/j.mattod.2018.01.035.

Liu, Chao, Jizi Liu, Jing Wang, Jiansheng Li, Rui Luo, Jinyou Shen, Xinyun Sun, Weiqing Han, and Lianjun Wang. 2018. "Electrospun Mulberry-Like Hierarchical Carbon Fiber Web for High-Performance Supercapacitors." *Journal of Colloid and Interface Science* 512 (February): 713–21. doi:10.1016/J.JCIS.2017.10.093.

Luo, Zhiqiang, Ting Yu, Jingzhi Shang, Yingying Wang, Sanhua Lim, Lei Liu, Gagik G. Gurzadyan, Zexiang Shen, and Jianyi Lin. 2011. "Large-Scale Synthesis of Bi-Layer Graphene in Strongly Coupled Stacking Order." *Advanced Functional Materials* 21 (5): 911–17.

Mafra, Daniela, Georgy Samsonidze, Leandro Malard, Daniel Cunha Elias, Juliana Brant, Flavio Plentz, Elmo Salomao Alves, and Marcos Assunção Pimenta. 2007. "Determination of LA and TO Phonon Dispersion Relations of Graphene near the Dirac Point by Double Resonance Raman Scattering." *Physical Review B* 76: 233407. doi:10.1103/PhysRevB.76.233407.

Malard, Leandro, Marcos Assunção Pimenta, Gene Dresselhaus, and Mildred Dresselhaus. 2009. "Raman Spectroscopy in Graphene." *Physics Reports* 473 (5–6): 51–87.

Manickam, Sivakumar, Kasturi Muthoosamy, Renu Geetha Bai, Ibrahim Babangida Abubakar, Surya Mudavasseril Sudheer, Lim Hongngee, Loh Hwei-San, Huang Nayming, and Chia Ch. 2015. "Exceedingly Biocompatible and Thin-Layered Reduced Graphene Oxide Nanosheets Using an Eco-Friendly Mushroom Extract Strategy." *International Journal of Nanomedicine* 10 (1): 1505. doi:10.2147/IJN.S75213.

Masikhwa, Tshifhiwa Moureen, Moshawe Jack Madito, Damilola Momodu, Abdulhakeem Bello, Julien Dangbegnon, and Ncholu Manyala. 2016. "High Electrochemical Performance of Hybrid Cobalt Oxyhydroxide/Nickel Foam Graphene." *Journal of Colloid and Interface Science* 484. doi:10.1016/j.jcis.2016.08.069.

Masikhwa, Tshifhiwa Moureen, Moshawe Jack Madito, Abdulhakeem Bello, Julien Dangbegnon, and Ncholu Manyala. 2017. "High Performance Asymmetric Supercapacitor Based on Molybdenum Disulphide/Graphene Foam and Activated Carbon from Expanded Graphite." *Journal of Colloid and Interface Science* 488: 155–65. doi:10.1016/j.jcis.2016.10.095.

Miller, John, Ronald Outlaw, and Brian Holloway. 2010. "Graphene Double-Layer Capacitor with Ac Line-Filtering Performance." *Science* 329 (5999): 1637–39. doi:10.1126/science.1194372.

Mirghni, Abdulmajid Abdallah, Moshawe Jack Madito, Tshifhiwa Moureen Masikhwa, Kabir O. Oyedotun, Abdulhakeem Bello, and Ncholu Manyala. 2017. "Hydrothermal Synthesis of Manganese Phosphate/ Graphene Foam Composite for Electrochemical Supercapacitor Applications." *Journal of Colloid and Interface Science*. doi:10.1016/j.jcis.2017.01.098.

Nair, Rahul, Peter Blake, Alexander Grigorenko, Kostya Sergeevich Novoselov, Tim Booth, Tobias Stauber, Nuno Peres, and Andrei Konstantinovich Geim. 2008. "Fine Structure Constant Defines Visual Transparency of Graphene." *Science* 320 (5881): 1308.

Naoi, Katsuhiko, and Patrice Simon. 2008. "New Materials and New Configurations for Advanced Electrochemical Capacitors." *Journal of the Electrochemical Society (JES)* April. http://oatao.univ-toulouse.fr/3934/1/Naoi_3934.pdf.

Ndiaye, Ndeye Maty, Balla Diop Ngom, Ndeye Fatau Sylla, Tshifhiwa Moureen Masikhwa, Moshawe Jack Madito, Damilola Momodu, Tshepo Ntsoane, and Ncholu Manyala. 2018. "Three Dimensional Vanadium Pentoxide/Graphene Foam Composite as Positive Electrode for High Performance Asymmetric Electrochemical Supercapacitor." *Journal of Colloid and Interface Science* 532 (December): 395–406. doi:10.1016/J.JCIS.2018.08.010.

Novoselov, Kostya Sergeevich, Andrei Konstantinovich Geim, Sergei Vladimirovich Morozov, Da Jiang, Mikhail Katsnelson, Irina Grigorieva, Sergey Dubonos, Alexander Firsov. 2005. "Two-Dimensional Gas of Massless Dirac Fermions in Graphene." *Nature* 438 (7065): 197–200. doi:10.1038/nature04233.

Ochai-Ejeh, Faith, Moshawe Jack Madito, Katlego Makgopa, Mologadi Nkiyase Rantho, Okikiola Olaniyan, and Ncholu Manyala. 2018. "Electrochemical Performance of Hybrid Supercapacitor Device Based on Birnessite-Type Manganese Oxide Decorated on Uncapped Carbon Nanotubes and Porous Activated Carbon Nanostructures." *Electrochimica Acta* 289 (November): 363–75. doi:10.1016/J. ELECTACTA.2018.09.032.

Paek, Seung-Min, Eun Joo Yoo, and Itaru Honma. 2009. "Enhanced Cyclic Performance and Lithium Storage Capacity of SnO_2/Graphene Nanoporous Electrodes with Three-Dimensionally Delaminated Flexible Structure." *Nano Letters* 9 (1): 72–75. doi:10.1021/nl802484w.

Park, Sungjin, and Rodney Ruoff. 2009. "Chemical Methods for the Production of Graphenes." *Nature Nanotechnology* 4 (4): 217–24. doi:10.1038/nnano.2009.58.

Pisana, Simone, Michele Lazzeri, Cinzia Casiraghi, Kostya Sergeevich Novoselov, Andrei Konstantinovich Geim, Andrea Ferrari, and Francesco Mauri. 2007. "Breakdown of the Adiabatic Born–Oppenheimer Approximation in Graphene." *Nature Materials* 6 (3): 198–201.

Rao, Chintamani Nagesa Ramachandra, Ajay Kumar Sood, Kota Surya Subrahmanyam, and Achutharao Govindaraj. 2009. "Graphene: The New Two-Dimensional Nanomaterial." *Angewandte Chemie (International Ed. in English)* 48 (42): 7752–77. doi:10.1002/anie.200901678.

Saito, Riichiro, Ado Jorio, Antonio Gomes Souza Filho, Gene Dresselhaus, Mildred Dresselhaus, and Marcos Assunção Pimenta. 2002. "Probing Phonon Dispersion Relations of Graphite by Double Resonance Raman Scattering." *Physical Review Letters* 88: 027401. doi:10.1103/ PhysRevLett.88.027401.

Schedin, Fredrik, Andrei Konstantinovich Geim, Sergei Vladimirovich Morozov, Edward Hill, Peter Blake, Mikhail Katsnelson, and Kostya Sergeevich Novoselov. 2007. "Detection of Individual Gas Molecules Adsorbed on Graphene." *Nature Materials* 6 (9): 652–55.

Simon, Patrice, and Yury Gogotsi. 2008. "Materials for Electrochemical Capacitors." *Nature Materials* 7 (11): 845–54. doi:10.1038/nmat2297.

Stankovich, Sasha, Richard Piner, Xinqi Chen, Nianqiang Wu, SonBinh Nguyen, and Rodney Ruoff. 2006. "Stable Aqueous Dispersions of Graphitic Nanoplatelets via the Reduction of Exfoliated Graphite Oxide in the Presence of Poly(Sodium 4-Styrenesulfonate)." *Journal of Material Chemistry* 16 (2): 155–58. doi:10.1039/B512799H.

Staudenmaier, L. 1898. "Verfahren Zur Darstellung Der Graphitsäure." *Berichte Der Deutschen Chemischen Gesellschaft* 31 (2): 1481–87. doi:10.1002/cber.18980310237.

Stoller, Meryl, Sungjin Park, Yanwu Zhu, Jinho An, and Rodney Ruoff. 2008. "Graphene-Based Ultracapacitors." *Nano Letters* 8 (10): 3498–3502. doi:10.1021/nl802558y.

Stoller, Meryl, and Rodney Ruoff. 2010. "Best Practice Methods for Determining an Electrode Material's Performance for Ultracapacitors." *Energy & Environmental Science* 3 (9): 1294–1301. doi:10.1039/ c0ee00074d.

Su, Dawei, Dong Han Seo, Yuhang Ju, ZhaoJun Han, Kostya Ostrikov, Shixue Dou, Hyo-Jun Ahn, Zhangquan Peng, and Guoxiu Wang. 2016. "Ruthenium Nanocrystal Decorated Vertical Graphene Nanosheets@ Ni Foam as Highly Efficient Cathode Catalysts for Lithium-Oxygen Batteries." *NPG Asia Materials*. doi:10.1038/am.2016.91.

Toupin, Mathieu, Thierry Brousse, and Daniel Be. 2004. "Charge Storage Mechanism of MnO_2 Electrode Used in Aqueous Electrochemical Capacitor." *Chemistry of Materials* 16 (9): 3184–90.

Tung, Vincent, Matthew Allen, Yang Yang, and Richard Kaner. 2009. "High-Throughput Solution Processing of Large-Scale Graphene." *Nature Nanotechnology* 4 (1): 25–29. doi:10.1038/nnano.2008.329.

Wan, Xiangjian, Yi Huang, and Yongsheng Chen. 2012. "Focusing on Energy and Optoelectronic Applications: A Journey for Graphene and Graphene Oxide at Large Scale." *Accounts of Chemical Research* 45 (4): 598–607. doi:10.1021/ar200229q.

Wang, Da-Wei, Feng Li, Jinping Zhao, Wencai Ren, Zhi-Gang Chen, Jun Tan, Zhong-Shuai Wu, Ian Gentle, Gao Qing Lu, and Hui-Ming Cheng. 2009. "Fabrication of Graphene/Polyaniline Composite Paper *via In Situ* Anodic Electropolymerization for High-Performance Flexible Electrode." *ACS Nano* 3 (7): 1745–52. doi:10.1021/nn900297m.

Wang, Guoping, Lei Zhang, and Jiujun Zhang. 2012. "A Review of Electrode Materials for Electrochemical Supercapacitors." *Chemical Society of Review* 41 (2): 797–828. doi:10.1039/C1CS15060J.

Wang, Huan-Wen, Zhong-Ai Hu, Yan-Qin Chang, Yan-Li Chen, Zi-Yu Zhang, Yu-Ying Yang, and Hong-Ying Wu. 2011. "Preparation of Reduced Graphene Oxide/Cobalt Oxide Composites and Their Enhanced Capacitive Behaviors by Homogeneous Incorporation of Reduced Graphene Oxide Sheets in Cobalt Oxide Matrix." *Materials Chemistry and Physics* 130 (1–2): 672–79. doi:10.1016/J. MATCHEMPHYS.2011.07.043.

Wang, Kai, Pu Zhao, Xiaomo Zhou, Haiping Wu, and Zhixiang Wei. 2011. "Flexible Supercapacitors Based on Cloth-Supported Electrodes of Conducting Polymer Nanowire Array/SWCNT Composites." *Journal of Materials Chemistry* 21 (41): 16373. doi:10.1039/c1jm13722k.

Wang, Shiyong, Li Ma, Mengyu Gan, Shenna Fu, Wenqin Dai, Tao Zhou, Xiaowu Sun, Huihui Wang, and Huining Wang. 2015. "Free-Standing 3D Graphene/Polyaniline Composite Film Electrodes for High-Performance Supercapacitors." *Journal of Power Sources* 299 (December): 347–55. doi:10.1016/j. jpowsour.2015.09.018.

Wang, Yang, Aurelien Du Pasquier, Derek Li, Paolina Atanassova, Scott Sawrey, and Miodrag Oljaca. 2018. "Electrochemical Double Layer Capacitors Containing Carbon Black Additives for Improved Capacitance and Cycle Life." *Carbon* 133 (July): 1–5. doi:10.1016/J.CARBON.2018.03.001.

Weingarth, Daniel, Heeju Noh, Annette Foelske-Schmitz, Alexander Wokaun, and Rüdiger Kötz. 2013. "A Reliable Determination Method of Stability Limits for Electrochemical Double Layer Capacitors." *Electrochimica Acta* 103 (July): 119–24. doi:10.1016/j.electacta.2013.04.057.

Wilkes, John. 2002. "A Short History of Ionic Liquids—From Molten Salts to Neoteric Solvents." *Green Chemistry* 4: 73–80. doi:10.1039/b110838g.

William Hummers, Jr., and Richard Offeman. 1958. "Preparation of Graphitic Oxide." *Journal of American Chemical Society* 80 (1937): 1339. doi:10.1021/ja01539a017.

Wu, Guan, Pengfeng Tan, Dongxing Wang, Zhe Li, Lu Peng, Ying Hu, Caifeng Wang, Wei Zhu, Su Chen, and Wei Chen. 2017. "High-Performance Supercapacitors Based on Electrochemical-Induced Vertical-Aligned Carbon Nanotubes and Polyaniline Nanocomposite Electrodes." *Scientific Reports* 7 (1): 43676. doi:10.1038/srep43676.

Wu, Yaping, Harry Chou, Hengxing Ji, Qingzhi Wu, Shanshan Chen, Wei Jiang, Yufeng Hao, et al. 2012. "Growth Mechanism and Controlled Synthesis of AB-Stacked Bilayer Graphene on Cu-Ni Alloy Foils." *ACS Nano* 6 (9): 7731–38.

Wu, Zhong-Shuai, Guangmin Zhou, Li-Chang Yin, Wencai Ren, Feng Li, and Hui-Ming Cheng. 2012. "Graphene/Metal Oxide Composite Electrode Materials for Energy Storage." *Nano Energy* 1 (1): 107–31. doi:10.1016/j.nanoen.2011.11.001.

Xu, Yuxi, Zhaoyang Lin, Xiaoqing Huang, Yang Wang, Yu Huang, and Xiangfeng Duan. 2013. "Functionalized Graphene Hydrogel-Based High-Performance Supercapacitors." *Advanced Materials* 25 (40): 5779–84. doi:10.1002/adma.201301928.

Yan, Jun, Tong Wei, Bo Shao, and Zhuangjun Fan. 2009. "Preparation of a Graphene Nanosheet / Polyaniline Composite with High Specific Capacitance." *Carbon* 48 (2): 487–93. doi:10.1016/j. carbon.2009.09.066.

Yan, Jun, Tong Wei, Bo Shao, Zhuangjun Fan, Weizhong Qian, Milin Zhang, and Fei Wei. 2010. "Preparation of a Graphene Nanosheet/Polyaniline Composite with High Specific Capacitance." *Carbon* 48 (2): 487–93. doi:10.1016/J.CARBON.2009.09.066.

Yang, Xiaowei, Chi Cheng, Yufei Wang, Ling Qiu, and Dan Li. 2013. "Liquid-Mediated Dense Integration of Graphene Materials for Compact Capacitive Energy Storage." *Science (New York, N.Y.)* 341 (6145): 534–37. doi:10.1126/science.1239089.

Zhai, Yunpu, Yuqian Dou, Dongyuan Zhao, Pasquale Fulvio, Richard Mayes, and Sheng Dai. 2011. "Carbon Materials for Chemical Capacitive Energy Storage." *Advanced Materials* 23 (42): 4828–50.

Zhang, Fan, Tengfei Zhang, Xi Yang, Long Zhang, Kai Leng, Yi Huang, and Yongsheng Chen. 2013. "A High-Performance Supercapacitor-Battery Hybrid Energy Storage Device Based on Graphene-Enhanced Electrode Materials with Ultrahigh Energy Density." *Energy & Environmental Science* 6 (5): 1623. doi:10.1039/c3ee40509e.

Zhang, Jing, Yao Yu, Lin Liu, and Yue Wu. 2013. "Graphene–Hollow PPy Sphere 3D-Nanoarchitecture with Enhanced Electrochemical Performance." *Nanoscale* 5 (7): 3052. doi:10.1039/c3nr33641g.

Zhi, Mingjia, Chengcheng Xiang, Jiangtian Li, Ming Li, and Nianqiang Wu. 2013. "Nanostructured Carbon-Metal Oxide Composite Electrodes for Supercapacitors: A Review." *Nanoscale* 5 (1): 72–88. doi:10.1039/c2nr32040a.

7

Transition Metal Oxide-Based Nanomaterials for High Energy and Power Density Supercapacitor

Raphael M. Obodo, Assumpta C. Nwanya, Tabassum Hassina, Mesfin A. Kebede, Ishaq Ahmad, M. Maaza, and Fabian I. Ezema

CONTENTS

7.1   Introduction 132
7.2   Supercapacitors 133
    7.2.1   Types of Supercapacitors 133
    7.2.2   Electrostatic Double-Layer Capacitor 134
    7.2.3   Pseudocapacitor 135
    7.2.4   Hybrid Supercapacitor 136
    7.2.5   Asymmetric Supercapacitor 136
    7.2.6   Composite Supercapacitor 136
    7.2.7   Battery Type Supercapacitor 136
7.3   Supercapacitors Electrode Materials 136
    7.3.1   Electrodes for Electrostatic Double-Layer Capacitors 136
    7.3.2   Activated Carbon and Carbon Nanomaterials 137
    7.3.3   Activated Carbon Fibres 137
    7.3.4   Carbon Acrogel 137
    7.3.5   Carbide-Derived Carbon 137
    7.3.6   Graphene 137
    7.3.7   Carbon Nanotubes 137
7.4   Electrodes for Pseudocapacitors 138
    7.4.1   Transition Metal Oxides and Hydroxides 138
    7.4.2   Ruthenium Oxide 139
    7.4.3   Manganese (IV) Oxide 140
    7.4.4   Cobalt Oxide/Hydroxide 140
    7.4.5   Iron 2 and 3 Oxides (Fe_2O_3 and Fe_3O_4) 141
    7.4.6   Nickel Oxide/Hydroxide 141
    7.4.7   Binary Metal Oxides 141
7.5   Transition Metal Phosphide Nanomaterials 141
7.6   Transition Metal Sulphides and Carbon Derivatives Nanomaterials 142
    7.6.1   Cobalt Sulphides 142
    7.6.2   Nickel Sulphides 142
7.7   Carbon Composite-Based Nanomaterials 143
    7.7.1   Carbon/Carbon Composites 143
    7.7.2   Carbon/Metal Oxides Composites 143
    7.7.3   Carbon/Conducting Polymer Composites 143
7.8   Applications of Supercapacitors 143
    7.8.1   Commercial Scale Applications of Supercapacitors 143
    7.8.2   Electronics 144
    7.8.3   Low-Power Equipment Power Buffer 144

    7.8.4   Voltage Stabilizer 144
    7.8.5   Streetlights... 144
    7.8.6   Medical... 144
    7.8.7   Transport 145
    7.8.8   Smart Devices Applications of Supercapacitors 145
7.9   Summary/Challenges 145
    7.9.1   Challenges Facing Supercapacitor Development... 145
    7.9.2   Recommendations 146
References... 146

7.1 Introduction

Recently, the demand for energy sustainability has progressively increased, and the high rate of the world energy depletion has also accelerated the efforts in finding and refining alternative energy resourceful devices. In terms of augmenting the existing storage devices, supercapacitors are superior to the battery technology, but have also been a fundamental concern to scientists recently. Energy storage is as important as energy production, hence, the need for effective storage systems especially in our modern electrical and electronic devices. To effectively manage our energy requirements nowadays, our present society wants small, light weight, inexpensive, and environmentally friendly energy storage systems (Meng et al. 2010). Presently, the battery is a widely used energy storage system, but its unassailability, high discharge rate, slow charge rate, and small lifespan limit its use in modern electronics devices (Meng et al. 2010). Supercapacitors have shown to be a better option for storage of energy because of their better energy and power density, light weight, quick charge rate, and long lifespan (Jayalakshmi and Balasubramanian 2008).

Supercapacitors are used in many applications, such as crossbreed vehicles, power gridlock, military services, and electronic devices like mobile phones, laptops, wristwatches, iPods, etc. (Wang, Lu et al. 2012). The geometric increase in world population has created a great deal of pressure on the current energy depletion. In order to contribute to the solution of the energy crisis, a convenient technology is required as a storage device. The struggle for a more efficient energy storage system based on the dimension of nanostructures is shown in Figure 7.1.

Carbon nanotubes (CNTs) are attractive electrode materials owing to their good power density, energy density, rate capability, specific capacitance, as well as cycling stability, which make them suitable for construction of 1D, 2D, 3D and its usage in supercapacitors electrode materials.

There has been various research work going on to increase the energy and power density of supercapacitors to enable them to function effectively as primary power storages and sources, as both are found in our minor and major gadgets. Figure 7.2 illustrates the steps, mechanisms, and ways of enhancing supercapacitors performance. The observance of these strategies and approaches could manifest in the effective replacement of battery technology.

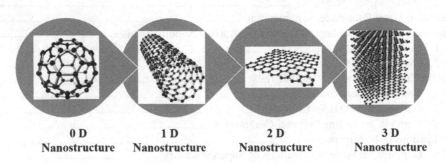

           0 D   **1 D**   **2 D**   **3 D**
  Nanostructure   **Nanostructure**   **Nanostructure**   **Nanostructure**

FIGURE 7.1   Stages in advancement of 0D to 3D used for supercapacitors. (Reproduced from Kiew, S.F. et al., *J. Control Release*, 226, 217–228, 2016. With permission.)

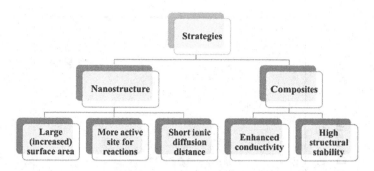

FIGURE 7.2 Strategies to assist increase in the performance of supercapacitor.

The aim of this chapter is to present the recent works on the use of transition metal compounds in supercapacitors and summarize the achievements, challenges, and possible solutions being attempted.

7.2 Supercapacitors

Supercapacitors most often known as electrochemical capacitors, also called supercap, ultracapacitor, or goldcap. They are devices capable of storing electrical energy (Xia, Tu, Zhang, et al. 2011). Supercapacitors have two electrodes divided by a separator with an electrolyte that links both electrodes ionically (Xin and Bingqing 2013) (Figure 7.3).

Electrodes are polarized when a voltage is applied, hence, various ions inside the electrolyte create a double layer of the electric field opposite to the polarity of the electrodes. Electrodes that are charged positively will form a shield of cations midway of electrode/electrolyte, with a balanced charge deposit of anions adsorbed onto the electrodes and vice versa as shown above. The type of electrode material and superficial shape determine the ions adsorption rate on the double layer and their impact on the overall capacitance of the supercapacitor (Figure 7.4).

7.2.1 Types of Supercapacitors

There are three main classes of supercapacitors, namely: (1) electric (electrostatic) double-layer supercapacitor, (2) pseudocapacitor (electrochemical) supercapacitors, and (3) hybrid supercapacitors.

Supercapacitors are categorized by their mechanism for storing and delivering charges. These include non-faradaic also known as electrostatic because no transfer of charge, faradaic, which involves chemical reaction, hence, transfer of charges, and hybrid, which is a combination of non-faradaic (electrostatic)

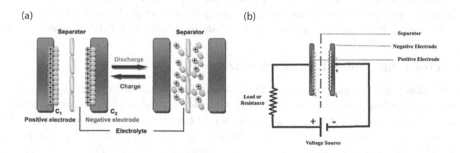

FIGURE 7.3 Schematic diagram of (a) charged and discharged supercapacitor (b) typical supercapacitor. (Reproduced from Xin, L. and Bingqing, W., *Nano Energy*, 2, 159–173, 2013. With permission.)

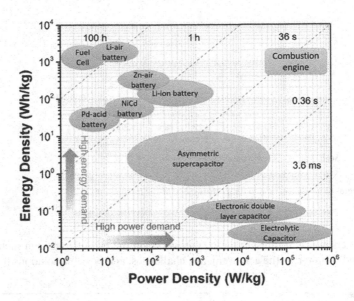

FIGURE 7.4   Ragone chart of energy storage and conversion devices. (Reproduced from Yuanlong, S. et al., *Chem. Rev.*, 118, 9233–80, 2018. With permission.)

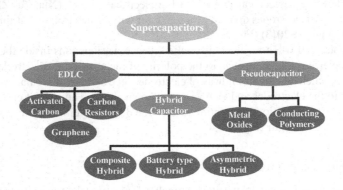

FIGURE 7.5   Classification of supercapacitors.

and faradaic (chemical reaction). However, non-faradaic mechanisms do not use chemicals, but charges are spread on the surface of the electrode by physical procedures that do not require making or breaking of chemical bonds, hence, described as electrostatic. Faradaic mechanisms involve redox reactions, which require the exchange of charges among electrolyte and electrode (Figure 7.5).

The supercapacitor performance compared to other energy storage alternative devices is shown in Table 7.1. The output power and energy density of various alternatives to supercapacitors are also shown.

7.2.2 Electrostatic Double-Layer Capacitor

Electrostatic double-layer capacitors (EDLCs) are produced from a combination of two carbon-based electrodes, an electrolyte, and a separator in between the two electrodes. EDLCs charges are stored non-faradaically (electrostatically), hence, no charge exchange (absorption or emission) between electrolyte and electrode. It has almost the same working principle with conventional capacitors. There is physical charge accumulation at the interface of the electrolyte and electrode (Conway 1999). The process of charging transports electrons from cathode to anode. This charging facilitates the exchange of anions towards the cathode, while cations move towards the anode in the electrolyte (Dubal, Kim and Lokhande 2012).

TABLE 7.1

Performance Evaluation of Supercapacitors and Various Alternatives

S/N	Energy Storage Systems	No. of Cycles	Maximum Weight	Specific Supercapacitance	Power Density Range (w/kg)	Energy Density Range (Wh/kg)
1.	Lead acid batteries	150–1500	> 10 kg	N/A	50–500	30–50
2.	Li-ion batteries	150–1500	1 g–10 kg	N/A	300–3000	100–180
3.	Supercapacitors	>50,000	1 g–230 g	100 mF–1500 F	10–120	1–10
4.	Orthodox capacitors	>100,000	1 g–10 kg	100 pF–202 mF	0.01–0.05	0.01–0.02

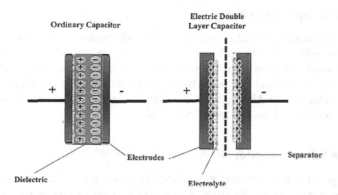

FIGURE 7.6 Conventional and electrochemical double-layer capacitor.

On the other hand, anions and cations travel in the opposite course during discharging (Conway 1999). EDLCs have a long lifespan because there is no exchange of charges between electrode and electrolyte, hence, no significant volume or morphology change of electrodes (Fan et al. 2014) (Figure 7.6).

The thickness of the separator is a major factor that governs the capacitance of conventional capacitors, while the thickness of the double layer at the electrode/electrolyte boundary decides the capacitance of EDLCs.

EDLCs have a bigger capacitance as compared to the conventional capacitor because the thickness of the double layer is very small compared to the separator (Amatucci et al. 2001). The shortness of the distance between the electrodes and increased area of electrode/electrolyte boundary in a conventional capacitor contributes to the added capacitance of EDLCs. A good electrolyte with an enhanced electrical conductivity in an EDLC reduces the internal resistance of electrodes, while good solubility of the electrolyte will raise the transmission of ions within the holes of electrodes and increases the capacitive competency of EDLCs (Amatucci et al. 2001).

7.2.3 Pseudocapacitor

Pseudocapacitors operate with oxidation and a reduction in faradaic reactions. Reversible and rapid faradaic reactions occur on the pseudocapacitor electrode materials when a potential is applied across its electrodes. This practical potential facilitates the creation of charges and activates charge movement between the double layers. An analogous charging and discharging practice also arises in battery operations.

Three major faradaic reactions occur in the electrodes which include: (1) oxidation and reduction reactions of transition metal oxides, (2) reversible electrochemical functionalization, and (3) reversible adsorption. It is observed that pseudocapacitors give bigger energy density and specific capacitance compared to EDLCs because the faradaic reactions take place on the surface and inner parts of the electrode materials (Vuorilehto and Nuutinen Bautzen 2014). Long cycling of pseudocapacitors causes instability when it operates on redox reactions. EDLCs give a higher power density than pseudocapacitors because non-faradaic routes are faster than faradaic sequences.

7.2.4 Hybrid Supercapacitor

A hybrid supercapacitor refers to a supercapacitor with a combination of pseudocapacitor and EDLC qualities. One of the electrodes exhibits electrostatic capacitive techniques, while the other shows the electrochemical capacitive method. EDLC has good cyclic stability, while a pseudocapacitor offers a better specific capacitance, and the two qualities are sandwiched in a hybrid supercapacitor.

A hybrid supercapacitor with good electrode combinations can deliver a high cell voltage, hence, an improved power and energy density performance. There is a growing interest in hybrid supercapacitors recently owing to their dual quality. There are three types of hybrid supercapacitor, namely: (1) asymmetric, (2) composite, and (3) battery-type.

7.2.5 Asymmetric Supercapacitor

This is a type of hybrid supercapacitor with two unlike electrodes, mostly a carbon material and a battery-like electrode with a faradaic performance. A good operational voltage window was obtainable due to it's an exceptional design which raises the energy density due to the oxidation and reduction reaction process at the electrochemical electrode (Burke 2007). Asymmetric supercapacitors are fabricated in a way that metal oxide or a conducting polymer (COP) serves as an anode, while carbon material takes a cathode function (Syarif, Tribidasari, and Wibowo 2012).

7.2.6 Composite Supercapacitor

Composite supercapacitors are fabricated by sandwiching a carbon material and an oxide of a transition metal or conductive polymer in one electrode. This technique upgrades one single electrode, combining chemical and physical charge storage machinery.

The carbon material provides a high specific surface area, which increases the capacitive double layer of charge while metal oxides or a COP increases the pseudocapacitive performance of the composite electrode (Halper and Ellenbogen 2006). Presently, we have two types of composites, namely: (1) binary composite, which uses two unlike materials to fabricate one electrode and (2) ternary composite that uses three diverse materials to produce one electrode.

7.2.7 Battery Type Supercapacitor

This type of supercapacitor resembles an asymmetric supercapacitor, but has unique features by integrating a supercapacitor electrode with the battery electrode as anode and cathode or vice versa. The enhanced properties of supercapacitors and batteries are inserted in one system which optimizes both qualities in one supercapacitor (Halper and Ellenbogen 2006).

7.3 Supercapacitors Electrode Materials

Electrode materials are tiny coatings that are electrically and functionally stable in a current conductive collector. The qualities of a good supercapacitor electrode include good conductivity, stable in a high temperature, chemically stable in an electrolyte (inert), resistant to corrosion, and an enhanced surface area per unit mass and volume. Supercapacitors electrode materials should also be readily available, non-toxic, and cost effective.

7.3.1 Electrodes for Electrostatic Double-Layer Capacitors

There are many factors considered when choosing a material for supercapacitors electrode fabrication. These factors include stability in chemical reactions and high temperatures, the capacitance per surface area, electrical conductivity, the percentage of surface area, charge storage performance, etc.

7.3.2 Activated Carbon and Carbon Nanomaterials

In the early days of EDLCs, the earliest electrode material used was activated carbon (AC) because it has a good surface area, enhanced electrical conductivity, and low cost (Syarif, Tribidasari, and Wibowo 2012). This electrode material can be fabricated by chemical or physical reactions from carbon compounds like coal, charcoal, wood, etc. Chemical fabrication of AC requires a lower temperature range of 400°C–700°C in combination with activation materials such as zinc chloride, phosphoric acid, potassium hydroxide, and sodium hydroxide. The physical preparation needs the burning of carbon compounds at a temperature range of 700°C–1200°C with the help of oxidizing agents, which includes CO_2, air, and steam (Pandolfo and Hollenkamp 2006). AC is a porous form of the carbon electrode material with a better specific surface area and less costly than other carbon products (Nesbitt and Sun 2004).

7.3.3 Activated Carbon Fibres

This electrode material is an offspring of AC after some purification. They have a small diameter of 10 μm, pore size dispersal simply arranged. Activated carbon fibre electrode materials have small electrical resistance along the fibre axis, hence, good contact to the collector. Due to the small arrangement of their microspores, activated carbon fibre electrodes mainly operate on double-layer capacitance with a little pseudocapacitance.

7.3.4 Carbon Aerogel

Carbon aerogel is a synthetic, weightless, and very permeable electrode material produced by inserting a gas inside an organic gel with a molten component. They are mostly produced by pyrolysis of resorcinol-formaldehyde aerogels (Fischer et al. 1997), and they are highly conductive compared to other AC electrodes. An aerogel electrode also absorbs high mechanical and vibrational steadfastness for a supercapacitor in use where a large vibrational environment is required.

7.3.5 Carbide-Derived Carbon

This carbide-derived carbon also known as tunable nanoporous carbon is an offspring of the carbide compound developed from carbide predecessors, which include binary silicon carbide and titanium carbide. These carbide materials are changed into pure carbon mainly by a physical change like a heat disintegration method or chemical halogenation technique (Korenblit et al. 2010). They have a high surface area and a variable pore radius that changes from micropores to mesopores to increase optimization of ions incarceration and intensify a pseudocapacitive technique by rapid electrochemical adsorption of hydrogen. Most carbide-derived carbon electrode materials have a unique pore arrangement, which gives them enhanced specific energy compared to normal ACs.

7.3.6 Graphene

Graphene is a sheet of refined graphite in which the atoms are densely and closely packed. These graphite atoms that formed graphene are arranged systematically and neatly in a hexagonal configuration (Palaniselvam and Baek 2015). They are also known as nanocomposite papers (Palaniselvam and Baek 2015).

Their unique properties such as high specific surface area, good electrical conductivity, and stable in chemical reactions (inertness) make them a good candidate for energy storage devices. Their high malleable characteristic gives graphene an edge over AC because it can be reformed into many shapes, as shown in Figure 7.7. Graphene sheets can also be used directly as electrode materials without collectors for transferable supercapacitor application (Shaijumon, Gowda, and Ajayan 2009).

7.3.7 Carbon Nanotubes

CNTs are also known as buckytubes. They are carbon nanostructured molecules with tubular shapes. They have made an astonishing contribution to the science and engineering of carbon electrode materials for supercapacitors. Its extraordinary characteristics of good mechanical shock absorbance, heat

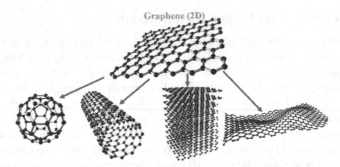

Fullerenes (0D)   Carbon nanotubes (1D) Graphite (3D)   Nanoribbon (3D)

FIGURE 7.7   Formation of graphene into various shapes. (Reproduced with permission from Kiew et al., Assessing Biocompatibility of graphene oxide-based nanostructure: A review, *J. Control Release*, 226, 217–228, 2016.)

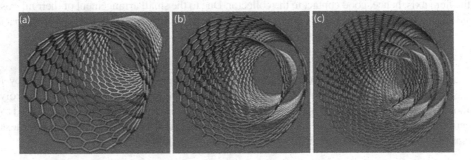

FIGURE 7.8   (a) Single-walled carbon nanotube. (b) Double-wall carbon nanotube. (c) Multi-walled carbon nanotube. (Reproduced Giulianini, M. and Motta, N., Polmer self-assembly on carbon nanotubes, in edited by Bellucci, *Self-assembly of Nanostructures: Lecture Notes in Nanoscale Science and Technology*, Springer, New York, NY, 2012. With permission.)

stability, nice pore structure, and good electrical conductivity increased the interest towards CNTs as supercapacitor electrode material (Jian et al. 2016). CNTs are generally divided into two categories, namely: (1) single-walled nanotubes and (2) multi-walled nanotubes. These two types of CNTs can be used as supercapacitors electrode material. Some typical examples are shown in Figure 7.8.

7.4   Electrodes for Pseudocapacitors

Materials used as electrodes for pseudocapacitors exhibit electrochemical supercapacitive behaviour. Their mechanisms of charge storage are mainly electron transfer instead of the accumulation of ions as in the electrochemical double layer. Pseudocapacitive capacitance is produced because of faradaic redox reactions taking place exclusively with the electrodes material (Zheng and Jow 1995).

7.4.1   Transition Metal Oxides and Hydroxides

Transition metal oxides and hydroxides are new generation materials used in the production of supercapacitor electrodes due to their excellently small resistance and good specific capacitance. These intrinsic properties of metal oxides and hydroxides make them suitable for fabrication of a supercapacitor with a reasonable power and energy density. Those commonly and widely used as materials for supercapacitor electrodes include RuO_2 (Zheng and Jow 1995), Co_3O_4 (Lin, Ritter, and Popov 1998), MnO_2 (Pang, Anderson, and Chapman 2000), NiO (Liu and Anderson 1996), and V_2O_5 (Reddy and Reddy 2006).

These metal oxides have variable oxidation states, which increase the capacitance of the supercapacitors. Recently, there was a growing interest in pseudocapacitive materials, such as transition metal oxides/hydroxides and conductive polymer instead of pure carbon materials (Wang et al. 2014) owing to their amazing properties. They are divided into three major groups, namely: (1) noble metal oxides/hydroxides which include RuO_2 (Chen et al. 2011) and IrO_2 (Wei et al. 2011), (2) low cost metal oxides/hydroxides, which include MnO_2 (Cheng et al. 2012), Co_3O_4 (Xia, Tu, Mai et al. 2011), NiO (Xia, Tu, Zhang et al. 2011), $Co(OH)_2$ (Jiang, Liu et al. 2011), and (3) binary transition metal oxides/hydroxides, such as $Ni(OH)_2$ (Xia, Tu, Mai, et al. 2011). Regardless of this growing interest, they have some weaknesses, such as poor cycle stability and insignificant power density, which are major hindrances to its usage nowadays. This small power density of metal oxides/hydroxides can be attributed to the poor electrical conductivity, which restricts the easy and fast rate of electron transfer. It has been reported individually that cycling instability in electrode materials often is caused by swelling and shrinking processes during charging and discharging cycles, which also affect the chemical and physical properties of the electrodes materials (Wei et al. 2011).

There are two major steps in solving these instability problems which include: (1) combining metal oxide/hydroxides with a conductive polymer and (2) application of good porous nanocomposites. The good electrical conductivity property of transition metal oxides/hydroxides helps to speed up reaction dynamics in electrodes materials (Xia, Tu, Mai et al. 2011). Their well nanostructured pores also absorb the stress resulting from the expanding and shrinking of the electrode. They have many ion adsorption ratio or active sites for the charge transfer reactions, which also reduces the route for the ion.

7.4.2 Ruthenium Oxide

RuO_2 has a high theoretical supercapacitance of approximately ~2000 F/g (Kim et al. 2015) and is one of the most ancient materials used for the production of supercapacitor electrodes. It is also known as one of the best electrode materials for supercapacitors because of its wide potential window, long lifespan, good electrical conductivity, high capability rate, and high reversibility of electrochemical reactions (Zhang et al. 2011). Equation 7.1 explains the reversible and fast pseudocapacitive performance of RuO_2 electrode material in an acidic electrolyte.

$$RuO_2 + xH^+ + xe^- \leftrightarrow RuO_{2-x}(OH)_x, \qquad (7.1)$$

where $0 \leq x \leq 2$.

The oxidation states of ruthenium in an acidic medium fluctuate from (II) to (IV), but adopt a different oxidation state in an alkaline medium. RuO_2 and carbon composite, when charged in an electrode material, oxidize to RuO_4^{2-}, RuO_4^-, and RuO_4, but these high oxidation numbers get reduced to RuO_2 when the electrode is discharged (Su et al. 2007). This explains why the supercapacitances of RuO_2 electrode materials are smaller than the theoretical value because the increase in crystallinity reduces supercapacitance due to power restrictions (Kim et al. 2015).

The supercapacitive performance of RuO_2 lies on the synthesis techniques, which include concentration, annealing temperature, crystallinity, particle size, etc. The presence of water in hydrous RuO_2 accelerates the diffusion of hydrogen ions in the electrode material. The liquid ruthenium oxide exhibits a high supercapacitance of 861 F/g and 900 F/g (Long et al. 1999), which are much higher compared to the anhydrous state. The decrease in water content decreases it from $RuO_2 \cdot 0.5H_2O$ to $RuO_2 \cdot 0.03H_2O$, hence, a decrease in supercapacitance from 720 to 19.2 F/g (Zheng 1995). Crystallinity is also a factor that affects the supercapacitive performance of RuO_2 (Pataki, Lokhande, and Joo 2009).

The exchange of ions and electrons has difficulty taking place because of a well crystallized and compactness of the RuO_2 structure. It causes a huge rise in electrochemical impedance and a reduction in the overall supercapacitive performance of RuO_2 electrodes material. The oxidation and reduction reactions of unrefined RuO_2 take place on its surface and also in the unpackaged of the materials. There has been a comparison of the performance of amorphous RuO_2 materials and its crystallized structures which favours the crystals (Kuratani, Kiyobayashi, and Kuriyama 2009).

The amorphous RuO_2 thin films deposited anodically on stainless steel substrates exhibit stable electrochemical supercapacitive properties, showing a maximum supercapacitance of 1190 F/g in H_2SO_4 (Cormier, Andrea, and Zhang 2011). Equally, annealing temperature improves the supercapacitive performance of RuO_2 because annealing at a high temperature enhances the crystallinity (Kim and Popov 2002).

7.4.3 Manganese (IV) Oxide

Recently, there has been a growing interest on the use of manganese (IV) oxide as an alternative electrode material for supercapacitors because of its exceptional properties. These include inexpensive, small toxicity, readily available, excellent supercapacitive performance, especially in aqueous electrolytes, and environmental friendliness (Zhao et al. 2013). The capacitance of manganese (IV) oxides comes mainly from pseudocapacitance.

Two major ways in which manganese (IV) oxide charge storage behaviour can be illustrated are shown in equations 7.2 and 7.3. The first is the addition of electrolyte cations (C^+ = H^+, Li^+, Na^+, and K^+) into the reactions mainstream of the electrode (Brousse et al. 2006).

$$MnO_2 + C^+ + e^- \leftrightarrow MnOOC \tag{7.2}$$

While the second equation shows adsorption of electrolyte cations on the MnO_2 electrode surface (Brousse et al. 2006):

$$(MnO)_{surface} + C^+ e^- \leftrightarrow (MnOOC)_{surface} \tag{7.3}$$

The two mechanisms involve a redox reaction between the III and IV oxidation states.

A MnO_2-based compound has numerous crystal structures, such as (α^-, β^-, γ^-, δ^-, and λ^-), which govern their electrochemical performance, especially when the dimensions of the bonds limit the interactions of cations (Donne, Hollenkamp, and Jones 2010).

7.4.4 Cobalt Oxide/Hydroxide

Cobalt is a transition metal with three typical polymorphs known as: (1) monoxide or cobaltous oxide (CoO), (2) cobaltic oxide (Co_2O_3), and (3) cobaltosic oxide or cobalt cobaltite (Co_3O_4) (Nwanya et al. 2017; Obodo et al. 2019).

Cobalt oxide (Co_3O_4) is also a cubic structure with an AB_2O_4 spinel configuration with a theoretical supercapacitance of about 3560 F/g (Meenakshi et al. 2012). It has been observed that annealing in oxygen can cause environmental changes from $Co(OH)_2$ to crystalline Co_3O_4. There two classes of $Co(OH)_2$, namely: (1) α-$Co(OH)_2$ and (2) β-$Co(OH)_2$, where α-$Co(OH)_2$ has a better supercapacitive compared to β-$Co(OH)$.

The reaction equations of a supercapacitive performance of Co_3O_4 are illustrated as follows in equations (7.4) and (7.5).

$$Co_3O_2 + OH^+ + H_2O \leftrightarrow 3CoOOH^+ + e^- \tag{7.4}$$

$$CoOOH + OH^- \leftrightarrow CoO_2 + H_2O + e^-. \tag{7.5}$$

Co_3O_4 is a p-type semiconductor and has a very small electronic and ionic conductivity (Fan et al. 2014). It exhibits both small rate capability and huge capacity change resulting in a short life cycle during pulverization (Shan and Gao 2007). A nanosize Co_3O_4 provides a larger supercapacitance than bulk Co_3O_4 due to its high surface area to volume ratio.

7.4.5 Iron 2 and 3 Oxides (Fe_2O_3 and Fe_3O_4)

Iron oxides have a huge electric conductivity compared to other metal oxides (Xiong et al. 2012), but their small supercapacitance limits their practical applications in supercapacitors owing to the high redox behaviour of iron.

7.4.6 Nickel Oxide/Hydroxide

Nickel oxide is a good electrodes material because of its high theoretical supercapacitance of 3750 F/g, low cost, and toxicity (Iwueke et al. 2015). Their high supercapacitive performance are explained by two reaction theories: (1) the first states that the energy storing process arises among NiO and NiOOH [equations (7.6) and (7.7)], and (2) the second one states that NiO changes to $Ni(OH)_2$ first in alkaline electrolyte before electrochemical reactions occur between $Ni(OH)_2$ and NiOOH [equations (7.8) and (7.9)] (Wu et al. 2009). The equations of reaction of the mechanisms are shown below:

$$NiO + OH^- \leftrightarrow NiOOH + e^- \tag{7.6}$$

$$NiO + H_2O \leftrightarrow NiOOH + H^- + e^-. \tag{7.7}$$

Or

$$Ni(OH)_2 \leftrightarrow NiOOH + H^+ + e^- \tag{7.8}$$

$$Ni(OH)_2 + OH^- \leftrightarrow NiOOH + H_2O + e^-. \tag{7.9}$$

However, the two theories finally agree that Ni^{2+} oxidizes to NiOOH by losing an electron, which resulted in a supercapacitive performance (Prasad and Miura 2004). The first theory surpasses, but the second theory is also useful. In alkaline medium, NiO combines with OH^- to produce $Ni(OH)_2$, which contributes to the increased capacitive performance.

There are some difficulties experienced with NiO supercapacitor electrode materials applications, which include low cycle stability and poor electronic conductivity (Offiah et al. 2014). In order to solve these problems, nanostructured NiO electrodes were employed because they can provide a large specific surface area, short diffusion, and transport paths. This way, the reaction kinetics can be facilitated (Zhou et al. 2015).

7.4.7 Binary Metal Oxides

There is a huge growing interest in binary metal oxides currently, such as $NiCo_2O_4$ (Wang et al. 2013), $NiFe_2O_4$ (Zhou et al. 2015), $CoFe_2O_4$ (Xiong et al. 2012), $ZnMnO_4$ (Kim et al. 2011), $ZnCo_2O_4$ (Karthikeyan et al. 2009), and $CoMn_2O_4$ (Yu et al. 2013). It is believed that binary metal oxides perform better electric conduction compared to single metal oxide due to their dual contributions to the overall capacitance. This double contribution of individual metal oxides leads to an increased electrochemical performance over a single metal oxide (Huang et al. 2013).

However, they show poor cycle performance because charge transfer resistance of the electrodes material increases in each cycle (Wang et al. 2013).

7.5 Transition Metal Phosphide Nanomaterials

The need for clean and renewable energy extends the search for effective materials to store energy for consumption since their applications in the supercapacitor, electronics, optical, and optoelectronic devices are viable. Transition metal phosphides have been recognized as a promising metal-based material for supercapacitors. The classes are not limited to Fe_2P, FeP, Co_2P, CoP, Ni_2P, $Ni_{12}P_5$, etc.

There are many ways to prepare phosphide compounds. One of which involves heating the metal and phosphorus in inert atmospheric or vacuum conditions. Normally, all metal phosphides and polyphosphides can be synthesized from elemental phosphorus and the desired metal element in stoichiometric quantities. However, the synthesis is somewhat difficult due to the numerous problems associated with it. For instance, the chemistry is completely exothermic, and the energy evolved in the form of heat is always enormous and hard to control. This, in turn, limits its applicability in supercapacitors.

7.6 Transition Metal Sulphides and Carbon Derivatives Nanomaterials

Recently, transition metal sulphides (TMSs) were contributing broadly in the development of supercapacitors, catalysts, solar cells, and electrodes material. TMSs have attracted researchers' interest because of their unique chemical and physical properties which prompted their usage in magnetic, catalytic, semiconducting, and optical materials (Kershaw, Susha, and Rogach 2013).

These TMSs that contribute in new modernizations include: cadmium sulphide (Behboudnia and Khanbabaee 2005), zinc sulphide (Behboudnia and Khanbabaee 2005), tin sulphide (Kristl, Gyergyek, and Kristl 2015), as well as copper sulphides (Xu, Wang, and Zhu 2006).

7.6.1 Cobalt Sulphides

Cobalt sulphides are commonly obtained from minerals containing cobalt compounds. Binary cobalt sulphides minerals include cattierite (CoS_2) and linnaeite (Co_3S_4). Cobalt sulphides occur in many stoichiometric and non-stoichiometric forms, such as Co_9S_8, CoS, Co_2S_3, and CoS_2 (Luo et al. 2014). Usually, the sulphides of cobalt are black in colour, insoluble in water, and semiconducting. Cobalt sulphides minerals are transformed to cobalt by burning and treatment with aqueous acid. The cobalt salts are further purified by precipitation when aqueous solutions of cobalt (II) ions are bubbled with hydrogen sulphides (H_2S). It has been reported that they have potential to be applied as catalysts (Patil et al. 2015), in water splitting (Feng et al. 2015), magnetic materials (Rumale et al. 2015), counter electrodes in solar cells (Congiu et al. 2015), lithium-ion battery anodes (Wang, Zou et al. 2015), and high-performance supercapacitors (You et al. 2015).

7.6.2 Nickel Sulphides

Nickel sulphides have numerous phases and stoichiometry, such as NiS (α-NiS and β-NiS), NiS_2, Ni_3S_2, Ni_3S_4, and Ni_7S_6 (Kullerud and Yund 1962). Nickel-based compounds possess exceptional chemical and physical properties that give nickel sulphides a wide range of applicability in many technological applications, such as catalysis, supercapacitors, energy conversion, and storage applications (Wang, Batter et al. 2015). Ni_3S_4 has been shown to be an efficient electrode material for supercapacitors high performance (Gupta and Fisher 2017) and lithium-ion batteries (Ji et al. 2016).

The electrochemical performance of NiS_x in pseudocapacitors is a reversible reaction between the Ni(II) and Ni(III) oxidation states in the company of a hydroxide ion as shown in equation (7.10) (Guan et al. 2017):

$$NiS_x + OH^- \leftrightarrow NiS_xOH + e^-. \tag{7.10}$$

Unfortunately, the growth and reduction of nickel sulphides occur during frequent charge and discharge cycles, which devalue the electrochemical properties of the electrodes and obstruct their practical applications (Justin and Rao 2010). To solve these problems, two possible approaches have commonly been used and that include nano structuring of the electrodes (A. Wang et al. 2013) and manufacturing composites with carbon-based nanomaterials (Liu, Duay, and Lee 2011).

7.7 Carbon Composite-Based Nanomaterials

Supercapacitor composite electrodes are manufactured by inserting pseudocapacitive active materials like metal oxide or conductive polymer in a carbon-based material.

7.7.1 Carbon/Carbon Composites

Carbon composite materials interwoven with one another produce larger capacitive performance because there is an increase in total specific surface area that is available for diffusion. This enhanced effective surface area in two integrated carbon materials creates higher power and energy density in supercapacitors when incorporated into the electrode materials (Liu, Duay, and Lee 2011).

7.7.2 Carbon/Metal Oxides Composites

A combination of carbon and metal oxide materials creates high specific capacitance, which is credited to the highly porous structure of metal oxides and high surface area of carbon materials. There are many advantages of carbon materials when they form composites with metal oxides, which include the carbon ability to be surface functionalized by a metal oxide and polymers or self-oxidation against surface impurities (Iro, Subramani, and Dash 2016).

Research has also concentrated on exploring the specific surface area of metal oxides through the ranked configuration of their composite with carbon nanomaterials (CNs) to enhance their usage as nanohybrid supercapacitor electrode materials. Carbon-supported metal oxide electrodes have become a current methodology for actualizing multi-functionalities, firmness, and fabricating new physicochemical nanomaterial properties due to the surface and interface enhanced qualities and the combined effect among constituents (Cozzoli, Pellegrino, and Manna 2006). The applications in the areas of supercapacitor electrodes, optoelectronics, and photovoltaic based on nanohybrids of carbon and metal have also increased.

7.7.3 Carbon/Conducting Polymer Composites

The combination of CNs and a conductive polymer like polyaniline (Downs et al. 1999), polypyrrole (Fan 1999), polythiophene (Kymakis and Amaratunga 2002), poly (3,4-ethylenedioxy thiophene) (Lefrant et al. 2005), poly (p-phenylene vinylene) (Curran et al. 1998), and poly (m-phenylene vinylene-co-2,5-dioctoxy-p-phenylene) (Panhuis et al. 2003) offers a technique to strengthen the polymer. It is also believed that the CNs functionalized the polymer or CN is a dopant to COPs, hence, a charge transfer occurs between the two constituents (Baibarac et al. 2004). The resulting composite always exhibits excellent performances for supercapacitor electrodes.

7.8 Applications of Supercapacitors

The recent increase in the population worldwide has intensified the need for increased energy in society, and the enlarged environmental pollution has increased the search for an improvement of a renewable and clean energy storage system, which the supercapacitor is displaying a motivating role.

A supercapacitor's amazing properties have increased its applications in numerous areas, for instance, a place where a large amount of power is required for a small time, numerous charge/discharge cycle, and a long lasting lifespan is needed. These enticing characteristics increased researchers' interest in the science and technology of supercapacitors and broadened its applications.

7.8.1 Commercial Scale Applications of Supercapacitors

Scientists and technological inventors need to produce supercapacitor units and structures that meet the very specific working desires of individual applications that serve societal interest and objectives, which will effectively and completely replace the battery.

7.8.2 Electronics

There is a need for a continuous power supply to electronic processors and their energy storage devices for constant operation even when the utility which serves as a primary power source cannot sustain the load and also support decoupling between the input and output converters. These storage devices that need a shock absorber include conventional capacitors or batteries because they possess many weaknesses, such as small energy storage capability, low transient response, and relatively high equivalent serial resistance in the case of the batteries. Supercapacitors have drastically reduced problems encountered in our traditional energy storage devices, such as reduced losses, improved power, energy densities, and low toxic compounds that are friendly to our environment, unlike batteries.

7.8.3 Low-Power Equipment Power Buffer

Power buffer equipment can be defined as a building block that can deliver a sustainable power to the load for a short period of time without upsetting its component or lifespan. The power buffer is not the main energy source, but works in combination with internal combustion engines or batteries. A power buffer could be supercapacitors, flywheels, or high power density batteries. They are less expensive and have a higher energy density, the power fluctuations are meaningfully reduced, and the dynamic response is enhanced.

7.8.4 Voltage Stabilizer

There are numerous voltage drops and deviations associated with electrical and electronic components, which include excess loads to the system, fast reclosing of circuit breakers, auto start, short circuiting, lightning strikes on overhead lines, etc. Supercapacitors can sustain voltage drops, as well as short voltage variations by moderating the effect of voltage variations within milliseconds, hence, stabilizing the continuous available voltage to the output or load.

7.8.5 Streetlights

Recently, the use of a long lifespan device called a light-emitting diode (LED) emerged. It provides a long lifespan to the lighting technology, which became helpful in operating flashlights and streetlights. The combination of a lithium-ion battery and supercapacitor form a perfect system for operating LEDs because the low power and energy of LEDs allow a hybrid of supercapacitor and battery to provide a longer duration of energy sources than using a battery alone.

It is also observed that LEDs with supercapacitor hybrid devices, such as flashlights, lanterns, streetlights, etc., have a longer lifespan and consume less energy and power compared to electric bulbs and halogen lamps. It is strongly recommended that producers of various lamps, bulbs, etc. now should use hybrid supercapacitor devices because their lifespan generally exceeds the lifespan of LEDs. Supercapacitor hybrid devices have shown excellent performance, which is 20 times greater in lifespan and 60 times faster recharging rate, hence, improving LED lighting products for numerous companies and homes with small or no maintenance (Nippon 2010).

7.8.6 Medical

Recently, scientists at the University of California, Los Angeles and the University of Connecticut produced a 'biological supercapacitor' capable of storing electrical energy inside the body using safe, nontoxic components, including the body's own fluids. The need for long lifespan power devices arises because patients with cardiac related diseases and other diseases that need implantable electronic medical devices end up requiring reimplantation when the battery runs down.

The researchers hope that the new supercapacitors will work with an internal energy harvester, which uses heat or movement, to provide reliable energy systems for implantable medical devices. Supercapacitors can also be used in defibrillators to enhance the beating of the heart for optimum performance.

7.8.7 Transport

The researchers' major concern in transport, recently, is how to reduce energy consumption and also decrease carbon (IV) oxide (CO_2) emissions to the environment, hence, the world rapidly moves towards greener methods of transportation. In the past, supercapacitors were used in motor startup (to start especially cold diesel engines) and large engines in tanks and submarines, where it moderates the battery for a higher voltage source. In 2003, Mannheim, Germany, used a model light-rail vehicle which has a micro train computer (MITRAC) energy saver system from Bombardier Transportation to store mechanical braking energy with a roof-mounted supercapacitor unit (Bogdan et al. 2013).

The performance of supercapacitors is also witnessed on a triple-hybrid forklift truck, which uses a battery and fuel cell as the main source of energy storage and also buffers power peaks using supercapacitors to store braking energy. This hybrid system provides the forklift with ultimate power of more than 30 kW. It has been reported that these triple-hybrid systems provide over 50% more energy savings than fuel cells or diesel engines (fuelcellworks.com).

Recently, a Toyota TS030 hybrid LMP1 racing car was manufactured under Le Mans prototype rules, which use a hybrid drivetrain with supercapacitors to generate energy (Hamilton 2009). Hybrid electric vehicles (HEVs) are also studied; reports show that 20% to 60% fuel reduction has been achieved by brake energy recovering in HEVs (Van 2009). A supercapacitor's remarkable properties, such as charging faster than a battery, electrical stability, a wide range of temperature absorbance, and longer lifespan make them suitable for use, however, volume, weight, and high cost reduce their usage.

7.8.8 Smart Devices Applications of Supercapacitors

Supercapacitors are used in smart devices, such as remote controllers and screwdrivers. A good example of this device is a cordless electric screwdriver, which uses supercapacitors for its energy storage and runs double the time compared to a battery-type screwdriver and can be fully charged within 90 seconds. It can maintain 85% of its charged power after three months if it is not used. Supercapacitors usage in smart devices is very economical and beneficial because they prolong the lifespan of the device and reduce the cost of maintenance.

7.9 Summary/Challenges

The discovery of supercapacitors as a substitute energy storage device appeared enticing because they possess numerous properties, which suit them for many applications. Some of these qualities include high power density, stability, and electrochemical properties. These good qualities suited them to be used in many devices that include: (1) emergency power supplies, (2) HEVs, (3) power systems, (4) streetlights, (5) battery buffers, etc.

7.9.1 Challenges Facing Supercapacitor Development

The main component of a supercapacitor is the electrode. Electrodes with sufficient electrical conductivity, adequate mechanical properties, and cost effectiveness should be researched and encouraged. Recently, carbon derivatives have shown to deliver high electrical conductivity, they should be modified into various types that can serve several multi-functionalities.

Separators are normally electrical insulators, ionically conducting, and withstand high mechanical stresses caused by expansion, compression, etc. It is the interface between the electrodes, but many separators nowadays hardly serve the multi-purpose function needed, hence, they must have excellent interfacial adhesion to provide sufficient mechanical stability in conjunction with multi-functional matrices for high performance supercapacitors.

The cost of supercapacitors are still very high because the technology required for their production is still sophisticated, hence, it is difficult for them to replace batteries effectively irrespective of their numerous advantages nowadays.

7.9.2 Recommendations

As presented in this review, an upgrade is required for the production techniques of supercapacitors to perfectly suit more applications requiring huge energy and power densities. Recently, scientists concentrated on numerous materials for electrodes, such as carbon materials, metal oxides, and COPs, henceforth, carbon materials, having a high specific surface area with a balanced distribution of pores were produced in 3D nanostructured electrode materials, even though their energy density and capacitances are still low and require an improvement.

COPs have a high specific capacitance, but their swelling and shrinking, when charging and discharging, respectively, lead to a small lifespan, which is a considerable threat to its applications. In regards to metal oxides, they possess excellent electrical conductivity, high theoretical capacitance, good thermal stability, high specific surface area, and enhanced diffusion of ions on top of the bulk of material, but have poor cycle stability and sophisticated fabrication method, hence, the need for more concerted effort for improvement.

We strongly recommend the following future prospects below:

1. There is a need for better techniques of fabricating supercapacitors that will control the issue of self-discharging.
2. Investigations should focus on simpler and affordable methods for the fabrication of supercapacitors to make it less expensive for the effective replacement of the battery.
3. Scientists should work on improving supercapacitors high cycling stability, long lifespan, and small charging time.
4. CNs need to be examined effectively because of their high specific surface area, which enhances supercapacitance.
5. Supercapacitors have huge power density, but suffer small energy density, hence, an invitation to increase their energy density for optimum performance.

REFERENCES

Amatucci, G. G., F. Badway, P. A. Du, and T. Zheng. 2001. "An Asymmetric Hybrid Nonaqueous Energy Storage Cell." *Journal of the Electrochemical Society* 148 (8): A930–39.

Baibarac, M., I. Baltog, C. Godon, S. Lefrant, and O. Chauvet. 2004. "Covalent Functionalization of Single-walled Carbon Nanotubes by Aniline Electrochemical Polymerization." *Carbon* 42: 3143.

Behboudnia, M., and M. Khanbabaee. 2005. "Conformational Study of CdS Nanoparticles Prepared by Ultrasonic Waves." *Colloids and Surfaces* A (290): 229–232.

Bogdan, V., T. Rochdi, L. Richard, and C. Gérard. 2013. "Implementation and Test of a Hybrid Storage System on an Electric Urban Bus." *Transportation Research Part C: Emerging Technologies* 30: 55–66.

Brousse, T., M. Toupin, R. Dugas, L. Athouel, O. Crosnier, and D. Belanger. 2006. "Crystalline MnO_2 as Possible Alternatives to Amorphous Compounds in Electrochemical Supercapacitors." *Journal of Electrochemical Society* 153: A2171–80.

Burke, A. 2007. "R&D Considerations for the Performance and Application of Electrochemical Capacitors." *Electochimica Acta* 53: 1083–91.

Chen, Y. M., J. H. Cai, Y. S. Huang, K. Y. Lee, and D. S. Tsai. 2011. "Preparation and Characterization of Iridium Dioxide–Carbon Nanotube Nanocomposites for Supercapacitors." *Journal of Nanotechnology* 22: 115706.

Cheng, Y. W., S. T. Lu, H. B. Zhang, C. V. Varanasi, and J. Liu. 2012. "Synergistic Effects from Graphene and Carbon Nanotubes Enable Flexible and Robust Electrodes for High-Performance Supercapacitors." *Nano Letters* 12: 4206–11.

Congiu, M., L. G. S. Albano, F. Decker, and C. F. O. Graeff. 2015. "Single Precursor Route to Efficient Cobalt Sulphide Counter Electrodes for Dye Sensitized Solar Cells." *Electrochimica Acta,* 151: 517–24.

Conway, B. E. 1999. *Electrochemical Supercapacitors: Scientific Fundamentals and Technological Applications (in German).* Berlin, Germany: Springer.

Cormier, Z. R., H. A. Andrea, and P. Zhang. 2011. "Temperature-Dependent Structure and Electrochemical Behavior of RuO_2/Carbon Nanocomposites." *Journal of Physical Chemistry C* 115: 19117–28.

Cozzoli, P. D., T. Pellegrino, and L. Manna. 2006. "Synthesis, Properties and Perspectives of Hybrid Nanocrystal Structures." *Chemical Society Reviews,* 35: 1195–208.

Curran, S. A., P. M. Ajayan, W. J. Blau, D. L. Carroll, J. N. Coleman, A. B. Dalton, A. P. Davey, et al. 1998. "A Composite from Poly(m-phenylenevinylene-co-2,5-dioctoxy-p-phenylenevinylene) and Carbon Nanotubes: A Novel Material for Molecular Optoelectronics." *Advanced Materials* 10: 1091.

Donne, S. W., A. F. Hollenkamp, and B. C. Jones. 2010. "Structure, Morphology and Electrochemical Behaviour of Manganese Oxides Prepared by Controlled Decomposition of Permanganate." *Journal of Power Sources* 195: 367–73.

Downs, C., J. Nuget, P. M. Ajayan, D. J. Duquette, and K. S. V. Santhanam. 1999. "Growth from Surface Methodology for the Fabrication of Functional Dual Phase Conducting Polymer Polypyrrole/Polycarbazole/Polythiophene." *Advanced Materials* 12: 1028.

Dubal, D. P., W. B. Kim, and C. D. Lokhande. 2012. "Galvanostatically Deposited Fe: MnO_2 Electrodes for Supercapacitor Application." *Journal of Physics and Chemistry of Solids* 73 (1): 18–24.

Fan, J., M. Wan, D. Zhu, B. Chang, Z. Pan, and S. Xie. 1999. "Synthesis, Characterizations, and Physical Properties of Carbon Nanotubes Coated by Conducting Polypyrrole." *Journal of Applied Polymer Science* 74: 2605.

Fan, S., L. Li, X. Wang, C. Gua, and J. Tu. 2014. "Metal Oxide/Hydroxide-Based Materials for Supercapacitors." *RSC Advances* 4: 41910–21.

Feng, L. L., G. D. Li, Y. P. Liu, P. Y. Wu, H. Chen, Y. Wang, Y. C. Zou, D. J. Wang, and X. X. Zou. 2015. "Mesoporous Nitrogen, Sulfur Co-Soped Carbon Dots/CoS Hybrid as an Efficient Electrocatalyst for Hydrogen Evolution." *ACS Applied Materials and Interfaces* 7: 980–88.

Fischer, U., R. Saliger, V. Bock, R. Petricevic, and J. Fricke. 1997. "Carbon Aerogels as Electrode Material in Supercapacitors." *Journal of Porous Materials* 4 (4): 281–85.

"Fuel Cell Works Supplemental News Page." Archived from the original on 21 May 2008. Retrieved 29 May 2013.

Giulianini, M., and N. Motta. 2012. "Polymer Self-Assembly on Carbon Nanotubes" In *Self-Assembly of Nanostructures,* edited by S. Bellucci, Lecture Notes in Nanoscale Science and Technology. New York, NY: Springer.

Guan, B., Y. Li, B. Yin, K. Liu, D. Wang, H. Zhang, and C. Cheng. 2017. "Synthesis of Hierarchical NiS Microflowers for High Performance Asymmetric Supercapacitor." *Journal of Chemical Engineering,* 308: 1165.

Gupta, R., and T. S. Fisher. 2017. "Scalable Coating of Single-Source Nickel Hexadecanethiolate Precursor on 3D Graphitic Petals for Asymmetric Supercapacitors." *Energy Technology* 5: 740–74.

Halper, M. S., and J. C. Ellenbogen. 2006. *Supercapacitors: A Brief Overview.* Virginia: The MITRE Corporation, McLean.

Hamilton, T. 2009. *Technology review.com.* MIT Technology Review.

Huang, L., D. C. Chen, Y. Ding, S. Feng, Z. L. Wang, and M. L. Liu. 2013. "Nickel–Cobalt Hydroxide Nanosheets Coated on $NiCo_2O_4$ Nanowires Grown on Carbon Fiber Paper for High-Performance Pseudocapacitors." *Nano Letters* 13: 3135–39.

Iro, Z. S., C. Subramani, and S. S. Dash. 2016. "A Brief Review on Electrode Materials for Supercapacitor." *International Journal of Electrochemical Science,* 11: 10628–43.

Iwueke, D. C., C. I. Amechi, A. C. ANwanya, A. B. C. Ekwealor, P. U. Asogwa, R. U. Osuji, M. Maaza, and F. I. Ezema. 2015. "A Novel Chemical Preparation of $Ni(OH)_2$/CuO Nanocomposite Thin Films for Supercapacitive Applications." *Journal of Materials Science: Materials in Electron* 26: 2236–2242.

Jayalakshmi, M., and K. Balasubramanian. 2008. "Simple Capacitors to Supercapacitors—An Overview." *International Journal of Electrochemical Science* 3: 1196–217.

Ji, S., L. Zhang, L. Yu, L. Xu, and J. Liu. 2016. "In Situ Carbon-coating and Ostwald Ripening-Based Route for Hollow Ni_3S_4@C Spheres with Superior Li-ion Storage Performances." *RSC Advances* 6: 101752–59.

Jian, X., S. Liu, Y. Gao, W. Tian, Z. Jiang, X. Xiao, H. Tang, and L. Yin. 2016. "Carbon-Based Electrode Materials for Supercapacitor: Progress, Challenges and Prospective Solutions." *Journal of Electrical Engineering,* 4: 75–87.

Justin, P., and G. R. Rao. 2010. "CoS Spheres for High-Rate Electrochemical Capacitive Energy Storage Application." *International Journal Hydrogen Energy* 35: 9709.

Karthikeyan, K., D. Kalpana, and N. Ranganathan. 2009. "Synthesis and Characterization of $ZnCo_2O_4$ Nanomaterial for Symmetric Supercapacitor Applications." *Ionics* 15: 107–10.

Kershaw, S. V., A. S. Susha, and A. L. Rogach. 2013. "Narrow Bandgap Colloidal Metal Chalcogenide Quantum Dots: Synthetic Methods, Heterostructures, Assemblies, Electronic and Infrared Optical Properties." *Chemical Society Reviews* 42: 3033–3087.

Kiew, S. F., L. V. Kiew, H. B. Lee, T. Imae, and L. Y. Chung. 2016. "Assessing Biocompatibility of Graphene Oxide-Based Nanostructure: A Review." *Journal of Control Release* 226: 217–228.

Kim, B. K., S. Sy, A. Yu, and J. Zhang. 2015. *Electrochemical Supercapacitors for Energy Storage and Conversion.* Hoboken, NJ: John Wiley & Sons.

Kim, H., and B. N. Popov. 2002. "Characterization of Hydrous Ruthenium Oxide/Carbon Nanocomposite Supercapacitors Prepared by a Colloidal Method." *Journal of Power Sources* 104: 52–61.

Kim, S. W., H. W. Lee, P. Muralidharan, D. H. Seo, W. S. Yoon, D. K. Kim, and K. Kang. 2011. "Electrochemical Performance and Ex Situ Analysis of ZnMn2O4 Nanowires as Anode Materials for Lithium Rechargeable Batteries." *Nano Research* 4: 505–10.

Korenblit, Y., M. Rose, E. Kockrick, L. Borchardt, A. Kvit, S. Kaskel, and G. Yushin. 2010. "High-Rate Electrochemical Capacitors Based on Ordered Mesoporous Silicon Carbide-Derived Carbon." *ACS Nano* 4 (3): 1337–44.

Kristl, M., S. Gyergyek, and M. Kristl. 2015. "Synthesis and Characterization of Nanosized Silver Chalcogenides Under Ultrasonic Irradiation." *Materials Express* 5: 359–66.

Kullerud, G., and R. A. Yund. 1962. "The Ni-S System and Related Minerals." *Journal of Petrology* 3: 126–75.

Kuratani, K., T. Kiyobayashi, and N. Kuriyama. 2009. "Influence of the Mesoporous Structure on Capacitance of the RuO$_2$ Electrode." *Journal of Power Sources* 189: 1284–91.

Kymakis, E., and G. A. J. Amaratunga. 2002. "Polymer Composites of Carbon Nanotubes Aligned by a Magnetic Field." *Applied Physics Letters,* 80: 112–88.

Lefrant, S., M. Baibarac, I. Baltog, T. Velula, J. Y. Mevellec, and O. C. Diam. 2005. "Raman and FTIR Spectroscopy as Valuable Tools for the Characterization of Polymer and Carbon Nanotube Based Composites." *Related Materials* 14: 867.

Lin, C., J. A. Ritter, and B. N. Popov. 1998. "Characterization of Sol-gel-derived Cobalt Oxide Xerogels as Electrochemical Capacitors." *Journal of Electrochemical Society* 145 (12): 4097–103.

Liu, K. C., and M. A. Anderson. 1996. "Porous Nickel Oxide/nickel Films for Electrochemical Capacitors." *Journal of Electrochemical Society* 143 (1): 124–30.

Liu, R., J. Duay, and S. B. Lee. 2011. "Heterogeneous Nanostructured Electrode Materials for Electrochemical Energy Storage." *Chemical Communications,* 47: 1384.

Long, J. W., K. E. Swider, C. I. Merzbacher, and D. R. Rolison. 1999. "Voltammetric Characterization of Ruthenium Oxide-Based Aerogels and Other RuO$_2$ Solids: The Nature of Capacitance in Nanostructured Materials." *Langmuir* 15: 780–85.

Luo, F., J. Li, H. Yuan, and D. Xiao. 2014. "Rapid Synthesis of Three-dimensional Flower-Like Cobalt Sulfide Hierarchitectures by Microwave Assisted Heating Method for High-performance Supercapacitors." *Electrochimica Acta* 123: 183–89.

Meenakshi, S. D., M. Rajarajan, S. Rajendran, Z. R. Kennedy, and G. B. Elixi. 2012. "Synthesis and Characterization of Magnesium Oxide Nanoparticles." *Nanotechnology* 50: 10618–20.

Meng, C., C. Liu, L. Chen, C. Hu, and S. Fan. 2010. "Highly Flexible and All-Solid-State Paperlike Polymer Supercapacitors." *Nano Letters* 10: 4025–31.

Nesbitt, C. C., and X. Sun. 2004. *Consolidated Amorphous Carbon Materials, Their Manufacture and Use.* Los Altos, CA: Reticle Inc.

Nippon, Chemi-Con. 2010. Nippon Chemi-Con, Stanley Electric and Tamura announce: Development of "Super CaLeCS," an environment-friendly EDLC-powered LED Street Lamp. Press Release, 30 March 2010.

Nwanya, A. C., S. U. Offiah, I. C. Amaechi, S. Agbo, S. C. Sone, B. T. Ezugwu, R. U. Osuji, M. Maaza, and F. I. Ezema. 2015. "Electrochromic and Electrochemical Supercapacitive Properties of Room Temperature PVP Capped Ni(OH)$_2$/NiO Thin Films. Electrode for Supercapacitive Applications." *Journal of Materials Science: Materials in Electron* 26: 2236–42.

Nwanya, D. Obi, R. U. Osuji, R. Bucher, M. Maaza, and F. I. Ezema. 2017. "Simple Chemical Route for Nanorod-like Cobalt Oxide Films for Electrochemical Energy Storage Applications." *Journal of Solid State Electrochem* 21: 2567–76.

Obodo, R. M., A. C. Nwanya, A. B. C. Ekwealor, I. Ahmad, T. Zhao, M. Maaza, and F. I. Ezema. 2019. "Influence of pH and Annealing on the Optical and Electrochemical Properties of Cobalt (III) Oxide (Co$_3$O$_4$) Thin Flms." *Surfaces and Interfaces* 16: 114–19.

Offiah, S. U., M. U. Nwanya, A. C. Nwodo, S. C. Ezugwu, S. N. Agbo, P. E. Ugwuoke, R. U. Osuji, and F. I. Ezema. 2014. "Ni(OH)₂/NiO Thin Films: Post-Thermal Treatments and Substrates Effects." *Optik* 125: 2905–08.

Palaniselvam, T., and J. B. Baek. 2015. "Graphene Based 2D-Materials for Supercapacitors." IOP Publishing Ltd. 2: 032002.

Pandolfo, A. G., and A. F. Hollenkamp. 2006. "Carbon Properties and Their Role in Supercapacitors." *Journal of Power Sources* 157 (1): 11–27.

Pang, S. C., M. A. Anderson, and T. W. Chapman. 2000. "Novel Electrode Materials for Thin-film Ultracapacitors: Comparison of Electrochemical Properties of Sol-Gel-Derived and Electrodeposited Manganese Dioxide." *Journal of Electrochemical Society* 147 (2): 444–50.

Panhuis, M. I. H., A. Maiti, A. B. van den Noort, A. Dalton, J. N. Coleman, B. McCarthy, and W. J. Blau. 2003. "Selective Interaction in a Polymer–Single-Wall Carbon Nanotube Composite." *Journal of Physical Chemistry B* 107: 478–482.

Pataki, V. D., C. D. Lokhande, and O. S. Joo. 2009. "Electrodeposited Ruthenium Oxide Thin Films for Supercapacitor: Effect of Surface Treatments." *Applied Surface Science* 255: 4192–96.

Patil, S. A., D. V. Shinde, I. Lim, K. Cho, S. S. Bhande, R. S. Mane, N. K. Shresta, J. K. Lee, T. K. Yoon, and S. H. Han. 2015. "Ultrathin CoS₂ Shells Anchored on Co₃O₄ Nanoneedles for Efficient Hydrogen Evolution Electrocatalysis." *Journal of Materials Chemistry* A (3): 7900–09.

Prasad, K. R., and N. Miura. 2004. "Formation of Nanostructure on the Nickel Metal Surface on Ionic Liquid under Anodizing." Applied Physics Letters 85: 4199–4201.

Reddy, R. N., and R. G. Reddy. 2006. "Porous Structured Vanadium Oxide Electrode Material for Electrochemical Capacitors." *Journal of Power Sources* 156 (2): 700–704.

Rumale, N., S. Arbuj, G. Umarji, M. Shinde, U. Mulik, P. Joy, and D. Amalnerkar. 2015. "Tuning Magnetic Behavior of Nanoscale Cobalt Sulfide and Its Nanocomposite with an Engineering Thermoplastic." *Journal of Electronic Materials* 2308–11.

Shaijumon, M. M., S. R. Gowda, and P. M. Ajayan. 2009. "Coaxial MnO₂/Carbon Nanotube Array Electrodes for High-Performance Lithium Batteries." *Nano Letters,* 104 (34): 13574–77.

Shan, Y., and L. Gao. 2007. "Formation and Characterization of Multi-walled Carbon Nanotubes/ Co₃O₄ Nanocomposites for Supercapacitors." *Materials Chemistry and Physics* 103: 206–10.

Su, Y. F., F. Wu, L. Y. Bao, and Z. H. Yang. 2007. "RuO₂/Activated Carbon Composites as a Positive Electrode in an Alkaline Electrochemical Capacitor." *New Carbon Materials* 22: 53–58.

Syarif, N., I. Tribidasari, and W. Wibowo. 2012. "Direct Synthesis Carbon/Metal Oxide Composites For." *International Transaction Journal of Engineering, Management, & Applied Science & Technology* 21 (3): 1906–9642.

Van, Bossche. 2009. *Vehicle Symposium.* Stavanger/Norway: Vehicle Symposium.

Vuorilehto, K., and M. Nuutinen Bautzen. 2014. "Supercapacitors Basics and Applications." *Skeleton Technologies,* 01: 23. http://www.skeletontechnologies.com.

Wang, A., H. Wang, S. Zhang, C. Mao, J. Song, H. Niu, B. Jin, and Y. Tian. 2013. "Controlled Synthesis of Nickel Sulfide/Graphene Oxide Nanocomposite for High-Performance Supercapacitor." *Applied Surface Science* 282: 704.

Wang, G., X. Lu, Y. Ling, T. Zhai, H. Wang, Y. Tong, and et al. 2012. "LiCl/PVA Gel Electrolyte Stabilizes Vanadium Oxide Nanowire Electrodes for Pseudocapacitors." *ACS Nano* 6: 10296–302.

Wang, K., H. P. Wu, Y. N. Meng, and Z. X. Wei. 2014. "Flexible All Solid-State Supercapacitors Based on Chemical Vapor Deposition Derived Graphene Fibers." *Journal of Small* 10: 14–31.

Wang, Q., R. Zou, W. Xia, J. Ma, B. Qiu, A. Mahmood, R. Zhao, Y. Yang, D. Xia, and Q. Xu. 2015. "Facile Synthesis of Ultrasmall CoS₂ Nanoparticles within Thin N-Doped Porous Carbon Shell for High Performance Lithium-Ion Batteries." *Small* 11: 2511–17.

Wang, X., B. Batter, Y. Xie, K. Pan, Y. Liao, C. Lv, M. Li, S. Sui, and H. Fu. 2015. "Highly Crystalline, Small Sized, Monodisperse α-NiS Nanocrystal Ink as an Efficient Counter Electrode for Dye-Sensitized Solar Cells." *Journal of Mater Chem A* 3: 15905–12.

Wang, Z., X. Zhang, Y. Li, Z. T. Liu, and Z. P. Hao. 2013. "Synthesis of Graphene–NiFe₂O₄ Nanocomposites and Their Electrochemical Capacitive Behavior." *Journal of Materials Chemistry A,* 1: 6393–6399.

Wei, W. F., X. W. Cui, W. X. Chen, and D. G. Ivey. 2011. "Manganese Oxide-Based Materials as Electrochemical Supercapacitor Electrodes." *Journal of Chemical Society Reviews* 2011(40): 1697–721.

Wu, X., H. Feng, A. Q. Zhang, L. Z. Wang, T. C. Xia, Y. Zhang, and Y. G. Gui. 2009. "Progress of Electrochemical Capacitor Electrode Materials: A Review." *International Journal of Hydrogen Energy* 34: 2467–70.

Xia, Xinhui, Jiangping Tu, Yongjin Mai, Rong Chen, Xiuli Wang, Changdong Gu, Xinbing Zhao. 2011. "Graphene Sheet and Porous NiO Hybrid Film for Supercapacitor Applications." *Journal of Chemistry* 17: 10898–905.

Xia, X. H., J. P. Tu, Y. J. Mai, X. L. Wang, C. D. Gu, and X. B. Zhao. 2011. "Self-Supported Hydrothermal Synthesized Hollow Co_3O_4 Nanowire Arrays with High Supercapacitor Capacitance." *Journal of Materials Chemistry* 21: 9319–25.

Xia, X. H., J. P. Tu, Y. Q. Zhang, Y. J. MaiX, L. Wang, C. D. GuX, and B. Zhao. 2011. "Three-Dimensional Porous Nano-Ni/Co(OH)$_2$ Nanoflake Composite Film: A Pseudocapacitive Material with Superior Performance." *Journal of Physical Chemistry C201111545* 115(45): 22662–68.

Xia, Xinhui, Yongqi Zhang, Dongliang Chao, Cao Guan, Yijun Zhang, Lu Li, Xiang Ge, Ignacio Mínguez Bacho, Jiangping Tu, and Hong Jin Fan. 2014. "Solution Synthesis of Metal Oxides for Electrochemical Energy Storage Applications." *Nanoscale* 6: 5008–5048.

Xin, L., and W. Bingqing. 2013. "Supercapacitors Based on Nanostructured Carbon." *Nano Energy* 2: 159–173.

Xiong, Q. Q., J. P. Tu, Y. Lu, J. Chen, Y. X. Yu, X. L. Wang, and C. D. Gu. 2012. "Three-Dimensional Porous Nano-Ni/Fe$_3$O$_4$ Composite Film: Enhanced Electrochemical Performance for Lithium-Ion Batteries." *Journal of Materials Chemistry* 22: 18639–45.

Xu, H., W. Wang, and W. Zhu. 2006. "Sonochemical Synthesis of Crystalline CuS Nanoplates via an In Situ Template Route." *Material Letters* 60: 2203–06.

You, B., N. Jiang, M. Sheng, and Y. Sun. 2015. "Microwave vs. Solvothermal Synthesis of Hollow Cobalt Sulfide Nanoprisms for Electrocatalytic Hydrogen Evolution and Supercapacitors." *Chemical Communications.* 51: 4252–55.

Yu, L., L. Zhang, H. B. Wu, G. Q. Zhang, and X. W. E. Lou. 2013. "Formation of Nickel Cobalt Sulfide Ball-in-Ball Hollow Spheres with Enhanced Electrochemical Pseudocapacitive Properties." *Energy Environmental Science* 6: 2664–71.

Yuanlong, Shao, F. El-Kady Maher, Sun Jingyu, et al. 2018. "Design and Mechanisms of Asymmetric Supercapacitors." *Chemical Review* 118: 9233–80.

Zhang, J. T., J. W. Jiang, H. L. Li, and X. S. Zhao. 2011. "A High-Performance Asymmetric Supercapacitor Fabricated with Graphene-Based Electrodes." *Energy Environmental Science* 4: 4009–15.

Zhao, Y., J. Liu, Y. Hu, H. Cheng, C. Hu, and C. Jiang. 2013. "Highly Compression-Tolerant Supercapacitor Based on Polypyrrole-Mediated Graphene Foam Electrodes." *Wiley Online Library* 177.

Zheng, J. P., P. J. Cygan, and T. R. Jow. 1995. "Hydrous Ruthenium Oxide as an Electrode Material for Electrochemical Capacitors." *Journal of Electrochemical Society.* 142: 2699–03.

Zheng, J. P., and T. R. Jow. 1995. "A New Charge Storage Mechanism for Electrochemical Capacitors." *Journal of Electrochemical Society. 1995* 142 (1): L6–8.

Zhou, J., Y. Huang, X. Cao, B. Ouyang, W. Sun, C. Tan, Y. Zhang, Q. Ma, S. Liang, Q. Yan, and H. Zhang. 2015. "Two-Dimensional NiCo$_2$O$_4$ Nanosheet-Coated Three-Dimensional Graphene Networks for High-Rate, Long-Cycle-Life Supercapacitors." *Nanoscale,* 7: 7035–39.

8

The Role of Modelling and Simulation in the Achievement of Next-Generation Electrochemical Capacitors

Innocent S. Ike, Iakovos J. Sigalas, Sunny E. Iyuke, and Egwu E. Kalu

CONTENTS

8.1   Introduction 151
8.2   Electrochemical Capacitors Modelling and Simulation 154
8.3   Modelling and Simulation: Key to Obtaining the Next-Generation Electrochemical Capacitors 155
8.4   Progress on Electrochemical Capacitors/Ultracapacitors Modelling and Simulation 156
        8.4.1   Electrical Behaviour Models/Dissipation Transmission Line Models 157
        8.4.2   Electrochemical Equations/Continuum Equations 158
        8.4.3   Quantum Equations of Ultracapacitors 159
        8.4.4   Atomistic Models (Monte Carlos Molecular Dynamics) 159
        8.4.5   Fractional-Order Models 161
        8.4.6   Clarified Systematic Models 161
8.5   Thermal Modelling of Electrochemical Capacitors/Ultracapacitors 163
8.6   Self-discharges and Charge Redistributions of Electrochemical Capacitors 163
8.7   Difficulties in Ultracapacitors/Electrochemical Capacitors Modelling 167
8.8   Summary 167
Acknowledgements 168
References 168

8.1 Introduction

Electrochemical capacitors (ECs) also called ultracapacitors or supercapacitors deliver stored energy very fast and have outstanding durability (long lifecycle), high efficiency, environmental friendliness, and safety compared with conventional capacitors and batteries (Ban et al. 2013). Reasonable improvements have been achieved in its materials and chemistry, while thorough studies have been channelled towards solving the challenges of device management. ECs are basically grouped into three types depending on their various ways of storing energy: electric double-layer capacitors (EDLCs), pseudocapacitors, and a hybrid type achieved by combining the EDLC and pseudocapacitor (Ike and Iyuke 2016; Ike 2017). EDLCs store electric charges physically by adsorption of ions onto electrode active sites. This charge storage mechanism is physical without chemical reactions (redox reactions) in the electric double layers formed around the electrode/electrolyte interface, as shown Figure 8.1, thus, the phenomenon is very reversible with very high cycle life (Frackowiak 2007; Pandolfo and Hollenkamp 2006; Pech et al. 2010; Simon and Gogotsi 2008; L. L. Zhang and Zhao 2009; Ike 2017).

Pseudocapacitors store energy via quick surface redox reactions, as well as possible ion intercalation in the electrode (Kötz and Carlen 2000; Simon and Gogotsi 2008; Yong Zhang et al. 2009; Kim, Augustyn, and Dunn 2012; Augustyn et al. 2013), as depicted in Figure 8.2. Some ECs employ quick reversible

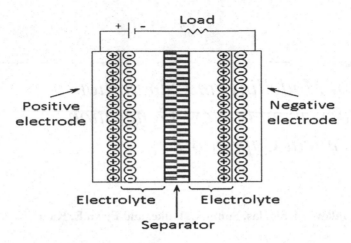

FIGURE 8.1 A diagrammatic representation of electric double-layer capacitors with the charge storage mechanism. (Reprinted with permission from Vangari, M. et al., *J. Energy Eng.*, 139, 72–79, 2013. Copyright 2011, American Society of Civil Engineers.)

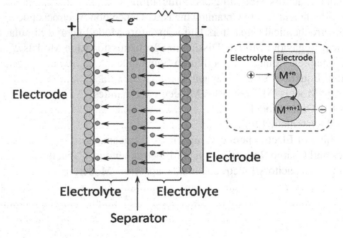

FIGURE 8.2 A diagrammatic representation of pseudocapacitors with the charge storage mechanism. (Reprinted with permission from Vangari, M. et al., *J. Energy Eng.*, 139, 72–79, 2013. Copyright 2011, American Society of Civil Engineers.)

reactions on the electrode surface, describing perfectly the pseudocapacitive performance, that usually outperforms those of carbon, which are mainly double layer charge storage. This explains why much attention has been channelled to these categories of systems. Thus, pseudocapacitors give more energy density, but have low cycle lives and low powers compared with the EDLCs (Yong Zhang et al. 2009; Conway, Birss, and Wojtowicz 1997).

Asymmetric or hybrid ECs appeared recently by combining pseudocapacitors and EDLCs (Simon and Gogotsi 2008; Long 2011; Lin and Wu 2011; Wang et al. 2013; Wang, Wang, and Xia 2005; Khomenko, Raymundo-Piñero, and Béguin 2006; Demarconnay, Raymundo-Piñero, and Béguin 2011; Brousse et al. 2007), as shown in Figure 8.3. In asymmetric/hybrid ECs, the energy storage in one electrode, usually carbon-based, is by physical means in the electric double layers, whereas the second electrode being redox-active stores the charge by reversible redox reactions (Simon and Gogotsi 2008; Long 2011; Lin and Wu 2011; Wang et al. 2013; Wang, Wang, and Xia 2005; Khomenko, Raymundo-Piñero, and Béguin 2006; Demarconnay, Raymundo-Piñero, and Béguin 2011; Brousse et al. 2007). The pseudocapacitive electrode accounts for the energy and the carbon-based electrode accounts for the power in asymmetric ECs (Simon and Gogotsi 2008; Long 2011; Lin and Wu 2011; Wang et al. 2013; Wang, Wang, and Xia 2005; Khomenko, Raymundo-Piñero, and Béguin 2006; Demarconnay, Raymundo-Piñero, and Béguin 2011; Brousse et al. 2007).

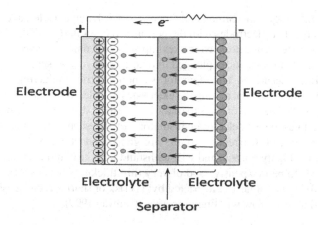

FIGURE 8.3 A diagrammatic representation of asymmetric or hybrid ECs with the charge storage mechanism. (Reprinted with permission from Vangari, M. et al., *J. Energy Eng.*, 139, 72–79, 2013. Copyright 2011, American Society of Civil Engineers.)

ECs are specifically suitable for applications that demand high power for a few seconds (Conway 1999; Miller and Burke 2008; Wu 2002; Simon and Burke 2008; Simon and Gogotsi 2008). When the ECs are charged, anions of electrolytes are adsorbed on a positively polarized electrode and cations on the negative one, leading to an electric double layer at each electrode. Separation of charge occurs on the incomplete or total dissociation into ions at the interface of electrode–electrolyte, creating what has been explained by Helmholtz (Beguin and Frackowiak 2013; Huang et al. 2008c, 2008d):

$$C = \frac{\varepsilon_0 \varepsilon_r A}{d} \text{ or } \frac{C}{A} = \frac{\varepsilon_0 \varepsilon_r A}{d},$$

(8.1)

where ε_r is the electrolyte dielectric constant, ε_0 is the dielectric constant of a vacuum, d is the active thickness of a double layer, and A is the area of electrode surface. The quantity of energy stored in an EC, (E) is related straight to the capacitance (C):

$$E = \frac{1}{2}CV^2.$$

(8.2)

The voltage during charging/discharging of the EC is determined by equivalent series resistance. Usually, power P is the rate of releasing energy and is basically given by:

$$P = IV = \frac{V^2}{4 \times ESR},$$

(8.3)

where I is applied current, V is applied voltage, and ESR is the equivalent series resistance of a capacitor's components to current flow (Ban et al. 2013; Burke 2000; Chu and Braatz 2002).

The energy density of ECs that store charges through both mechanisms are higher than those of convectional capacitors. The name supercapacitors or ultracapacitors emanated from the high charge that can be stored and the rate at which the charges are stored or released in this type of energy storage devices. The high prospects of ECs in engineering applications with increasing power demands have recently attracted serious attention of researches.

The high capacitance of ECs is essential due to the electrodes high specific surface area, which is attained by utilizing very porous electrodes (Simon and Gogotsi 2008; Kroupa, Offer, and Kosek 2016). These pores are basically categorized based on the size range, namely: micropores (<2 nm), mesopores

(2–50 nm), or macropores (>50 nm) (Simon and Gogotsi 2008; Kroupa, Offer, and Kosek 2016). A standard EC cell is made up of two electrodes with the porous separator in the middle, all having electrolytes within their pores and current collectors, which seal up the cell from both ends.

ECs basically consist of two electrodes plunged into an electrolyte and separated by a non-conductive, semipermeable membrane separator. The separator is a non-electric conducting membrane that permits only the movement of an ionic current among the electrodes in order to store the charge and energy in the device (Stoller et al. 2008). The aim of the separator is to restrict direct movement of electrons between the electrodes, that is, it has to be an electronic insulator, but also should permit easier movement of ions to complete the electric circuit. Current collectors are good conductors. A desired separator has to be electrochemically stable, highly porous, and will withstand high heat and chemical resistance (Augusta et al. 2013). Also, it has to be inexpensive and of very thin thickness because it is not active in capacitors (Augusta et al. 2013). Its thinness is restricted by: (1) risk of short circuiting (Richner 2001) and (2) mechanical strength that will allow winding processes (Conway 1999).

8.2 Electrochemical Capacitors Modelling and Simulation

Modelling of ECs is an art of developing mathematical descriptions of EC features and the variables of phenomena taking place in the device, while considering the physical, electrical, and the electrochemical variables of electrode materials, electrolytes, as well as the components' design. Features of ultracapacitors and variables of phenomena taking place in it are closely related and depend on these variables of the device components and design (Ike 2017; Kazaryan et al. 2006; Ike, Sigalas, and Iyuke 2016). Thus, modelling of an electrochemical capacitor for evaluation of the device capacity and every functioning features is inconceivable if due consideration of the component variables were ignored. The majority of works on modelling and simulation of ECs did not fully account for the components physical, electrical, and structural features, which resulted in a variation in outcomes of practical ECs variables in theories and experiments (Kazaryan et al. 2006). Many of the existing models ignored unwanted reactions at the interface and derived them as if the EDL is the only way to store energy (Vazquez et al. 2010). They are also not capable to accurately estimate elements like temperature change within the device, field at interface, and spatial voltage variation for random charging/discharging current ranges. Those models that considered these side reactions (Devan, Subramanian, and White 2004; Kim and Popov 2003; Lin, Popov, and Ploehn 2002), oversimplified the reactions by ignoring the temperature effects, the influence of pores, and its various sizes were not considered in a qualitative and rigorous way. For electrochemical capacitors or ultracapacitors systems, modelling is very important for design estimation, supervisory state, and management synthesis. The exactness of a model which is at most a substitute for practical arrangements is dependent on the suppositions and needs, thus it has to be developed based on a given motive (Ike 2017; Bolwig et al. 2019; Muzaffar et al. 2019; Zhang et al. 2018). Consequently, many EC models have been presented in the literature for various motives, such as highlighting electrical working, thermal working, self-discharge, aging simulation, etc. (Ike 2017; Bolwig et al. 2019; Muzaffar et al. 2019; Zhang et al. 2018).

Models in which their derivations are based on physics electrochemistry's fundamental principles have the capacity to predict temperature inside the device and also the electric field at the interface in practical situations. This will give more realistic estimations and promotes the ECs process design higher than every electrical energy storage gadget in terms of keeping the voltage for a given loading spread. This in turn aids in the production of porous structures, leading to a higher power (Pech et al. 2010; Ike and Iyuke 2016; Ike 2017; Ike et al. 2015; Ike, Sigalas, and Iyuke 2016, 2017b, 2017c). The purposes of EC modelling are to build a theory for the determination and enhancement of the variables of various kinds and configurations with good accounts of properties of electrodes, separators, and the designs and also to provide via modelling an excellent alternative to the time and cost demanding experiments by giving blueprints for the design and optimization of ECs for numerous applications. Again, modelling helps to optimize the variables of ultracapacitors under the type and quantity of electrodes conductivity, width, porosity, and capacitance, electrolyte's conductivity, separator parameters, and charge and discharge currents (Ike and Iyuke 2016; Ike 2017; Ike et al. 2015; Ike, Sigalas, and Iyuke 2016, 2017b, 2017c).

8.3 Modelling and Simulation: Key to Obtaining the Next-Generation Electrochemical Capacitors

Quantitative modelling has been employed to predict the theoretical constraints of the electrochemical capacitor's energy and power densities. Experimental and theoretical researches to estimate the constraining factors that stops ECs from achieving their theoretical maximum have given novel insights into methods to attain optimized capacitor design. Although great efforts and interest have been directed towards the development of electrodes with better properties to enhance the energy storage capacity, models proposed and recommended that the ECs energy density limitations are mainly due to ion concentration and the electrolyte's voltage collapse (Zheng 2003, 2005; Zheng, Huang, and Jow 1997). Further researches also proposed that the power density of ECs could equally be restricted by the electrolyte employed (Pell and Conway 2001; Conway and Pell 2003). Therefore, results from several studies insist that electrodes and electrolytes' optimization are crucial in obtaining high energy and power very close to theoretical values of electrochemical capacitors.

Mathematical modelling and simulation obviously remain crucial in achieving and designing the new-generation electrochemical capacitors. Different models of different ultracapacitors have been presented in the literature (Lin et al. 1999; Srinivasan and Weidner 1999; Lin, Popov, and Ploehn 2002; Kazaryan et al. 2006; Staser and Weidner 2014; Verbrugge and Liu 2005; Srinivasan, Weidner, and White 2000; Somasundaram, Birgersson, and Mujumdar 2011; Sikha, White, and Popov 2005; Pillay and Newman 1996; Fan and White 1991; Kim and Popov 2003), but virtually none was derived and solved based on mechanism(s) of self-discharge(s), until very recently some groups have presented it (Ike 2017; Ike, Sigalas, and Iyuke 2016, 2017a, 2017b, 2017c, 2018). Upon the level of experimental breakthrough in ECs, mathematical models that moved past equivalent circuits and combined the mechanics of charging, charge losses using physical and chemical properties, and processes of solvents and ions in nano-pores and charge transports in electrodes are very important. Again, careful physical models and precise numerical models will obviously enhance the optimization of electrodes' form, shape, or structure, and recognition of a perfect electrolyte.

Modelling and simulation have greatly helped in the provision of expressions and guidelines to find the practical thickness of electrodes, optimum charging current, and material utilization in cells with a given practical conductivity of electrodes and electrolytes for new-generation ECs (Ike, Sigalas, and Iyuke 2017c). The high performance ECs are achievable because of the guidelines obtained via modelling since this will eliminate an electrode's underutilization, increased inefficiencies, and potential drops (Ike, Sigalas, and Iyuke 2017c). Optimizing electrode and electrolyte morphology experimentally is mainly through the trial and error method, which is very strenuous due to the numerous parameters and contesting processes required. This makes modelling and simulation very inevitable.

Theoretical equations for various performance variables of different ultracapacitors were developed and used to optimize their performance, while considering states of electrodes and electrolytes employing optimal coefficients related to battery-type material K_{BMopt} and electrolyte material K_{Eopt}. The dependence of various ECs performances on electrode and electrolyte fabricating conditions, such as mass ratio of electrode, electrode type, materials reaction, and potential window of electrolyte and capacitance were shown from modelling and simulation (Ike 2017; Ike, Sigalas, and Iyuke 2017b; Choi and Park 2014). Modelling and simulation give absolute comprehension of the influences of various operating conditions, constructional design parameters, and self-discharge on capacitor's performance. Thus, these are guidelines to obtain an optimal design and fabrication of new-generation ECs (Ike, Sigalas, and Iyuke 2016). Having perfect knowledge of the effects of each structural parameter, such as electrode thickness, porosity, separator thickness and porosity, electrode and electrolyte effective conductivities, and operation conditions like charging current densities and times is inevitable in design and fabrication of electrochemical capacitors with high performance. Modelling provides an overall framework that will allow for the different factors which affect the EC performance and implementation of the EDL differential capacitance, while considering capacitance that depends on either potential or concentration as presented by different researchers in recent times (Khomenko et al. 2006; Simon and Burke 2008).

Through modelling and simulation, the electrochemical capacitor models, which incorporated self-discharge mechanisms and are used to determine the minimum impurity or redox species concentration and the optimum entire width of separator and anode, have been presented (Ike 2017; Ike, Sigalas, and Iyuke 2017a, 2018). Also, a guideline to determine optimum design configurations and operating conditions for optimal performance of electrochemical capacitors has been created, and by using a typical length scale[w] over which the liquid potential fall occurs, effective electrode thickness and electrode utilization (Ike, Sigalas, and Iyuke 2016, 2017b, 2017c; Ike 2017) have been obtained via modelling as well. Further, a theoretical guideline using optimum ratios for design and fabrication of symmetric EDLCs that operate as asymmetric capacitors was reported (Ike 2017).

Useful guidelines and requirements for the determination of optimum ratios, proper organic electrolyte for optimal performance, and the entire design and fabrication of electrochemical energy storage devices with enhanced energy and power with a reduction in device mass and volume (Ike, Sigalas, and Iyuke 2016, 2017b, 2017c; Ike 2017) were presented. These guidelines and requirements aid in the production of asymmetric ultracapacitors with the optimum battery-type mass ratio, potential range ratio, maximum potential range ratio, and ratio of capacitance for capacitor-type (Ike, Sigalas, and Iyuke 2016, 2017b, 2017c; Ike 2017).

Molecular dynamics (MDs) simulation results have the capacity to give the blueprints for optimizing energy density within electrodes of carbon material at a minute extent. In order to obtain high energy, optimization of the EDL capacitance and quantum capacitance concurrently is very important, and this is realizable through a MDs simulation (Bo et al. 2018). The approaches of enhancing the ECs' power through MD simulations could be classified into intensification of the ion dynamics and reduction of the ion track (Bo et al. 2018). To get the ECs with high power densities, the great/severe ion dynamics and shortened ion track are achievable by regulating the electrode formations and surface features to obtain an optimum via simulation before actual synthesis and fabrication (Bo et al. 2018). Future studies on improving the performances of ECs will obviously gain from the endorsed electrode formations and surface features obtained from MD simulations.

Modelling and simulation are vital in synthesizing advanced electrodes and electrolyte materials with enhanced properties for fabrication of new-generation ultracapacitors. Development of different types of advanced electrode active materials and new cell architectures via simulation has highly improved the performance of electrochemical energy storage devices over the past decade (Noori et al. 2019). Tian et al. (2018) discovered through density functional theory (DFT) calculations that the interstitial carbon atoms increase the adsorption energy of B and N atoms on the Co(111) surface, thereby reducing the diffusion activation energy and, in turn, enhancing the nucleation and growth of 2D h-BN electrode materials. Modelling has played a vital role in the discovery, synthesis, optimization, and fabrication of numerous new electrode and electrolyte materials, which appreciably helped in the improvement of the ultracapacitors performance (Tie et al. 2018; Li et al. 2019; Kumar et al. 2019; Menzel, Frackowiak, and Fic 2019; Boota and Gogotsi 2019; Liu et al. 2019; Tian et al. 2019; Etsuro Iwama et al. 2019; Melchior et al. 2018; Doyoung Kim et al. 2019; Mirzaeian et al. 2017; Yu et al. 2015; Krummacher et al. 2018; Balducci 2016; Shabangoli et al. 2018; Mei et al. 2018).

8.4 Progress on Electrochemical Capacitors/Ultracapacitors Modelling and Simulation

Theoretical expressions for electrochemical capacitors originate from the indigenous Helmholtz equation to mean-field continuum equations, external curvature-based post-Helmholtz equations, and the current simulations in minute levels. Real-life expressions of ECs are derived by employing a high stage of progression in traditional and discrete molecular change approaches and parallel high-performance computation. Helmholtz (Wang and Pilon 2011) uncovered the EDL process and related it employing an expression where every charge was taken to be adsorbed at the surface of an electrode, which resembles the ordinary dielectric capacitor structure (Masliyah and Bhattacharjee 2006).

Mathematical modelling is an ordinary structure, usually mathematics, which researchers and engineers employ in description and analysis of systems. Modelling and simulation is very important for blueprint estimation, situation observation, and formation of ECs systems control. Every model which is a representation of real-life systems is derived for a given purpose, thus, its accuracy is dependent upon the assumptions and requirements for it. Various models are derived for several purposes, such as highlighting electrical conduct, thermal conduct, self-discharge, decaying simulation, etc. The most frequently employed models are electrical behaviour modelling/dissipative transmission line equations, electrochemical equations, atomistic equations (Monte Carlo, molecular dynamics), quantum equations (ab initio quantum chemistry and DFT), fractional-order models, and simple analytical equations. The electrochemical models can highlight only the actual reaction phenomenon within the ECs without considering coupled partial differential equations, and as such, they only have high accuracy without computational efficiency (Ike and Iyuke 2016; Ike 2017; Zhang et al. 2018; Ike et al. 2015). Each model having been derived to achieve different goals has various merits and demerits associated to it.

8.4.1 Electrical Behaviour Models/Dissipation Transmission Line Models

This category of models uses parameterized RC (resistor capacitor) systems to imitate the ECs electrical behaviour. They are simple and easy to implement because they employ simple differential equations in derivations of the expressions (Hu, Li, and Peng 2012). The basic electrical behaviour of ultracapacitors presented in literature is equivalent-resistor linked with a capacitor in a series (Kim et al. 2011), where the ultracapacitor and series resistor represent the canonical capacitance effect and the overall resistance of ECs, respectively. Spyker and Nelms (2000) introduced an additional parallel resistor, which represents the self-discharge process to create a classical equivalent circuit equation. Zubieta and Bonert (2000) presented an equation which has three RC arms: instant arm, prolonged arm, and long-term arm for the purpose of power electronic applications. Every branch presents the EC characteristics on a different time scale. In order to extract the model parameters, Rajani, Pandya, and Shah (2016) reported a new average point method. Other groups developed analogous models with various characterization methods (Faranda 2010; Weddell et al. 2011; Logerais et al. 2015; Chai and Zhang 2015).

Specifically, Liu et al. (2013) studied the effect of temperature on ultracapacitor model variables and developed a three-arm model, which is dependent on temperature. Zhang and Yang (2011) utilized a changing resistor to describe a self-discharge phenomenon in the three-arm equation, whereas Buller et al. (2002) suggested a dynamic expression employing electrochemical impedance spectroscopy in the frequency realm. The model has a sequence resistor, bulk capacitor, and two parallel RC systems. Zhang et al. (2018) used the expanded Kalman filter to retrieve the expression variables within dynamic propelling cycles (Zhang et al. 2014), while Musolino, Piegari, and Tironi (2013), with the aim of explaining the entire EC frequency-range feature, utilized a dynamic expression to substitute the instant arm of the three-arm expression, and as well added an aligned escape resistor to create a joined EC equation.

Gualous et al. (2003) experimentally studied EC serial resistance and capacitance changes and developed an equivalent circuit model with parameters dependent on temperature. Transmission line models which account for transient and a long-term feature were launched in order to estimate the spread-out capacitance and electrolyte resistance decided by the porous electrodes. This model's complication depends usually on the number of the RC systems utilized (Rafik et al. 2007; Torregrossa et al. 2014). In general, multiplying the RC systems favours a model's fidelity, but with low computational efficiency. Rizoug, Bartholomeus, and Moigne (2010) used a hybrid approach made up of the frequency method and a temporal method to describe the transmission line model. Dougal, Gao, and Liu (2004) obtained automatic model order choice of the transmission, the model relied on the simulation time step through numerical approach, thereby generating better flexible and a computationally efficient model. Research has shown that the dynamic model gives the best all-inclusive performance (Zhang et al. 2015). Parvini et al. (2016) reported an electrical and thermal model, which is computationally efficient to consider their coupling influence for a cylindrical EC device. This category of model could only successively constitute EC dynamics over a time range of several seconds, and this greatly restricts its real-life applications.

Ban et al. (2013) developed an experimentally validated straightforward model which consists of voltage-independent parallel leakage systems and electrochemical disintegration to interpret the ECs behaviours, while charging and discharging, in a bid to obtain a fundamental comprehension of the functioning of capacitors during operations. This simulated parameter value which accounted for parallel leakage and solvent decomposition can be used to predict the self-discharge, shelf-life, performance/efficiency, and the limiting workable cell voltage of ECs. The models are utilized in the evaluation and identification of ultracapacitors and also present insights into the devices charging–discharging behaviour. Again, the model is functional in acquiring the important EC parameters like equivalent series resistance and capacitance based on reported charging–discharging curves.

However, a small disagreement which could be attributed to experimentally insignificant factors like the influences of evolution of gases, dissolution of current collectors, crystallization of electrolyte in the separator, packaging quality, etc. precisely existed on discharge curves. The main difficulty in this category of equations are their little calibre for states of functioning, which are at a variance with the ones employed in acquiring the variables (Diab et al. 2009). Also they are incapable of predictions when the time is extended and heat produced inside the device obviously changes the electrical behaviours. These setbacks are basically because they were not developed from physics principles and are only correct where temperature growth is very negligible. Thus, predictions using these models are not realistic and do not depict the behaviour of real electrochemical capacitors.

The terminal voltage grows in a non-linear manner because of the restriction of ionic motion through the electrodes pore structure, while moving down into regions that reside deep into a pore. Thus, models that describe dynamic and steady-state reaction of the electrolyte portion are valuable because that mass region is presented with a common resistor of the first order. The foremost models of porous electrodes which considered EDLC like a transmission line and electrolyte resistance were presented by de Levie (1963). Zubieta and Bonert (2000) equally proposed a common equation, which considered capacitance and most linear changes with tension. Dougal, Gao, and Liu (2004) and Belhachemi, Rael, and Davat (2000) utilized related expressions with capacitance that have a constant part C_o and straightway voltage dependent part $C_v = K_V \cdot U$, where K_V is a coefficient that relies on the technology employed (Kurzweil and Fischle 2004).

Zaccagnini et al. (2019) presented a model built upon the Poisson-Nernst-Planck equation using adsorbing electrodes and utilized it to fit the voltammogram from experiments. Results from the model and data from experiments agreed well, showing the adsorption process is important in an ultracapacitors electric response to an outside electric field. Thus, development of a physical equation which has the capacity to interpret an ultracapacitors electric response to an outside field by accounting for the electrodes ionic adsorption is inevitable. Several researchers recently presented models in this category, which were built from different approaches for different purposes (Batchelor-McAuley, Ngamchuea, and Compton 2018; Bukola and Creager 2019; Fellows et al. 2019; Raicopol et al. 2019).

8.4.2 Electrochemical Equations/Continuum Equations

The Helmholtz model was adjusted more to consider the mobility of ions in the electrolyte due to electrostatic forces and diffusion by Gouy (Wang and Pilon 2011) and Chapman (Gouy 1910). Stern (Conway 1999) then integrated the Helmholtz equation and the Gouy-Chapman equation and divided the EDL into two identifiable different surfaces known as the Stern surface (Helmholtz surface) and the diffuse surface (Gouy-Chapman surface) (Grahame 1947). The Stern layer considers a given ion's absorption on the surface of an electrode, and the diffuse surface integrates the Gouy-Chapman equation (Grahame 1947). The overall capacitance of EDLs are considered as a link of the Stern surface serially and the diffuse surface capacitances. An impractical model which considers ions as the point charge by neglecting their concrete size was acquired by developing the Poisson-Boltzmann expression. The assumption is only true when the concentration of ions and electrical potential are low (Bagotsky 2006; Bard and Faulkner 2001). Bikerman (1942) redeveloped the Poisson-Boltzmann equation by introducing the definite ion size effect in equilibrium situations, while the electrolytic ions are considered to have similar valence with different sizes. Verbrugge and Liu (2005) on the grounds of a dilute-solution hypothesis and examination of electrode pores presented a one-dimensional mathematical model in one domain. The ultracapacitor was considered like a continuum body having similar and isotropic concrete features. Allu et al. (2014) enlarged the equation to a three-realm equation through consistent derivation of the electrode-electrolyte process. This demonstrated the need to express the benefits of the non-uniform geometric layout, charge movement, the associated achievement in higher extents, introduction of spatio-secular changes, aniso-tropic concrete features, as well as the upstream variables into estimations. Wang and Pilon (2013) presented a three-dimensional (3D) ultracapacitor equation which captured the 3D electrode form, shape, or structure, definite size of ion, and the dialectic permittivity which field-depends on the electrolyte. Precisely, a common pair of border situations were formulated to explain the Stern surface conduct without its reproduction in the calculation realm.

8.4.3 Quantum Equations of Ultracapacitors

Special simulations of characteristics like an electrolyte with negligible vapour pressure, complete functional groups, and potential range have justified the recent research interest in ionic liquids. Extensive efforts have been directed towards studies on how the capacitance of an ionic liquid electrolyte depends on pore sizes in a mini-nanometres range. Kondrat and Kornyshev (2011) described an unexpected increase as imaginary forces that restrict the repulsion of the same charged counter-ions in a small pore equation of charged ions in a slit pore of metals. De-en and Jianzhong (Jiang and Wu 2013) dealt with the issues of the interface of electrode and electrolyte at microscopic point and pore size effects on capacitance via classical DFT (c-DFT). They also demonstrated that the c-DFT method has a large capacity to present microscopic understandings with small molecular coverage and computation cost, as well as considering a wider pore size range in comparison to larger ionic size scales.

Some researchers recently presented the basic aims and usefulness of c-DFT in electrolyte domains in reviews (Wu 2006; Wu and Li 2007). They showed that the mathematical environment of a c-DFT is similar to a familiar electronic DFT (Hohenberg and Kohn 1964). A c-DFT of electrolytes showed that free energy does not depend on the density of electrons, but on the localized ion densities and solvent molecules. The simple nature of the equation equally helps in discovering the genesis of the oscillation of capacitance in an ionic liquid electrolyte and offers a complete comprehension of the main variables in a large environmental span (Jiang, Meng, and Wu 2011; Wu et al. 2011). c-DFT is anticipated to have the capacity to predict more compounded types of capacitance-potential plots with the phenomena of charging and the variations in shape with size of pores because the ionic liquid equation depicts practical situations (Xing et al. 2013).

Models of EDLC in the past are not relevant because the models were built by dividing pore size settings into different size ranges, such as below 1, 1–2, 2–5 nm, etc. (Huang, Sumpter, and Meunier 2008a, 2008b), which produces a constant variation of structure and capacitance of EDL, whereas pore size changes with c-DFT use. c-DFT considers the very crucial physics of the system, as is evident in its good agreement with experimental results, but it has not handled the issues of the solvent molecules size, polarity of solvent, concentrations of ions, different sizes of ions, and the electrode's form (Feng, Jiang, and Cummings 2012; Nguyen et al. 2013). These parameters are crucial in enhancing the understanding of EDLCs with organic electrolytes (Pizio, Sokołowski, and Sokołowska 2012). Experimentalists employ the c-DFT approach to estimate pore size dispersal and surface area of actual porous materials (Ravikovitch and Neimark 2006; Gor et al. 2012). It can also take into account modelling challenges like confirming a predicted relationship between the structure and capacitance of actual material and a given electrolyte. c-DFTs still do not have the capacity to completely account for rigorous porous design, porous electrodes, as well as the prediction of perfect electrolytes for optimal energy and power densities of actual EDLCs.

8.4.4 Atomistic Models (Monte Carlos Molecular Dynamics)

Molecular modelling is highly inevitable for the optimal blueprint and fabrication of ultracapacitors. It's advantage is that it paves the way for the prediction of phenomena which are not spontaneously noticeable experimentally, for example, changes of an electrode's pattern due to polarization (Conway 1999) and an ion's orientation in an electrolyte (Chaban and Kalugin 2009) are hardly evaluated from an experiment and are crucial conditions that influence the capacitance of electrodes. The given pore geometries and sizes have to be employed simultaneously with the electrolyte as design considerations

(Masliyah and Bhattacharjee 2006; Grahame 1947; Varghese, Wang, and Pilon 2011). Models considering ions of different kinetic sizes during charging and discharging while functioning are developed for given electrolytes, and the optimum pore size in each electrode should be different (Woo et al. 2008; Wang and Pilon 2013; Freund 2002). The introduction of organic solvents into room temperature ionic-liquid electrolytes (RTILs) will increase their electric conductivity by employing the optimal concentration of the electrolyte (Qiao and Aluru 2003; Thompson 2003). An issue about this optimization was a change to the effect of the pore size on the power density (Qiao and Aluru 2004). An ultracapacitor's storable energy density could be enhanced by employing pores with diameters that are similar to the ion's diameter (Masliyah and Bhattacharjee 2006; Zhang and Zhao 2009), whereas the power densities are mitigated by the high resistance to diffusion due to the use of pores with small diameters.

The comprehension of equilibrium and dynamic processes displaying out in ultracapacitors on an atomic level was improved by molecular modelling, though the accuracy greatly depends on the genuineness of the force fields employed to describe the fluid phase molecular behaviours, as well as the modelled electrode's fields. The most acceptable molecular simulation approach is the Monte Carlo simulation, which depends on statistical mechanics through essential choice of the molecular system's stage space and MDs (Johnson and Newman 1971; Bertrand et al. 2010). MD simulations assess the Newton's motion models for different molecular processes, and their features were averaged within a short simulation period to evaluate the system's features. MD simulations also have the capacity to predict the diffusivity of ions (Schiffer et al. 2006). A group of researchers performed reliable classical evaluations of EDLs at an ionic liquid and metallic boundary (Kislenko, Samoylov, and Amirov 2009). Chaban and Kalugin (2009) carried out simulations on carbon nanotube electrodes using solid electrolytes. The MD simulation gives a platform to successfully characterize the electrolyte's ionic subsystem, without an insight into the electrode's electron subsystem. MD simulation has been widely employed in modelling electro-osmotic flow because it gives the most fundamental and alterable approach for analysing molecular behaviours (Freund 2002; Qiao and Aluru 2003; Kim and Darve 2006; Wu and Qiao 2011; Qiao and Aluru 2004; Thompson 2003) and also handles higher charge densities that are greatly important in EDLCs (Cui and Cochran 2002; Feng et al. 2010; Wander and Shuford 2010). The challenges associated to it are computational cost, which renders the approach unrealistic in terms of time and also length scales achievable in different usages (Lankin, Norman, and Stegailov 2010).

DFT research presented a bell-shaped plot resulting from withdrawal of co-ion and counter-ion's insertion without an increase in capacitance (Jiang and Wu 2013). This result from DFT computations was anticipated because of untangled RTIL ions in the bounded common equation and chosen pore size. An exohedral model of ECs was presented by Huang et al. (Silva et al. 2008) for spherical and cylindrical electrodes in an attempt to describe the size and relative capacitance of either onion-like carbons (OLCs) or carbon nanotubes (CNTs). Great attention has been channelled towards demonstrating that several C-V curves were captured in devices using planar electrodes (Kornyshev 2007; Lockett et al. 2010; Baldelli 2008; Trulsson et al. 2010; Feng, Zhang, and Qiao 2009). Some theories and experiments demonstrated that capacitance increases as temperature rises (Lockett et al. 2008; Silva et al. 2008), whereas others reported that it reduces as temperature grows (Alam et al. 2007; Vatamanu, Borodin, and Smith 2010). Some other groups presented at least a rigorous relationship between capacitance and temperature (Holovko et al. 2001; Vatamanu, Borodin, and Smith 2010; Boda, Henderson, and Chan 1999; Boda and Henderson 2000). Lin et al. (2011) recently reported that capacitance of ultracapacitors with OLC electrodes and RTIL electrolytes increases as the temperatures are increased, while those with RTIL electrolytes and vertically aligned CNT electrodes do not depend on temperature. MD simulations were used to present the expressions relating the ultracapacitors RTIL electrolytes and OLC/CNT electrodes with temperature, so as to understand how temperature depends on capacitance (Feng et al. 2013; Li et al. 2012).

Boda and Henderson (2000) presented a bell-shaped curve of capacitance-temperature relationships presented through Monte Carlo estimations. Thus, in-depth researches are required for the purpose of validating the description of temperature dependence of capacitances and the fundamental mechanisms resulting in the phenomena. Feng et al. (2013) reported that EDL capacitance around

CNT electrodes does not entirely depend on temperature within the range of 260 K and 400 K, which comfortably agreed with what Lin et al. (2011) obtained. It was, however, noticed that the capacitors' power, rate of charging, and discharging strongly depend on the mobility of ions and not the ion's packing, which can hardly be estimated from acquired capacitance-temperature dependence. Nano-pores can store more charge compared with planar electrodes because the general charge on the electrode by a given area of ions, not overshadowed, is in equilibrium (Merlet et al. 2012). The maximum storable energy depends on the diameter of pores, and the effective diameter of pores increases as the voltage increases (Kondrat et al. 2012).

Simulations of more rigorous structures of electrodes, such as a 3D, hierarchical pore system, which are not in the literature, need to be integrated into molecular expressions in order to increase the level of reality and get a platform to improve capacitor performance (Feng et al. 2011; Frackowiak, Lota, and Pernak 2005). Also, pseudocapacitors and hybrid ultracapacitors employing battery-type and carbon-based electrodes are areas of studies which attract reasonable awareness, even though real expressions of these devices are currently restricted to small simulations (Xing et al. 2012; Vatamanu, Borodin, and Smith 2012) and ab initio calculations (Ozoliņš, Zhou, and Asta 2013). Derivation of effective equations along these lines for precise recognition of conditions and materials which have great merits in ultracapacitor performance are inevitable.

8.4.5 Fractional-Order Models

Fractional-order calculus has been initiated so as to improve the correctness of ECs modelling and the applications (Wang et al. 2015; Xu et al. 2013). The introduced fractional-order models are made up of differential equations, which have the powerful capacity to express the EC dynamics, compared with the dissipation transmission line models. Concerning fractional-order equations, Riu, Retiere, and Linzen (2004) presented a novel work with the inclusion of a half-order equation for ultracapacitors with the capacity to credibly constitute the EC behaviour with reduction in computational implication. Martynyuk and Ortigueira (2015) reported a fractional-degree equation where the variables were recognized by the impedance facts. A non-linear fractional-degree equation was developed through frequency analysis by Bertrand et al. (2010). Martín et al. (2008) also reported a function-based Havriliak–Negami relaxation with the aim to obtain a perfect fitting to the entire frequency span, though its accuracy is greatly dependent on the equation that is represented in terms of parameters that rely heavily on the exactness and accessibility to the ECs impedance spectra that are only obtained from a laboratory. Dzielinski, Sarwas, and Sierociuk (2010), contrary to other existing fractional-order models, proposed a model with parameters that were recognized based on time-domain facts obtained via steady current charge experiments. Also, Freeborn, Maundy, and Elwakil (2013) evaluated the ECs impedance variables of the fractional-order equation through step voltage reaction instead of direct impedance quantification. The accuracy of the equation might be affected greatly if subjected to non-laboratory circumstances under varying loading conditions because parameters of the model could be easily affected by differing states of the variable. Gabano, Poinot, and Kanoun (2015), via a cubic spine interpolation approach, initiated a fractional uninterrupted linear variable-changing equation from a group of localized recognized fractional impedance models. The functional voltage-dependent non-linear characteristic of ultracapacitors was examined, achieving greater precision and toughness.

8.4.6 Clarified Systematic Models

Models in the clarified systematic category are built upon the electrochemistry and physics fundamentals of electrochemical capacitors. This type employed algebraic and differential equations to characterize the system's determining features of the charged species movement and reaction rate (Devan, Subramanian, and White 2004; Lin, Popov, and Ploehn 2002; Johnson and Newman 1971; Bertrand et al. 2010; Dunn and Newman 2000; Doyle, Fuller, and Newman 1993). The ECs are categorized into two types: EDLC that stores a charge by a physical process and pseudocapacitor, which stores its energy through a chemical processes (redox reactions). Many sections of ECs, such as side reaction effects (Johnson and Newman 1971), pore structure of electrodes (Bertrand et al. 2010; Dunn and Newman 2000), and evaluation of the given energy and power (Doyle, Fuller, and Newman 1993), were examined based on stated suppositions. Karthik et al. (2006) derived a lessened equation of ultracapacitors that were analysed with scaling reasoning, calibrated, and validated with results from experiments, while ignoring the equation of energy change and heat production. This was done on the assumption that the system is isothermal and temperature influence and side reactions are negligible.

Several studies concerning ultracapacitors were focused on comprehension of the effects of the electrode's widths and pore structure on performance parameters. Researches towards the development of pseudocapacitors were highly concentrated on producing several electrode materials, such as oxides of metal (Kazaryan et al. 2006; Simon and Gogotsi 2008), and other researches were concentrated on comprehension of the charging and discharging behaviour of these redox pair electrodes (Conway 1999). ECs in which two electrodes have the same mechanism of charge storage are known as symmetric capacitors (Kazaryan et al. 2006), while if the two electrodes are of different mechanisms, the device is called asymmetric ultracapacitors.

Srinivasan and Weidner (Lin, Popov, and Ploehn 2002; Srinivasan and Weidner 1999; Staser and Weidner 2014) presented a systematic solution of electrode voltage in terms of discharge current. The discharge of a redox pair electrode is mainly determined by the diffusion of a movable ion passed the film. The kinetic equation of the interface was developed earlier (Dunn and Newman 2000; Doyle, Fuller, and Newman 1993) and John et al. (Staser and Weidner 2014) broadened it to hybrid ECs. The redox pair electrode equation is all-inclusive and has the capacity to explain every electrode, provided that variables like initial concentration of ions and diffusivity are given. The quantity of diffusivity earlier presented by experimental approaches built on the constant diffusivity assumption (Belhachemi, Rael, and Davat 2000) were employed. The capacity of the systematic expression to explain the ultracapacitors' electrical behaviour using partial differential equations that characterize the phenomena inside the device is a major merit over other types of models. This approach is more amenable in allowing other variables and groups of constituting expressions and does not depend a large amount on empirical results. Nevertheless, this class of model makes it very strenuous to evaluate the aging processes of ultracapacitor equations that have a similar systematic modelling method as reported in the work of Doyle, Fuller, and Newman (1993). This was primarily because of the difference in electrolyte phase at electrode and electrolyte surfaces and the introduction of heating parameters while developing the systematic equation.

Kazaryan et al. (2006) developed models capable of estimating, controlling, and improving the performance parameters and other parameters essential for the safety and functioning of various types and designs of heterogeneous ultracapacitors, while accounting for properties of electrode materials, their designs, and separator's structure. They showed that the majority of energy losses were due to the device component's polarization resistance, while the remaining portion was due to electrode depolarization. It was also discovered that the ultracapacitors' energy efficiency with charging and discharging depends a great deal on effective conductivity, thickness of electrodes, and the charging/discharging currents. Growth in effective conductivity decreases the charge losses due to polarization and depolarization non-linearly when the device is charged and discharged.

Chang, Dawson, and Lian (2011) derived a simple planar EC model which showed the importance of polarization density, electrolyte solution, and surface contact area in model development, because a majority of the charges are piled up in a small area that is dependent on variables, such as concentrations of species, impurities, electric, and thermal fields. Varghese, Wang, and Pilon (2011) presented a 3D equation of ultracapacitors using cylindrical electrodes with mesopores to examine the influences of the electrolyte's features on the diffuse layer capacitance, electrolyte permittivity, pore radius, porosity, and ion's practical active diameter. They demonstrated that a reduction in the ion practical diameter and pore radius yielded an enormous rise in diffuse layer mass capacitance. Ganesh and Sanjeev (Gualous et al. 2003) presented 1D and 2D equations of ultracapacitors without Faradic reaction from the basic principles of ionic movement via isotropic movement features to describe the regain of potential while relaxed at the end of discharge or charge, and how it depends on current and concentration, and the device failure when charged with high currents. The EDL capacitances from the electrolyte side and the electrode side are serially joined to each other, and wall pore variables (Conway, Niu, and Pell 2003) play a key role in variations in electrode overall capacitance.

Mathematical expressions and theoretical bases for designing and optimizing ultracapacitors with enhanced energy, power, efficiency, retention capacity, and cycle-life under very practical suppositions have been presented (Ike 2017). Also, guidelines and requirements for design and fabrication of EC components and entire cells with outstanding performance and a reduction in device mass and volume have been reported (Ike 2017). Different ECs have to be designed and fabricated on the basis of the developed theoretical guidelines and requirements and their performance variables carefully evaluated. This knowledge acquired from modelling and simulation will certainly aid in bridging the gap between theoretical and experimental results for practical engineering purposes.

8.5 Thermal Modelling of Electrochemical Capacitors/Ultracapacitors

A reasonable quantity of heat resulting in considerable temperature changes might be produced within the ECs when charged/discharged at a high rate, not minding the device's low internal resistance (Miller 2006). This might have great effects in the ECs' aging and functioning processes that are temperature allergic (Liu, Verbrugge, and Soukiazian 2006). Precise determination of an ultracapacitor's heating situation is very crucial in the design of its temperature controlling systems, especially when modifying the parameters of electrical circuit models that are dependent on temperature and in assessment of the aging level (Weddell et al. 2011). Schiffer, Linzen, and Sauer (2006) examined through experiment the heating behaviour of ECs and noticed that heat production was due to unchangeable Joule heat and heat productions caused by entropy variation by ion movement between conditions of discharging and charging. Dandeville et al. (2011) developed ECs heat profiles that are dependent on time, through a calorimetric method and confirmed Schiffer et al.'s conclusion. Different models which predict EC temperature performance could be commonly categorized into two groups: basic principle equations and detailed equations. The basic principle equations focus on the application of partial differential equations that are normally analysed through numerically discrete approaches for a description of the ECs thermal dynamics. Gualous, Louahlia, and Gallay (2011) presented EC heat expression and analysed it via a finite difference approach to estimate the temperature variation in relation to time and distance. Likewise, Wang et al. (2013) proposed a 3D finite element heat equation of stackable ultracapacitors to examine the temperature field inside the device via a steady current charging and discharging experiment. d'Entremont (d'Entremont and Pilon 2014) developed on fundamental principles a controlling energy expression in addition to the extended Poisson-Nernst-Planck expression to reproduce electronic movement in binary electrolyte. The model captured the ion diffusion and steric influences and entropy variations on the changeable and unchangeable heat production. Several other researches proposed different detailed models to describe ultracapacitor thermal dynamics. Al Sakka et al. (2009) reported a heating model and thermal control of an ultracapacitor cell using a heating-electric analogy, resulting in a very intuitive estimation. Berrueta et al. (2014) also assembled an electrical circuit equation employing a heat equation to describe ECs behaviour while accounting for temperature influences. A joined electro-thermal equation was validated using a series of experiments depicting loading situations in practical uses. Sarwar et al. (2016) also presented an electro-thermal equation to forecast ultracapacitors' electrical and heating functioning subject to a wide scope of operating conditions. The electrical model was firmly joined to the high-accuracy heat equation that accounted for material geometries, heating features, and air gaps. The thermal production and movement expression was employed in temperature change predictions inside the device under different conditions.

8.6 Self-discharges and Charge Redistributions of Electrochemical Capacitors

ECs experience self-discharge while fully charged because of greater free energy as the thermodynamic pushing force compared with when fully discharged (Conway 1999). Ultracapacitors' self-discharge is an essential area to examine while designing ECs, especially for long-term static power supply uses

(Liu, Pell, and Conway 1997). There are pseudo-pushing forces which push open-circuit voltage reduction because Gibbs free energy of charged ECs is more than that of the discharged ones. This is called a 'self-discharge' process, and the open-circuit voltage might reduce up to 60% over many weeks, thereby restricting greatly the ECs' power-delivery capacity. The voltage decay in fully charged ECs at the rest state results in loss of energy due to self-discharge effects (Conway 1999). A amount of self-discharge in a capacitor will certainly limit its applications, especially when used for critical purposes (Simon and Gogotsi 2008; Zhang, Cao, and Yang 2009). Several groups have recognized and focused their work on the understanding of the self-discharge mechanisms. It was shown that the degree of oxygen retention by physical adsorption/introduction of acid functional groups on electrode surfaces affects the rate of self-discharge in ultracapacitors (Conway 1999; Yoshida, Tanahashi, and Nishino 1990). The increased concentration of complexes on electrode surfaces will catalyze the rate of self-discharge, since complexes increase reactions in carbon materials (Morimoto et al. 1996; Yoshida, Tanahashi, and Nishino 1990; Pandolfo and Hollenkamp 2006). Considerable suppression of current leakages and self-discharge were discovered by the removal of the attached functional groups from the surface of electrodes through a thermal application in suitable environments (Yoshida, Tanahashi, and Nishino 1990; Pandolfo and Hollenkamp 2006).

Basic self-discharge of pseudocapacitive substances is different from carbon substances, which exhibit double-layer capacitance. The charge transfer phenomenon in pseudocapacitive materials is Faradaic and consists of various oxidation states that might correspond to different solid phases (Conway, Pell, and Liu 1997). The potential decay over time of fully charged pseudocapacitors due to the diffusion-influenced self-discharge process is usually constant if the materials have high surface area (Hsieh and Teng 2002), thereby yielding a relaxed state of oxidation (Conway 1999). The higher surface area increases the active prismatic sites, which quickens the self-discharge rate (Skipworth and Donne 2007). The impurity of metallics which are present in electrolytes affects the self-discharge process greatly (Andreas, Lussier, and Oickle 2009; Kazaryan, Litvinenko, and Kharisov 2008; Kazaryan et al. 2007a). It has been presented that ions of Fe are more effective in causing self-discharge than other common metallic impurities, because Fe_2^+ ions are oxidised at the anode and Fe_3^+ ions are reduced at the cathode (Kazaryan, Litvinenko, and Kharisov 2008). Another method to examine self-discharge is the movement of Nernst capacities for water decomposition (Pillay and Newman 1996). The effects of charge redistribution on the self-discharge curve of ultracapacitors with carbon electrodes were studied by some groups (Black and Andreas 2009; Kaus, Kowal, and Sauer 2010; Black and Andreas 2010). The application of carbon electrodes with high surface area generates pores of different sizes in the range of micro to macro level, causing the uneven charging/discharging of electrodes along pore walls, thereby yielding a voltage gradient that appears as self-discharge after charging. Ultracapacitors have not replaced batteries as a power source purely because of their inability to retain energy due to increased self-discharge rate (Demarconnay, Raymundo-Piñero, and Béguin 2010; Zhang et al. 2008; Diederich et al. 1999).

Examination of self-discharge principles in the past was to develop acceptable and satisfactory equations of the self-discharge process. Ricketts and Ton-That (2000) presented that ultracapacitors' self-discharge has two main mechanisms, namely: diffusion of ions and current leakages. Ion diffusion that certainly produces charge redistribution comes from the instantaneous difficulty in accessing pores that are very small and deep in the electrodes while devices are charged or discharged (Graydon, Panjehshahi, and Kirk 2014). Black and Andreas (2009) presented from their examination on the influence of charge redistribution on self-discharge that the effect is purely as a result of the redistribution and not actual self-discharge. Niu, Conway, and Pell (2004) demonstrated that reduction in the potential of charged ultracapacitors was strongly influenced by initial potential and displayed an exponential relationship with time. This proved the influence of the charge redistribution is greatly dependent on the electrodes polarization potential. Yang and Zhang (2011) produced self-discharge in the form of changeable resistance in an EC equivalent circuit equation and developed the model to be related to voltage and not time. The derivation permitted the ultracapacitors' equation to be satisfactory for real-life uses as wireless sensors. Kaus, Kowal, and Sauer (2010) examined the key factors affecting self-discharge experimentally. Kowal et al. (2011) later reported an elaborate analysis of likely factors that affect self-discharge and explained their function using exponential relationships. Diab et al. (2009) took advantage of a parameterized equivalent circuit equation to describe ultracapacitors self-discharge and emphasised the leaking currents and ion diffusion at the electrode and electrolyte interface.

Ultracapacitors self-discharge has not been fully presented in review articles, which is very essential because it poses a severe restriction to their practical applicability as energy storage systems (Hou et al. 2008; Wang, Zhang, and Zhang 2012; Conway, Pell, and Liu 1997; Pell et al. 1999; Ricketts and Ton-That 2000; Niu, Conway, and Pell 2004; Kazaryan et al. 2007b; Kazaryan, Litvinenko, and Kharisov 2008; Kim et al. 2004; Black and Andreas 2009; Andreas, Lussier, and Oickle 2009; Ishimoto et al. 2009; Kaus, Kowal, and Sauer 2010; Kowal et al. 2011; Yang and Zhang 2011; Zhang et al. 2011; Pell and Conway 2001; Xu et al. 2008). Thus, acquiring a detailed understanding of the entire mechanism of ultracapacitors self-discharge is paramount to overcoming the challenges of self-discharge. A detailed examination of factors which determine the amount of storable energy at a given temperature (Andreas, Lussier, and Oickle 2009; Kowal et al. 2011) and the genesis of charge redistribution on self-discharge have previously been carried out (Black and Andreas 2009). Kaus et al. (2010) and Conway, Pell, and Liu (1997) presented equations to evaluate self-discharge distribution based on three possible mechanisms.

Many groups have in recent times focused on self-discharge examinations (Pillay and Newman 1996; Johnson and Newman 1971; Bertrand et al. 2010; Dunn and Newman 2000; Doyle, Fuller, and Newman 1993; Chang, Dawson, and Lian 2011; Gualous et al. 2003; Kötz, Hahn, and Gallay 2006; Rafik et al. 2007; Guillemet, Scudeller, and Brousse 2006; Schiffer, Linzen, and Sauer 2006; Hou et al. 2008; Wang, Zhang, and Zhang 2012; Conway, Pell, and Liu 1997; Pell et al. 1999; Ricketts and Ton-That 2000; Niu, Conway, and Pell 2004; Kim et al. 2004; Kazaryan et al. 2007b; Kazaryan, Litvinenko, and Kharisov 2008; Black and Andreas 2009; Andreas, Lussier, and Oickle 2009; Ishimoto et al. 2009). Any faults resulting from the EC production obviously cause little leakage in the electrodes (Conway 1999), leading to a perceivable self-discharge effect. Q. Zhang et al. (2014) presented that simple modification of the surface composition of the single-walled carbon nanotube (SWNT) electrodes of supercapacitors has the capacity to suppress the self-discharge. Chen et al. (2014) reported that the electrochemical capacitors' quick self-discharge was quenched successfully by employing an ion exchange membrane separator or $CuSO_4$ effective electrolyte, as shown in Figure 8.4. This gave a clue to a design of ultracapacitors with enhanced capacitance and energy retention.

Tevi et al. (2013) demonstrated from an experiment that using a thin-blocking layer coated onto electrodes improved the reduced self-discharge with a reduction in specific capacitance. Tevi and Takshi (2015) presented a simple expression with quantum mechanical and electrochemical processes to describe the ECs' discharge curve and energy using blocking layers of different widths, as depicted in

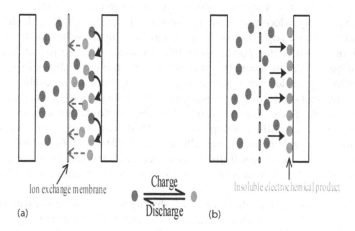

Ion exchange membrane

Charge
Discharge

Insoluble electrochemical product

(a)   (b)

FIGURE 8.4 Two master plans for preventing the defection of the mobile electrolyte among electrodes: (a) utilizing ion-interchange layer separator and (b) using a peculiar mobile electrolyte which transform into insoluble sorts on charge phenomenon. (Chen, L. et al., *Energy Environ. Sci.*, 7, 1750–1759, 2014. Reproduced by permission of the Royal Society of Chemistry.)

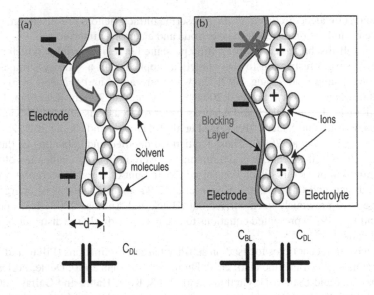

FIGURE 8.5 Diagrammatic illustration of the electrode-electrolyte interface in electrochemical capacitors: (a) chemical reaction at electrode surface leads to loss of charge and (b) implementation of thin stopping membrane reduces rate of migration and improves energy stocking potential. (Reprinted from *J. Power Sour.*, Tevi, T. et al., 241, Application of poly (p-phenylene oxide) as blocking layer to reduce self-discharge in supercapacitors, 589–596, Copyright 2013, with permission from Elsevier.)

Figure 8.5. It was observed that the model was capable of explaining their earlier experimental results (Tevi et al. 2013), and this plays a vital role in choosing the optimum width and the blocking layer materials to minimize leakages in specific utilizations. Zhang et al. (2014) developed mechanisms of self-discharge for ECs built with SWNTs and demonstrated that the divided potential was as a result of functional groups which cause relatively weak bonding among ions and the electrode-charged surface, thereby promoting the ionic diffusion out of Helmholtz layers.

Several researchers (Conway 1999; Nohara et al. 2006; Shinyama et al. 2006) examined the self-discharge features of ultracapacitors built with polymer hydrogel electrolyte using potassium poly acrylate and an aqueous solution of potassium hydroxide. Potential side reactions are mainly (Ishikawa et al. 1994; Conway 1999): (i) shuttle effects due to impurities in carbon material, (ii) advancement of H_2 and O_2, and (iii) electrolyte decomposition into constituent, etc. Shinyama et al. (2006) demonstrated experimentally that losses of charge and energy were satisfactorily quenched by the restriction of redox shuttle substances in the sulfonated polyolefin (S-PO) separator until the redox shuttle concentration grew higher as compared with the strength of the separator. Chen et al. (2014), in a bid to stop mobile electrolyte transfer among the electrodes, proposed that: (1) employing an ion-exchange separator has the capacity to suppress the mobile electrolyte transfer; and (2) employing a special mobile electrolyte which changes into a non-soluble species that are adsorbed onto an electrode during electrochemical reaction could stop transfer. Wang et al. (2014) developed hybrid ECs that store their energy by both Faradaic and capacitive means using a soluble redox species. Chun et al. (2015) recently presented an aqueous redox-improved ultracapacitor that had little self-discharge effect without an ion-exchange separator. This was achievable because of the adsorption of soluble redox species on the electrode, which was different from capacitors that employ inert electrolyte.

The development of absolute models of ECs self-discharge through an equivalent circuit yields a suitable simulation result that enhances the evaluation of self-discharge and energy losses in different devices. Models of ECs which incorporated different self-discharge mechanisms and are employed in the determination of the minimum impurity or redox species concentration and the optimum total thickness of separator and anode for self-discharge suppression have been presented (Ike 2017; Ike, Sigalas, and Iyuke 2017a, 2018). No single self-discharge mechanism has the capacity to fully describe the practical phenomena of ECs' self-discharge, particularly if the capacitance is high. Thus, a hybrid approach of ultracapacitors self-discharge is inevitable. Overall, novel self-discharge theories for optimal designs, operation parameters, enhanced energy capacity, and new-generation ultracapacitors, subject to various self-discharge mechanisms are suitable.

8.7 Difficulties in Ultracapacitors/Electrochemical Capacitors Modelling

Upon several researches to improve the performance of ECs, there are limited prospects to understand the theoretical aspects of energy density determination based on the electrode electrochemical active material. The majority of researches that achieved the most suitable conditions for best performance values were based on trial and error, which is not reliable and effective. Thus, it is very challenging to predict the performance of ultracapacitors. Intuitive prediction is very challenging because of the numerous parameters to be examined with numerous competitive processes involved. Also, experimental methods usually consume time and are expensive. Physical modelling of ECs is very demanding due to the many and closely related interfacial and transport processes, such as charge movement inside electrodes and electrode pore structure as a result of field and concentration changes, ionic steric repelling, and heat production occurring simultaneously.

The ultracapacitors' charging and discharging process needs elaborate knowledge of the effects of device geometry on its operation, effects of electrons and ions on electric potential, as well as those of temperature and field at the electrodes surface. ECs modelling has to account for all components of the device: electrodes, electrolyte, separator, and the current collectors. Every sector is derived distinctly with due account of the inhomogeneous pore formations with various interfaces. It is usually important to begin with a simplified system and gradually extend to complex equations so as to explain and designate the complicated system. The compound interactions between the device's components and their effects on performance parameters make successful optimization of EDLCs very difficult. The numerous factors to be accounted for in an ultracapacitors design makes the modelling very challenging. Such factors are an effect of the adsorption of ions on the electrodes surface, electron kinetics, transportation of solvated ions into pores of varying forms and sizes, and the electrolytes.

Although models of RTILs in planar electrodes successfully explained the difference in results of changing capacitance from experiments, it is difficult for it to be employed in an EDLC's design because the preferable porous electrodes with high surface area are governed by various physics.

All existing models cannot be employed in the examination of the effects of capacitor design and the heat production at steady current charging or discharging. Thus, it is very difficult to generate a blueprint of device heat managements so as to maintain a constant operating temperature at suitable window. Equations that modify the suppositions of the available systematic equations that are built upon and account for all expressions with full consideration of the ECs components' main variables in a meticulous manner are inevitable, but highly challenging.

8.8 Summary

Modelling and simulation play very crucial roles in the successful development of a novel EC with enhanced energy and power densities. Thus, there is a need for mathematical theory that extends beyond equivalent circuits to integrate charge mechanics, charge losses, solvents and ions mechanisms in electrode materials, and the charge movements in pores of electrodes. The development of ultracapacitor models from the basics of electrochemistry and physics is therefore very important. These theoretical models form the foundation for the new-generation ultracapacitors' design, optimization, and fabrication subject to practical assumptions. Another suitable approach is to develop a novel widespread theory of the ultracapacitors self-discharge and obtain an improved technology for the manufacturing of different types of capacitors with effective designs and enhanced storable energy and operation variables, under various mechanisms of self-discharge. Because no single self-discharge mechanism has the capacity to fully describe the practical processes of ECs' self-discharge, a hybrid mechanism of self-discharge is inevitable for the achievement of a new generation.

A thermal model with the capability to evaluate the spatial and instantaneous changes in various heat production rates and temperatures within ultracapacitors from basic rules with due consideration of the unchangeable Joule heating and all changeable heat production rates caused by the migration of ions, steric influences, and entropy of mixing variations will be produced. Thus, ECs' equations with a self-discharge term and the thermal equation provide an inside temperature curve, which helps in the design of electrode materials, estimation of charging and discharging temperatures, as well as creating heat control systems in available and novel designs. Mathematical expressions that relax the supposition of the existing systematic equations and consider the production of heat, charge redistribution, self-discharge, energy and charge losses, and the ECs components' main parameters in a rigorous way is inevitable. Modelling and simulation certainly aid in the development of several new advanced materials and cell architectures, as well as in the synthesis of different new electrolyte materials with enhanced properties for fabrication of new-generation ultracapacitors. DFT calculations and other types of modelling play vital roles in the discovery, synthesis, optimization, and fabrication of numerous new electrode and electrolyte materials, which can help in the new-generation electrochemical capacitors with improved performance.

Modelling and simulation are therefore very crucial in making excellent progress in areas of electrode material synthesis, optimization and fabrication, electrolyte synthesis and optimizations, separators synthesis, and fabrications. Subsequently, an optimized design of the new-generation ultra-capacitors with enhanced performance is achievable by employing the electrochemical capacitors blueprint obtained from improved electrodes, electrolytes, and separators acquired via modelling and simulation.

ACKNOWLEDGEMENTS

The financial assistance of the National Research Foundation (NRF) and DST/NRF Centre of Excellence in Strong Materials (COE-SM), University of the Witwatersrand, Private Bag 3, Johannesburg 2050, South Africa, towards this research is hereby acknowledged. Opinions expressed and conclusions arrived at are those of the authors and are not necessarily to be attributed to the NRF, DST/NRF Centre of Excellence in Strong Materials, and African Centre of Excellence in Future Energies and Electrochemical Systems (ACE-FUELS), Federal University of Technology, Owerri, Nigeria.

REFERENCES

Al Sakka, Monzer, Hamid Gualous, Joeri Van Mierlo, and Hasan Culcu. 2009. "Thermal Modeling and Heat Management of Supercapacitor Modules for Vehicle Applications." *Journal of Power Sources* 194 (2): 581–87. doi:10.1016/j.jpowsour.2009.06.038.

Alam, Muhammad Tanzirul, Md. Mominul Islam, Takeyoshi Okajima, and Takeo Ohsaka. 2007. "Measurements of Differential Capacitance at Mercury/Room-Temperature Ionic Liquids Interfaces." *The Journal of Physical Chemistry C* 111 (49): 18326–33. doi:10.1021/jp0758081.

Bard, Allen J., and Larry R. Faulkner. 2001. *Electrochemical Methods: Fundamentals and Applications.* 2nd edition. https://www.wiley.com/en-us/Electrochemical+Methods%3A+Fundamentals+and+Applications%2C+2nd+Edition-p-9780471043720.

Allu, S., B. Velamur Asokan, W. A. Shelton, B. Philip, and S. Pannala. 2014. "A Generalized Multi-Dimensional Mathematical Model for Charging and Discharging Processes in a Supercapacitor." *Journal of Power Sources* 256 (June): 369–82. doi:10.1016/j.jpowsour.2014.01.054.

Andreas, Heather A., Kate Lussier, and Alicia M. Oickle. 2009. "Effect of Fe-Contamination on Rate of Self-Discharge in Carbon-Based Aqueous Electrochemical Capacitors." *Journal of Power Sources* 187 (1): 275–83. doi:10.1016/j.jpowsour.2008.10.096.

Augustyn, Veronica, Jérémy Come, Michael A. Lowe, Jong Woung Kim, Pierre-Louis Taberna, Sarah H. Tolbert, Héctor D. Abruña, Patrice Simon, and Bruce Dunn. 2013. "High-Rate Electrochemical Energy Storage through Li+ Intercalation Pseudocapacitance." *Nature Materials* 12 (6): 518–22. doi:10.1038/nmat3601.

Bagotsky, V. S. 2006. *Fundamentals of Electrochemistry (Bagotsky, Wiley 2006, ISBN 0471700584) | Electrochemistry | Dissociation (Chemistry).* 2nd edition. https://www.scribd.com/document/52981827/Fundamentals-of-Electrochemistry-Bagotsky-Wiley-2006-ISBN-0471700584.

Baldelli, Steven. 2008. "Surface Structure at the Ionic Liquid–Electrified Metal Interface." *Accounts of Chemical Research* 41 (3): 421–31. doi:10.1021/ar700185h.

Balducci, A. 2016. "Electrolytes for High Voltage Electrochemical Double Layer Capacitors: A Perspective Article." *Journal of Power Sources* 326 (September): 534–40. doi:10.1016/j.jpowsour.2016.05.029.

Ban, Shuai, Jiujun Zhang, Lei Zhang, Ken Tsay, Datong Song, and Xinfu Zou. 2013. "Charging and Discharging Electrochemical Supercapacitors in the Presence of Both Parallel Leakage Process and Electrochemical Decomposition of Solvent." *Electrochimica Acta* 90 (February): 542–49. doi:10.1016/j.electacta.2012.12.056.

Batchelor-McAuley, Christopher, Kamonwad Ngamchuea, and Richard G. Compton. 2018. "Simulated Low-Support Voltammetry: Deviations from Ohm's Law." *Journal of Electroanalytical Chemistry* 830–831 (December): 88–94. doi:10.1016/j.jelechem.2018.10.032.

Beguin, Francois, and Elzbieta Frackowiak. 2013. *Supercapacitors: Materials, Systems and Applications.* Hoboken: John Wiley & Sons.

Belhachemi, F., S. Rael, and B. Davat. 2000. "A Physical Based Model of Power Electric Double-Layer Supercapacitors." In *Conference Record of the 2000 IEEE Industry Applications Conference, 2000,* 5:3069–76. doi:10.1109/IAS.2000.882604.

Berrueta, Alberto, Idoia San Martín, Andoni Hernández, Alfredo Ursúa, and Pablo Sanchis. 2014. "Electro-Thermal Modelling of a Supercapacitor and Experimental Validation." *Journal of Power Sources* 259 (August): 154–65. doi:10.1016/j.jpowsour.2014.02.089.

Bertrand, N., J. Sabatier, O. Briat, and J.-M. Vinassa. 2010. "Embedded Fractional Nonlinear Supercapacitor Model and Its Parametric Estimation Method." *IEEE Transactions on Industrial Electronics* 57 (12): 3991–4000. doi:10.1109/TIE.2010.2076307.

Bikerman, J. J. 1942. "XXXIX. Structure and Capacity of Electrical Double Layer." *The London, Edinburgh, and Dublin Philosophical Magazine and Journal of Science* 33 (220): 384–97. doi:10.1080/14786444208520813.

Black, Jennifer, and Heather A. Andreas. 2009. "Effects of Charge Redistribution on Self-Discharge of Electrochemical Capacitors." *Electrochimica Acta* 54 (13): 3568–74. doi:10.1016/j.electacta.2009.01.019.

Black, Jennifer, and Heather A. Andreas. 2010. "Prediction of the Self-Discharge Profile of an Electrochemical Capacitor Electrode in the Presence of Both Activation-Controlled Discharge and Charge Redistribution." *Journal of Power Sources* 195 (3): 929–35. doi:10.1016/j.jpowsour.2009.08.040.

Bo, Zheng, Changwen Li, Huachao Yang, Kostya Ostrikov, Jianhua Yan, and Kefa Cen. 2018. "Design of Supercapacitor Electrodes Using Molecular Dynamics Simulations." *Nano-Micro Letters* 10 (2): 33. doi:10.1007/s40820-018-0188-2.

Boda, Dezsö, and Douglas Henderson. 2000. "The Capacitance of the Solvent Primitive Model Double Layer at Low Effective Temperatures." *The Journal of Chemical Physics* 112 (20): 8934–38. doi:10.1063/1.481507.

Boda, Dezsö, Douglas Henderson, and Kwong-Yu Chan. 1999. "Monte Carlo Study of the Capacitance of the Double Layer in a Model Molten Salt." *The Journal of Chemical Physics* 110 (11): 5346–50. doi:10.1063/1.478429.

Bolwig, Simon, Gatis Bazbauers, Antje Klitkou, Peter D. Lund, Andra Blumberga, Armands Gravelsins, and Dagnija Blumberga. 2019. "Review of Modelling Energy Transitions Pathways with Application to Energy System Flexibility." *Renewable and Sustainable Energy Reviews* 101 (March): 440–52. doi:10.1016/j.rser.2018.11.019.

Boota, Muhammad, and Yury Gogotsi. 2019. "MXene—Conducting Polymer Asymmetric Pseudocapacitors." *Advanced Energy Materials* 9 (7): 1802917. doi:10.1002/aenm.201802917.

Brousse, Thierry, Pierre-Louis Taberna, Olivier Crosnier, Romain Dugas, Philippe Guillemet, Yves Scudeller, Yingke Zhou, Frédéric Favier, Daniel Bélanger, and Patrice Simon. 2007. "Long-Term Cycling Behavior of Asymmetric Activated Carbon/MnO_2 Aqueous Electrochemical Supercapacitor." *Journal of Power Sources* 173 (1): 633–41. doi:10.1016/j.jpowsour.2007.04.074.

Bukola, Saheed, and Stephen E. Creager. 2019. "A Charge-Transfer Resistance Model and Arrhenius Activation Analysis for Hydrogen Ion Transmission across Single-Layer Graphene." *Electrochimica Acta* 296 (February): 1–7. doi:10.1016/j.electacta.2018.11.005.

Buller, S., E. Karden, D. Kok, and R. W. De Doncker. 2002. "Modeling the Dynamic Behavior of Supercapacitors Using Impedance Spectroscopy." *IEEE Transactions on Industry Applications* 38 (6): 1622–26. doi:10.1109/TIA.2002.804762.

Burke, Andrew. 2000. "Ultracapacitors: Why, How, and Where Is the Technology." *Journal of Power Sources* 91 (1): 37–50. doi:10.1016/S0378-7753(00)00485-7.

Chaban, V. V., and O. N. Kalugin. 2009. "Structure and Dynamics in Methanol and Its Lithium Ion Solution Confined by Carbon Nanotubes." *Journal of Molecular Liquids*, Special Issue of contributions to the International Conference: Modern Physical Chemistry for Advanced Materials, 145 (3): 145–51. doi:10.1016/j.molliq.2008.06.003.

Chai, Ruizhi, and Ying Zhang. 2015. "A Practical Supercapacitor Model for Power Management in Wireless Sensor Nodes." *IEEE Transactions on Power Electronics* 30 (12): 6720–30. doi:10.1109/tpel.2014.2387113.

Chang, Jin Hyun, F. P. Dawson, and K. K. Lian. 2011. "A First Principles Approach to Develop a Dynamic Model of Electrochemical Capacitors." *IEEE Transactions on Power Electronics* 26 (12): 3472–80. doi:10.1109/TPEL.2011.2161096.

Chen, Libin, Hua Bai, Zhifeng Huang, and Lei Li. 2014. "Mechanism Investigation and Suppression of Self-Discharge in Active Electrolyte Enhanced Supercapacitors." *Energy Environmental Science* 7 (5): 1750–59. doi:10.1039/C4EE00002A.

Choi, Hong Soo, and Chong Rae Park. 2014. "Theoretical Guidelines to Designing High Performance Energy Storage Device Based on Hybridization of Lithium-Ion Battery and Supercapacitor." *Journal of Power Sources* 259 (August): 1–14. doi:10.1016/j.jpowsour.2014.02.001.

Chu, Andrew, and Paul Braatz. 2002. "Comparison of Commercial Supercapacitors and High-Power Lithium-Ion Batteries for Power-Assist Applications in Hybrid Electric Vehicles: I. Initial Characterization." *Journal of Power Sources* 112 (1): 236–46. doi:10.1016/S0378-7753(02)00364-6.

Chun, Sang-Eun, Brian Evanko, Xingfeng Wang, David Vonlanthen, Xiulei Ji, Galen D. Stucky, and Shannon W. Boettcher. 2015. "Design of Aqueous Redox-Enhanced Electrochemical Capacitors with High Specific Energies and Slow Self-Discharge." *Nature Communications* 6: 7818. doi:10.1038/ncomms8818.

Conway, B. E. 1999. *Electrochemical Supercapacitors: Scientific Fundamentals and Technological Applications.* New York: Kluwer Academic/Plenum.

Conway, B. E., V. Birss, and J. Wojtowicz. 1997. "The Role and Utilization of Pseudocapacitance for Energy Storage by Supercapacitors." *Journal of Power Sources* 66 (1–2): 1–14. doi:10.1016/S0378-7753(96)02474-3.

Conway, B. E., J. Niu, and W. G. Pell. 2003. "Seminar on Double Layer Capacitor and Hybrid Energy Storage Devices." In *Proceedings of the 13th International Seminar on Double Layer Capacitor and Hybrid Energy Storage Devices*, Deerfield Beach, FL.

Conway, B. E., and W. G. Pell. 2003. "Double-Layer and Pseudocapacitance Types of Electrochemical Capacitors and Their Applications to the Development of Hybrid Devices." *Journal of Solid State Electrochemistry* 7 (9): 637–44. doi:10.1007/s10008-003-0395-7.

Conway, Brian E., W. G. Pell, and T.-C. Liu. 1997. "Diagnostic Analyses for Mechanisms of Self-Discharge of Electrochemical Capacitors and Batteries." *Journal of Power Sources* 65 (1–2): 53–59. doi:10.1016/S0378-7753(97)02468-3.

Cui, S. T., and H. D. Cochran. 2002. "Molecular Dynamics Simulation of Interfacial Electrolyte Behaviors in Nanoscale Cylindrical Pores." *The Journal of Chemical Physics* 117 (12): 5850–54. doi:10.1063/1.1501585.

Dandeville, Y., Ph. Guillemet, Y. Scudeller, O. Crosnier, L. Athouel, and Th. Brousse. 2011. "Measuring Time-Dependent Heat Profiles of Aqueous Electrochemical Capacitors under Cycling." *Thermochimica Acta* 526 (1–2): 1–8. doi:10.1016/j.tca.2011.07.027.

Demarconnay, L., E. Raymundo-Piñero, and F. Béguin. 2010. "A Symmetric Carbon/Carbon Supercapacitor Operating at 1.6 V by Using a Neutral Aqueous Solution." *Electrochemistry Communications* 12 (10): 1275–78. doi:10.1016/j.elecom.2010.06.036.

Demarconnay, L., E. Raymundo-Piñero, and F. Béguin. 2011. "Adjustment of Electrodes Potential Window in an Asymmetric Carbon/MnO_2 Supercapacitor." *Journal of Power Sources* 196 (1): 580–86. doi:10.1016/j.jpowsour.2010.06.013.

Devan, Sheba, Venkat R. Subramanian, and R. E. White. 2004. "Analytical Solution for the Impedance of a Porous Electrode." *Journal of the Electrochemical Society* 151 (6): A905–13. doi:10.1149/1.1739218.

Diab, Y., P. Venet, H. Gualous, and G. Rojat. 2009. "Self-Discharge Characterization and Modeling of Electrochemical Capacitor Used for Power Electronics Applications." *IEEE Transactions on Power Electronics* 24 (2): 510–17. doi:10.1109/TPEL.2008.2007116.

Diederich, L., E. Barborini, P. Piseri, A. Podesta, P. Milani, A. Schneuwly, and R. Gallay. 1999. "Supercapacitors Based on Nanostructured Carbon Electrodes Grown by Cluster-Beam Deposition." *Applied Physics Letters* 75 (17): 2662–64. doi:10.1063/1.125111.

Dougal, R. A., L. Gao, and S. Liu. 2004. "Ultracapacitor Model with Automatic Order Selection and Capacity Scaling for Dynamic System Simulation." *Journal of Power Sources* 126 (1–2): 250–57. doi:10.1016/j.jpowsour.2003.08.031.

Doyle, Marc, Thomas F. Fuller, and John Newman. 1993. "Modeling of Galvanostatic Charge and Discharge of the Lithium/Polymer/Insertion Cell." *Journal of the Electrochemical Society* 140 (6): 1526–33. doi:10.1149/1.2221597.

Dunn, Darryl, and John Newman. 2000. "Predictions of Specific Energies and Specific Powers of Double-Layer Capacitors Using a Simplified Model." *Journal of the Electrochemical Society* 147 (3): 820–30. doi:10.1149/1.1393278.

Dzielinski, Andrzej, Grzegorz Sarwas, and Dominik Sierociuk. 2010. "Time Domain Validation of Ultracapacitor Fractional Order Model." In *49th IEEE Conference on Decision and Control (CDC)*, 3730–35. Atlanta, GA: IEEE. doi:10.1109/CDC.2010.5717093.

Entremont, Anna d', and Laurent Pilon. 2014. "First-Principles Thermal Modeling of Electric Double Layer Capacitors under Constant-Current Cycling." *Journal of Power Sources* 246 (January): 887–98. doi:10.1016/j.jpowsour.2013.08.024.

Etsuro Iwama, Tsukasa Ueda, Yoko Ishihara, Kenji Ohshima, Wako Naoi, McMahon Thomas Homer Reid, and Katsuhiko Naoi. 2019. "High-Voltage Operation of Li4Ti5O12/AC Hybrid Supercapacitor Cell in Carbonate and Sulfone Electrolytes: Gas Generation and Its Characterization." *Electrochimica Acta* 301 (April): 312–18. doi:10.1016/j.electacta.2019.01.088.

Fan, Deyuan, and Ralph E. White. 1991. "Mathematical Modeling of a Nickel-Cadmium Battery Effects of Intercalation and Oxygen Reactions." *Journal of the Electrochemical Society* 138 (10): 2952–60. doi:10.1149/1.2085347.

Faranda, R. 2010. "A New Parameters Identification Procedure for Simplified Double Layer Capacitor Two-Branch Model." *Electric Power Systems Research* 80 (4): 363–71. doi:10.1016/j.epsr.2009.10.024.

Fellows, Hannah M., Marveh Forghani, Olivier Crosnier, and Scott W. Donne. 2019. "Modelling Voltametric Data from Electrochemical Capacitors." *Journal of Power Sources* 417 (March): 193–206. doi:10.1016/j.jpowsour.2018.11.078.

Feng, G., J. S. Zhang, and R. Qiao. 2009. "Microstructure and Capacitance of the Electrical Double Layers at the Interface of Ionic Liquids and Planar Electrodes." *The Journal of Physical Chemistry C* 113 (11): 4549–59. doi:10.1021/jp809900w.

Feng, Guang, Jingsong Huang, Bobby G. Sumpter, Vincent Meunier, and Rui Qiao. 2011. "A 'Counter-Charge Layer in Generalized Solvents' Framework for Electrical Double Layers in Neat and Hybrid Ionic Liquid Electrolytes." *Physical Chemistry Chemical Physics* 13 (32): 14723. doi:10.1039/c1cp21428d.

Feng, Guang, De-en Jiang, and Peter T. Cummings. 2012. "Curvature Effect on the Capacitance of Electric Double Layers at Ionic Liquid/Onion-Like Carbon Interfaces." *Journal of Chemical Theory and Computation* 8 (3): 1058–63. doi:10.1021/ct200914j.

Feng, Guang, Song Li, Jennifer S. Atchison, Volker Presser, and Peter T. Cummings. 2013. "Molecular Insights into Carbon Nanotube Supercapacitors: Capacitance Independent of Voltage and Temperature." *The Journal of Physical Chemistry C* 117 (18): 9178–86. doi:10.1021/jp403547k.

Feng, Guang, Rui Qiao, Jingsong Huang, Bobby G. Sumpter, and Vincent Meunier. 2010. "Ion Distribution in Electrified Micropores and Its Role in the Anomalous Enhancement of Capacitance." *ACS Nano* 4 (4): 2382–90. doi:10.1021/nn100126w.

Frackowiak, Elzbieta. 2007. "Carbon Materials for Supercapacitor Application." *Physical Chemistry Chemical Physics* 9 (15): 1774–85. doi:10.1039/B618139M.

Frackowiak, Elzbieta, Grzegorz Lota, and Juliusz Pernak. 2005. "Room-Temperature Phosphonium Ionic Liquids for Supercapacitor Application." *Applied Physics Letters* 86 (16): 164104. doi:10.1063/1.1906320.

Freeborn, Todd J., Brent Maundy, and Ahmed S. Elwakil. 2013. "Measurement of Supercapacitor Fractional-Order Model Parameters from Voltage-Excited Step Response." *IEEE Journal on Emerging and Selected Topics in Circuits and Systems* 3 (3): 367–76. doi:10.1109/JETCAS.2013.2271433.

Freund, Jonathan B. 2002. "Electro-Osmosis in a Nanometer-Scale Channel Studied by Atomistic Simulation." *The Journal of Chemical Physics* 116 (5): 2194–2200. doi:10.1063/1.1431543.

Gabano, Jean-Denis, Thierry Poinot, and Houcem Kanoun. 2015. "LPV Continuous Fractional Modeling Applied to Ultracapacitor Impedance Identification." *Control Engineering Practice* 45 (December): 86–97. doi:10.1016/j.conengprac.2015.09.001.

Gor, Gennady Yu., Matthias Thommes, Katie A. Cychosz, and Alexander V. Neimark. 2012. "Quenched Solid Density Functional Theory Method for Characterization of Mesoporous Carbons by Nitrogen Adsorption." *Carbon* 50 (4): 1583–90. doi:10.1016/j.carbon.2011.11.037.

Gouy, G. 1910. "Sur La Constitution de La Charge Electrique a La Surface d'un Electrolyte." *J. Phys. (France)* 9.

Grahame, David C. 1947. "The Electrical Double Layer and the Theory of Electrocapillarity." *Chemical Reviews* 41 (3): 441–501. doi:10.1021/cr60130a002.

Graydon, John W., Milad Panjehshahi, and Donald W. Kirk. 2014. "Charge Redistribution and Ionic Mobility in the Micropores of Supercapacitors." *Journal of Power Sources* 245 (January): 822–29. doi:10.1016/j.jpowsour.2013.07.036.

Gualous, H., D. Bouquain, A. Berthon, and J. M. Kauffmann. 2003. "Experimental Study of Supercapacitor Serial Resistance and Capacitance Variations with Temperature." *Journal of Power Sources* 123 (1): 86–93. doi:10.1016/S0378-7753(03)00527-5.

Gualous, H., H. Louahlia, and R. Gallay. 2011. "Supercapacitor Characterization and Thermal Modelling with Reversible and Irreversible Heat Effect." *IEEE Transactions on Power Electronics* 26 (11): 3402–9. doi:10.1109/TPEL.2011.2145422.

Guillemet, Ph., Y. Scudeller, and Th. Brousse. 2006. "Multi-Level Reduced-Order Thermal Modeling of Electrochemical Capacitors." *Journal of Power Sources* 157 (1): 630–40. doi:10.1016/j.jpowsour.2005.07.072.

Hohenberg, P., and W. Kohn. 1964. "Inhomogeneous Electron Gas." *Physical Review* 136 (3B): B864–71. doi:10.1103/PhysRev.136.B864.

Holovko, Myroslav, Vitalyj Kapko, Douglas Henderson, and Dezsö Boda. 2001. "On the Influence of Ionic Association on the Capacitance of an Electrical Double Layer." *Chemical Physics Letters* 341 (3–4): 363–68. doi:10.1016/S0009-2614(01)00505-X.

Hou, Chia-Hung, Patricia Taboada-Serrano, Sotira Yiacoumi, and Costas Tsouris. 2008. "Monte Carlo Simulation of Electrical Double-Layer Formation from Mixtures of Electrolytes Inside Nanopores." *The Journal of Chemical Physics* 128 (4): 044705. doi:10.1063/1.2824957.

Hsieh, Chien-To, and Hsisheng Teng. 2002. "Influence of Oxygen Treatment on Electric Double-Layer Capacitance of Activated Carbon Fabrics." *Carbon* 40 (5): 667–74. doi:10.1016/S0008-6223(01)00182-8.

Hu, X., S. Li, and H. Peng. 2012. "A Comparative Study of Equivalent Circuit Models for Li-Ion Batteries." *Journal of Power Sources* 198 (January): 359–67. doi:10.1016/j.jpowsour.2011.10.013.

Huang, Jingsong, Bobby G. Sumpter, and Vincent Meunier. 2008a. "A Universal Model for Nanoporous Carbon Supercapacitors Applicable to Diverse Pore Regimes, Carbon Materials, and Electrolytes." *Chemistry – A European Journal* 14 (22): 6614–26. doi:10.1002/chem.200800639.

Huang, Jingsong, Bobby G. Sumpter, and Vincent Meunier. 2008b. "Theoretical Model for Nanoporous Carbon Supercapacitors." *Angewandte Chemie International Edition* 47 (3): 520–24. doi:10.1002/anie.200703864.

Huang, Jingsong, Bobby G. Sumpter, and Vincent Meunier. 2008c. "Theoretical Model for Nanoporous Carbon Supercapacitors." *Angewandte Chemie* 120 (3): 530–34. doi:10.1002/ange.200703864.

Huang, Jingsong, Bobby G. Sumpter, and Vincent Meunier. 2008d. "A Universal Model for Nanoporous Carbon Supercapacitors Applicable to Diverse Pore Regimes, Carbon Materials, and Electrolytes." *Chemistry – A European Journal* 14 (22): 6614–26. doi:10.1002/chem.200800639.

Ike, Innocent S. 2017. "Mathematical Modelling, Design, and Optimization of Electrochemical Capacitor Cells." Thesis. http://wiredspace.wits.ac.za/handle/10539/24089.

Ike, Innocent S., and Sunny Iyuke. 2016. "Mathematical Modelling and Simulation of Supercapacitors." In *Nanomaterials in Advanced Batteries and Supercapacitors*, edited by Kenneth I. Ozoemena and Shaowei Chen, 515–62. Nanostructure Science and Technology. Springer International Publishing. doi:10.1007/978-3-319-26082-2_15.

Ike, Innocent S., Iakovos Sigalas, and Sunny Iyuke. 2018. "The Effects of Self-Discharge on the Performance of Asymmetric/Hybrid Electrochemical Capacitors with Redox-Active Electrolytes: Insights from Modeling and Simulation." *Journal of Electronic Materials* 47 (1): 470–92. doi:10.1007/s11664-017-5796-y.

Ike, Innocent S., Iakovos Sigalas, and Sunny E. Iyuke. 2016. "The Influences of Operating Conditions and Design Configurations on the Performance of Symmetric Electrochemical Capacitors." *Physical Chemistry Chemical Physics* 18 (41): 28626–47. doi:10.1039/C6CP06214H.

Ike, Innocent S., Iakovos Sigalas, and Sunny E. Iyuke. 2017a. "The Effects of Self-Discharge on the Performance of Symmetric Electric Double-Layer Capacitors and Active Electrolyte-Enhanced Supercapacitors: Insights from Modeling and Simulation." *Journal of Electronic Materials* 46 (2): 1163–89. doi:10.1007/s11664-016-5053-9.

Ike, Innocent S., Iakovos Sigalas, and Sunny E. Iyuke. 2017b. "Optimization of Design Parameters and Operating Conditions of Electrochemical Capacitors for High Energy and Power Performance." *Journal of Electronic Materials* 46 (3): 1692–1713. doi:10.1007/s11664-016-5213-y.

Ike, Innocent S., Iakovos Sigalas, and Sunny E. Iyuke. 2017c. "Modelling and Optimization of Electrodes Utilization in Symmetric Electrochemical Capacitors for High Energy and Power." *Journal of Energy Storage* 12 (August): 261–75. doi:10.1016/j.est.2017.05.006.

Ike, Innocent S., Iakovos Sigalas, Sunny Iyuke, and Kenneth I. Ozoemena. 2015. "An Overview of Mathematical Modeling of Electrochemical Supercapacitors/Ultracapacitors." *Journal of Power Sources* 273 (January): 264–77. doi:10.1016/j.jpowsour.2014.09.071.

Ishikawa, Masashi, Masayuki Morita, Mitsuo Ihara, and Yoshiharu Matsuda. 1994. "Electric Double-Layer Capacitor Composed of Activated Carbon Fiber Cloth Electrodes and Solid Polymer Electrolytes Containing Alkylammonium Salts." *Journal of the Electrochemical Society* 141 (7): 1730–34. doi:10.1149/1.2054995.

Ishimoto, Shuichi, Yuichiro Asakawa, Masanori Shinya, and Katsuhiko Naoi. 2009. "Degradation Responses of Activated-Carbon-Based EDLCs for Higher Voltage Operation and Their Factors." *Journal of the Electrochemical Society* 156 (7): A563–71. doi:10.1149/1.3126423.

Jeffrey W. Long, and Daniel Bélanger. 2011. "Asymmetric Electrochemical Capacitors—Stretching the Limits of Aqueous Electrolytes." *MRS Bulletin* 36 (7): 513–22. doi:10.1557/mrs.2011.137.

Jiang, De-en, Dong Meng, and Jianzhong Wu. 2011. "Density Functional Theory for Differential Capacitance of Planar Electric Double Layers in Ionic Liquids." *Chemical Physics Letters* 504 (4–6): 153–58. doi:10.1016/j.cplett.2011.01.072.

Jiang, De-en, and Jianzhong Wu. 2013. "Microscopic Insights into the Electrochemical Behavior of Nonaqueous Electrolytes in Electric Double-Layer Capacitors." *The Journal of Physical Chemistry Letters* 4 (8): 1260–67. doi:10.1021/jz4002967.

Johnson, A. M., and John Newman. 1971. "Desalting by Means of Porous Carbon Electrodes." *Journal of The Electrochemical Society* 118 (3): 510–17. doi:10.1149/1.2408094.

Kaus, Maximilian, Julia Kowal, and Dirk Uwe Sauer. 2010. "Modelling the Effects of Charge Redistribution during Self-Discharge of Supercapacitors." *Electrochimica Acta* 55 (25): 7516–23. doi:10.1016/j.electacta.2010.01.002.

Kazaryan, S. A., G. G. Kharisov, S. V. Litvinenko, and V. I. Kogan. 2007a. "Self-Discharge Related to Iron Ions and Its Effect on the Parameters of HES PbO_2 | H_2SO_4 | C Systems." *Journal of the Electrochemical Society* 154 (8): A751–59. doi:10.1149/1.2742335.

Kazaryan, S. A., G. G. Kharisov, S. V. Litvinenko, and V. I. Kogan. 2007b. "Erratum: Self-Discharge Related to Iron Ions and Its Effect on the Parameters of HES PbO_2 | H_2SO_4 | C Systems [J. Electrochem. Soc., 154, A751 (2007)]." *Journal of the Electrochemical Society* 154 (11): S18. doi:10.1149/1.2775570.

Kazaryan, S. A., S. V. Litvinenko, and G. G. Kharisov. 2008. "Self-Discharge of Heterogeneous Electrochemical Supercapacitor of PbO_2 | H_2SO_4 | C Related to Manganese and Titanium Ions." *Journal of the Electrochemical Society* 155 (6): A464–73. doi:10.1149/1.2904456.

Kazaryan, S. A., S. N. Razumov, S. V. Litvinenko, G. G. Kharisov, and V. I. Kogan. 2006. "Mathematical Model of Heterogeneous Electrochemical Capacitors and Calculation of Their Parameters." *Journal of the Electrochemical Society* 153 (9): A1655–71. doi:10.1149/1.2212057.

Khomenko, V., E. Raymundo-Piñero, and F. Béguin. 2006. "Optimisation of an Asymmetric Manganese Oxide/Activated Carbon Capacitor Working at 2 V in Aqueous Medium." *Journal of Power Sources* 153 (1): 183–90. doi:10.1016/j.jpowsour.2005.03.210.

Kim, Daejoong, and Eric Darve. 2006. "Molecular Dynamics Simulation of Electro-Osmotic Flows in Rough Wall Nanochannels." *Physical Review E* 73 (5): 051203. doi:10.1103/PhysRevE.73.051203.

Kim, Doyoung, Keunsik Lee, Meeree Kim, Yongshin Kim, and Hyoyoung Lee. 2019. "Carbon-Based Asymmetric Capacitor for High-Performance Energy Storage Devices." *Electrochimica Acta* 300 (March): 461–69. doi:10.1016/j.electacta.2019.01.141.

Kim, Hansung, and Branko N. Popov. 2003. "A Mathematical Model of Oxide/Carbon Composite Electrode for Supercapacitors." *Journal of the Electrochemical Society* 150 (9): A1153–60. doi:10.1149/1.1593039.

Kim, Jong Woung, Veronica Augustyn, and Bruce Dunn. 2012. "The Effect of Crystallinity on the Rapid Pseudocapacitive Response of Nb_2O_5." *Advanced Energy Materials* 2 (1): 141–48. doi:10.1002/aenm.201100494.

Kim, Kyong-Min, Jin-Woo Hur, Se-Il Jung, and An-Soo Kang. 2004. "Electrochemical Characteristics of Activated Carbon/Ppy Electrode Combined with P(VdF-Co-HFP)/PVP for EDLC." *Electrochimica Acta* 50 (2–3): 863–72. doi:10.1016/j.electacta.2004.02.059.

Kim, Sang-Hyun, Woojin Choi, Kyo-Bum Lee, and Sewan Choi. 2011. "Advanced Dynamic Simulation of Supercapacitors Considering Parameter Variation and Self-Discharge." *IEEE Transactions on Power Electronics* 26 (11): 3377–85. doi:10.1109/TPEL.2011.2136388.

Kislenko, Sergey A., Igor S. Samoylov, and Ravil H. Amirov. 2009. "Molecular Dynamics Simulation of the Electrochemical Interface between a Graphite Surface and the Ionic Liquid [BMIM][PF6]." *Physical Chemistry Chemical Physics* 11 (27): 5584–90. doi:10.1039/B823189C.

Kondrat, S., and A. Kornyshev. 2011. "Superionic State in Double-Layer Capacitors with Nanoporous Electrodes." *Journal of Physics: Condensed Matter* 23 (2): 022201. doi:10.1088/0953-8984/23/2/022201.

Kondrat, S., C. R. Pérez, V. Presser, Y. Gogotsi, and A. A. Kornyshev. 2012. "Effect of Pore Size and Its Dispersity on the Energy Storage in Nanoporous Supercapacitors." *Energy & Environmental Science* 5 (4): 6474. doi:10.1039/c2ee03092f.

Kornyshev, Alexei A. 2007. "Double-Layer in Ionic Liquids: Paradigm Change?" *The Journal of Physical Chemistry B* 111 (20): 5545–57. doi:10.1021/jp067857o.

Kötz, R., and M. Carlen. 2000. "Principles and Applications of Electrochemical Capacitors." *Electrochimica Acta* 45 (15–16): 2483–98. doi:10.1016/S0013-4686(00)00354-6.

Kötz, R., M. Hahn, and R. Gallay. 2006. "Temperature Behavior and Impedance Fundamentals of Supercapacitors." *Journal of Power Sources* 154 (2): 550–55. doi:10.1016/j.jpowsour.2005.10.048.

Kowal, Julia, Esin Avaroglu, Fahmi Chamekh, Armands Šenfelds, Tjark Thien, Dhanny Wijaya, and Dirk Uwe Sauer. 2011. "Detailed Analysis of the Self-Discharge of Supercapacitors." *Journal of Power Sources* 196 (1): 573–79. doi:10.1016/j.jpowsour.2009.12.028.

Kroupa, Martin, Gregory J. Offer, and Juraj Kosek. 2016. "Modelling of Supercapacitors: Factors Influencing Performance." *Journal of the Electrochemical Society* 163 (10): A2475–87. doi:10.1149/2.0081613jes.

Krummacher, J., C. Schütter, L. H. Hess, and A. Balducci. 2018. "Non-Aqueous Electrolytes for Electrochemical Capacitors." *Current Opinion in Electrochemistry* 9 (June): 64–69. doi:10.1016/j.coelec.2018.03.036.

Kumar, Yogesh, Sangeeta Rawal, Bhawana Joshi, and S. A. Hashmi. 2019. "Background, Fundamental Understanding and Progress in Electrochemical Capacitors." *Journal of Solid State Electrochemistry* 23 (3): 667–92. doi:10.1007/s10008-018-4160-3.

Kurzweil, P., and H.-J. Fischle. 2004. "A New Monitoring Method for Electrochemical Aggregates by Impedance Spectroscopy." *Journal of Power Sources* 127 (1–2): 331–40. doi:10.1016/j.jpowsour.2003.09.030.

Lankin, A. V., G. E. Norman, and V. V. Stegailov. 2010. "Atomistic Simulation of the Interaction of an Electrolyte with Graphite Nanostructures in Perspective Supercapacitors." *High Temperature* 48 (6): 837–45. doi:10.1134/S0018151X10060106.

de Levie, R. 1963. "On Porous Electrodes in Electrolyte Solutions: I. Capacitance Effects." *Electrochimica Acta* 8 (10): 751–80. doi:10.1016/0013-4686(63)80042-0.

Li, Song, Guang Feng, Pasquale F. Fulvio, Patrick C. Hillesheim, Chen Liao, Sheng Dai, and Peter T. Cummings. 2012. "Molecular Dynamics Simulation Study of the Capacitive Performance of a Binary Mixture of Ionic Liquids Near an Onion-Like Carbon Electrode." *The Journal of Physical Chemistry Letters* 3 (17): 2465–69. doi:10.1021/jz3009387.

Li, Ziyu, Kaichang Kou, Jing Xue, Chen Pan, and Guanglei Wu. 2019. "Study of Triazine-Based-Polyimides Composites Working as Gel Polymer Electrolytes in ITO-Glass Based Capacitor Devices." *Journal of Materials Science: Materials in Electronics* 30 (4): 3426–31. doi:10.1007/s10854-018-00617-x.

Lin, Changqing, Branko N. Popov, and Harry J. Ploehn. 2002. "Modeling the Effects of Electrode Composition and Pore Structure on the Performance of Electrochemical Capacitors." *Journal of the Electrochemical Society* 149 (2): A167–75. doi:10.1149/1.1431575.

Lin, Chuan, James A. Ritter, Branko N. Popov, and Ralph E. White. 1999. "A Mathematical Model of an Electrochemical Capacitor with Double-Layer and Faradaic Processes." *Journal of the Electrochemical Society* 146 (9): 3168–75. doi:10.1149/1.1392450.

Lin, Rongying, Pierre-Louis Taberna, Sébastien Fantini, Volker Presser, Carlos R. Pérez, François Malbosc, Nalin L. Rupesinghe, Kenneth B. K. Teo, Yury Gogotsi, and Patrice Simon. 2011. "Capacitive Energy Storage from −50 to 100 °C Using an Ionic Liquid Electrolyte." *The Journal of Physical Chemistry Letters* 2 (19): 2396–2401. doi:10.1021/jz201065t.

Lin, Yen-Po, and Nae-Lih Wu. 2011. "Characterization of $MnFe_2O_4/LiMn_2O_4$ Aqueous Asymmetric Supercapacitor." *Journal of Power Sources* 196 (2): 851–54. doi:10.1016/j.jpowsour.2010.07.066.

Liu, K., C. Zhu, R. Lu, and C. C. Chan. 2013. "Improved Study of Temperature Dependence Equivalent Circuit Model for Supercapacitors." *IEEE Transactions on Plasma Science* 41 (5): 1267–71. doi:10.1109/TPS.2013.2251363.

Liu, Liying, Xuehang Wang, Vladimir Izotov, Dmytro Havrykov, Illia Koltsov, Wei Han, Yulia Zozulya, et al. 2019. "Capacitance of Coarse-Grained Carbon Electrodes with Thickness up to 800 Mm." *Electrochimica Acta* 302 (April): 38–44. doi:10.1016/j.electacta.2019.02.004.

Liu, Ping, Mark Verbrugge, and Souren Soukiazian. 2006. "Influence of Temperature and Electrolyte on the Performance of Activated-Carbon Supercapacitors." *Journal of Power Sources* 156 (2): 712–18. doi:10.1016/j.jpowsour.2005.05.055.

Liu, Tongchang, W. G. Pell, and B. E. Conway. 1997. "Self-Discharge and Potential Recovery Phenomena at Thermally and Electrochemically Prepared RuO_2 Supercapacitor Electrodes." *Electrochimica Acta* 42 (23–24): 3541–52. doi:10.1016/S0013-4686(97)81190-5.

Logerais, P. O., M. A. Camara, O. Riou, A. Djellad, A. Omeiri, F. Delaleux, and J. F. Durastanti. 2015. "Modeling of a Supercapacitor with a Multibranch Circuit." *International Journal of Hydrogen Energy* 40 (39): 13725–36. doi:10.1016/j.ijhydene.2015.06.037.

Lockett, Vera, Mike Horne, Rossen Sedev, Theo Rodopoulos, and John Ralston. 2010. "Differential Capacitance of the Double Layer at the Electrode/Ionic Liquids Interface." *Physical Chemistry Chemical Physics* 12 (39): 12499–512. doi:10.1039/C0CP00170H.

Lockett, Vera, Rossen Sedev, John Ralston, Mike Horne, and Theo Rodopoulos. 2008. "Differential Capacitance of the Electrical Double Layer in Imidazolium-Based Ionic Liquids: Influence of Potential, Cation Size, and Temperature." *The Journal of Physical Chemistry C* 112 (19): 7486–95. doi:10.1021/jp7100732.

Martín, Rodolfo, Jose J. Quintana, Alejandro Ramos, and Ignacio de la Nuez. 2008. "Modeling of Electrochemical Double Layer Capacitors by Means of Fractional Impedance." *Journal of Computational and Nonlinear Dynamics* 3 (2): 021303–021303–6. doi:10.1115/1.2833909.

Martynyuk, Valeriy, and Manuel Ortigueira. 2015. "Fractional Model of an Electrochemical Capacitor." *Signal Processing* 107 (February): 355–60. doi:10.1016/j.sigpro.2014.02.021.

Masliyah, J. H., and Bhattacharjee, S. 2006. *Electrokinetic and Colloid Transport Phenomena.* Hoboken, NJ: John Wiley & Sons.

Mei, Bing-Ang, Jonathan Lau, Terri Lin, Sarah H. Tolbert, Bruce S. Dunn, and Laurent Pilon. 2018. "Physical Interpretations of Electrochemical Impedance Spectroscopy of Redox Active Electrodes for Electrical Energy Storage." *The Journal of Physical Chemistry C.* doi:10.1021/acs.jpcc.8b05241.

Melchior, Sharona A., Kumar Raju, Innocent S. Ike, Rudolph M. Erasmus, Guy Kabongo, Iakovos Sigalas, Sunny E. Iyuke, and Kenneth I. Ozoemena. 2018. "High-Voltage Symmetric Supercapacitor Based on 2D Titanium Carbide (MXene, Ti2CTx)/Carbon Nanosphere Composites in a Neutral Aqueous Electrolyte." *Journal of the Electrochemical Society* 165 (3): A501–11. doi:10.1149/2.0401803jes.

Menzel, Jakub, Elzbieta Frackowiak, and Krzysztof Fic. 2019. "Electrochemical Capacitor with Water-Based Electrolyte Operating at Wide Temperature Range." *Journal of Power Sources* 414 (February): 183–91. doi:10.1016/j.jpowsour.2018.12.080.

Merlet, Céline, Benjamin Rotenberg, Paul A. Madden, Pierre-Louis Taberna, Patrice Simon, Yury Gogotsi, and Mathieu Salanne. 2012. "On the Molecular Origin of Supercapacitance in Nanoporous Carbon Electrodes." *Nature Materials* 11 (4): 306–10. doi:10.1038/nmat3260.

Miller, J. R, and A. F. Burke. 2008. "Electrochemical Capacitors: Challenges and Opportunities for Real-World Applications." *The Electrochemical Society Interface* 17 (1): 53–57.

Miller, John R. 2006. "Electrochemical Capacitor Thermal Management Issues at High-Rate Cycling." *Electrochimica Acta* 52 (4): 1703–8. doi:10.1016/j.electacta.2006.02.056.

Mirzaeian, Mojtaba, Qaisar Abbas, Abraham Ogwu, Peter Hall, Mark Goldin, Marjan Mirzaeian, and Hassan Fathinejad Jirandehi. 2017. "Electrode and Electrolyte Materials for Electrochemical Capacitors." *International Journal of Hydrogen Energy* 42 (40): 25565–87. doi:10.1016/j.ijhydene.2017.04.241.

Morimoto, T., K. Hiratsuka, Y. Sanada, and K. Kurihara. 1996. "Electric Double-Layer Capacitor Using Organic Electrolyte." *Journal of Power Sources* 60 (2): 239–47. doi:10.1016/S0378-7753(96)80017-6.

Musolino, V., L. Piegari, and E. Tironi. 2013. "New Full-Frequency-Range Supercapacitor Model with Easy Identification Procedure." *IEEE Transactions on Industrial Electronics* 60 (1): 112–20. doi:10.1109/TIE.2012.2187412.

Muzaffar, Aqib, M. Basheer Ahamed, Kalim Deshmukh, and Jagannathan Thirumalai. 2019. "A Review on Recent Advances in Hybrid Supercapacitors: Design, Fabrication and Applications." *Renewable and Sustainable Energy Reviews* 101 (March): 123–45. doi:10.1016/j.rser.2018.10.026.

Nguyen, Phuong T. M., Chunyan Fan, D. D. Do, and D. Nicholson. 2013. "On the Cavitation-Like Pore Blocking in Ink-Bottle Pore: Evolution of Hysteresis Loop with Neck Size." *The Journal of Physical Chemistry C* 117 (10): 5475–84. doi:10.1021/jp4002912.

Niu, Jianjun, Brian E. Conway, and Wendy G. Pell. 2004. "Comparative Studies of Self-Discharge by Potential Decay and Float-Current Measurements at C Double-Layer Capacitor and Battery Electrodes." *Journal of Power Sources* 135 (1–2): 332–43. doi:10.1016/j.jpowsour.2004.03.068.

Nohara, S., H. Wada, N. Furukawa, H. Inoue, and C. Iwakura. 2006. "Self-Discharge Characteristics of an Electric Double-Layer Capacitor with Polymer Hydrogel Electrolyte." *Research on Chemical Intermediates* 32: 491–96.

Noori, Abolhassan, Maher F. El-Kady, Mohammad S. Rahmanifar, Richard B. Kaner, and Mir F. Mousavi. 2019. "Towards Establishing Standard Performance Metrics for Batteries, Supercapacitors and Beyond." *Chemical Society Reviews* 48 (5): 1272–1341. doi:10.1039/C8CS00581H.

Ozoliņš, Vidvuds, Fei Zhou, and Mark Asta. 2013. "Ruthenia-Based Electrochemical Supercapacitors: Insights from First-Principles Calculations." *Accounts of Chemical Research* 46 (5): 1084–93. doi:10.1021/ar3002987.

Pandolfo, A. G., and A. F. Hollenkamp. 2006. "Carbon Properties and Their Role in Supercapacitors." *Journal of Power Sources* 157 (1): 11–27. doi:10.1016/j.jpowsour.2006.02.065.

Parvini, Y., J. B. Siegel, A. G. Stefanopoulou, and A. Vahidi. 2016. "Supercapacitor Electrical and Thermal Modeling, Identification, and Validation for a Wide Range of Temperature and Power Applications." *IEEE Transactions on Industrial Electronics* 63 (3): 1574–85. doi:10.1109/TIE.2015.2494868.

Pech, David, Magali Brunet, Hugo Durou, Peihua Huang, Vadym Mochalin, Yury Gogotsi, Pierre-Louis Taberna, and Patrice Simon. 2010. "Ultrahigh-Power Micrometre-Sized Supercapacitors Based on Onion-like Carbon." *Nature Nanotechnology* 5 (9): 651–54. doi:10.1038/nnano.2010.162.

Pell, Wendy G., and Brian E. Conway. 2001. "Voltammetry at a de Levie Brush Electrode as a Model for Electrochemical Supercapacitor Behaviour." *Journal of Electroanalytical Chemistry* 500 (1–2): 121–33. doi:10.1016/S0022-0728(00)00423-X.

Pell, Wendy G., and Brian E. Conway. 2001. "Analysis of Power Limitations at Porous Supercapacitor Electrodes under Cyclic Voltammetry Modulation and Dc Charge." *Journal of Power Sources* 96 (1): 57–67. doi:10.1016/S0378-7753(00)00682-0.

Pell, Wendy G., Brian E. Conway, William A. Adams, and Julio de Oliveira. 1999. "Electrochemical Efficiency in Multiple Discharge/Recharge Cycling of Supercapacitors in Hybrid EV Applications." *Journal of Power Sources* 80 (1–2): 134–41. doi:10.1016/S0378-7753(98)00257-2.

Pillay, Bavanethan, and John Newman. 1996. "The Influence of Side Reactions on the Performance of Electrochemical Double-Layer Capacitors." *Journal of the Electrochemical Society* 143 (6): 1806–14. doi:10.1149/1.1836908.

Pizio, O., S. Sokołowski, and Z. Sokołowska. 2012. "Electric Double Layer Capacitance of Restricted Primitive Model for an Ionic Fluid in Slit-Like Nanopores: A Density Functional Approach." *The Journal of Chemical Physics* 137 (23): 234705. doi:10.1063/1.4771919.

Qiao, R., and N. R. Aluru. 2003. "Ion Concentrations and Velocity Profiles in Nanochannel Electroosmotic Flows." *The Journal of Chemical Physics* 118 (10): 4692–4701.

Qiao, R., and N. R. Aluru. 2004. "Charge Inversion and Flow Reversal in a Nanochannel Electro-Osmotic Flow." *Physical Review Letters* 92 (19): 198301. doi:10.1103/PhysRevLett.92.198301.

Rafik, F., H. Gualous, R. Gallay, A. Crausaz, and A. Berthon. 2007. "Frequency, Thermal and Voltage Supercapacitor Characterization and Modeling." *Journal of Power Sources* 165 (2): 928–34. doi:10.1016/j.jpowsour.2006.12.021.

Raicopol, M., C. Dascalu, C. Devan, A. L. Alexe-Ionescu, and G. Barbero. 2019. "A Simple Model of Ac Hopping Surface Conductivity in Ionic Liquids." *Electrochemistry Communications* 100 (March): 16–19. doi:10.1016/j.elecom.2019.01.010.

Rajani, Sachin Vrajlal, Vivek J. Pandya, and Varsha A. Shah. 2016. "Experimental Validation of the Ultracapacitor Parameters Using the Method of Averaging for Photovoltaic Applications." *Journal of Energy Storage* 5 (February): 120–26. doi:10.1016/j.est.2015.12.002.

Ravikovitch, Peter I., and Alexander V. Neimark. 2006. "Density Functional Theory Model of Adsorption on Amorphous and Microporous Silica Materials." *Langmuir* 22 (26): 11171–79. doi:10.1021/la0616146.

Richner, R. P. 2001. "Entwicklung Neuartig Gebundener Kohlenstoff-Materiaien Fur Electrische Dop-Pelschichtkondensatorelektronden." Swiss Federal Institute of Technology Zurich.

Ricketts, B. W., and C. Ton-That. 2000. "Self-Discharge of Carbon-Based Supercapacitors with Organic Electrolytes." *Journal of Power Sources* 89 (1): 64–69. doi:10.1016/S0378-7753(00)00387-6.

Riu, Delphine M., Nicolas Retiere, and Dirk Linzen. 2004. "Half-Order Modelling of Supercapacitors." *Conference Record of the 2004 IEEE Industry Applications Conference, 2004. 39th IAS Annual Meeting.* 4: 2550–54 vol.4.

Rizoug, N., P. Bartholomeus, and P. Le Moigne. 2010. "Modeling and Characterizing Supercapacitors Using an Online Method." *IEEE Transactions on Industrial Electronics* 57 (12): 3980–90. doi:10.1109/TIE.2010.2042418.

Sarwar, Wasim, Monica Marinescu, Nick Green, Nigel Taylor, and Gregory Offer. 2016. "Electrochemical Double Layer Capacitor Electro-Thermal Modelling." *Journal of Energy Storage* 5 (February): 10–24. doi:10.1016/j.est.2015.11.001.

Schiffer, Julia, Dirk Linzen, and Dirk Uwe Sauer. 2006. "Heat Generation in Double Layer Capacitors." *Journal of Power Sources* 160 (1): 765–72. doi:10.1016/j.jpowsour.2005.12.070.

Shabangoli, Yasin, Mohammad S. Rahmanifar, Maher F. El-Kady, Abolhassan Noori, Mir F. Mousavi, and Richard B. Kaner. 2018. "Thionine Functionalized 3D Graphene Aerogel: Combining Simplicity and Efficiency in Fabrication of a Metal-Free Redox Supercapacitor." *Advanced Energy Materials* 8 (34): 1802869. doi:10.1002/aenm.201802869.

Shinyama, K., H. Nakamura, I. Yonezu, S. Matsuta, R. Maeda, Y. Harada, and T. Nohma. 2006. "Effect of Separators on the Self-Discharge Reaction in Nickel–Metal Hydride Batteries." *Research on Chemical Intermediates* 32: 447–52.

Sikha, Godfrey, Ralph E. White, and Branko N. Popov. 2005. "A Mathematical Model for a Lithium-Ion Battery/Electrochemical Capacitor Hybrid System." *Journal of the Electrochemical Society* 152 (8): A1682–93. doi:10.1149/1.1940749.

Silva, Fernando, Cristiana Gomes, Marta Figueiredo, Renata Costa, Ana Martins, and Carlos M. Pereira. 2008. "The Electrical Double Layer at the [BMIM][PF6] Ionic Liquid/Electrode Interface – Effect of Temperature on the Differential Capacitance." *Journal of Electroanalytical Chemistry* 622 (2): 153–60. doi:10.1016/j.jelechem.2008.05.014.

Simon, P., and A. F. Burke. 2008. "Nanostructured Carbons: Double-Layer Capacitance and More." *Electrochemical Society Interface* 17: 38–43.

Simon, Patrice, and Yury Gogotsi. 2008. "Materials for Electrochemical Capacitors." *Nature Materials* 7 (11): 845–54. doi:10.1038/nmat2297.

Skipworth, Ewen, and Scott W. Donne. 2007. "Role of Graphite in Self-Discharge of Nickel(III) Oxyhydroxide." *Journal of Power Sources*, Hybrid Electric Vehicles, 174 (1): 186–90. doi:10.1016/j.jpowsour.2007.07.078.

Somasundaram, Karthik, Erik Birgersson, and Arun Sadashiv Mujumdar. 2011. "Analysis of a Model for an Electrochemical Capacitor." *Journal of the Electrochemical Society* 158 (11): A1220–30. doi:10.1149/2.062111jes.

Spyker, R. L., and R. M. Nelms. 2000. "Classical Equivalent Circuit Parameters for a Double-Layer Capacitor." *IEEE Transactions on Aerospace and Electronic Systems* 36 (3): 829–36. doi:10.1109/7.869502.

Srinivasan, Venkat, and John W. Weidner. 1999. "Mathematical Modeling of Electrochemical Capacitors." *Journal of the Electrochemical Society* 146 (5): 1650–58. doi:10.1149/1.1391821.

Srinivasan, Venkat, John W. Weidner, and Ralph E. White. 2000. "Mathematical Models of the Nickel Hydroxide Active Material." *Journal of Solid State Electrochemistry* 4 (7): 367–82. doi:10.1007/s100080000107.

Staser, John A., and John W. Weidner. 2014. "Mathematical Modeling of Hybrid Asymmetric Electrochemical Capacitors." *Journal of the Electrochemical Society* 161 (8): E3267–75. doi:10.1149/2.031408jes.

Stoller, Meryl D., Sungjin Park, Yanwu Zhu, Jinho An, and Rodney S. Ruoff. 2008. "Graphene-Based Ultracapacitors." *Nano Letters* 8 (10): 3498–3502. doi:10.1021/nl802558y.

Tevi, Tete, and Arash Takshi. 2015. "Modeling and Simulation Study of the Self-Discharge in Supercapacitors in Presence of a Blocking Layer." *Journal of Power Sources* 273 (January): 857–62. doi:10.1016/j. jpowsour.2014.09.133.

Tevi, Tete, Houman Yaghoubi, Jing Wang, and Arash Takshi. 2013. "Application of Poly (p-Phenylene Oxide) as Blocking Layer to Reduce Self-Discharge in Supercapacitors." *Journal of Power Sources* 241 (November): 589–96. doi:10.1016/j.jpowsour.2013.04.150.

Thompson, Aidan P. 2003. "Nonequilibrium Molecular Dynamics Simulation of Electro-Osmotic Flow in a Charged Nanopore." *The Journal of Chemical Physics* 119 (14): 7503–11. doi:10.1063/1.1609194.

Tian, Hao, Alireza Khanaki, Protik Das, Renjing Zheng, Zhenjun Cui, Yanwei He, Wenhao Shi, Zhongguang Xu, Roger Lake, and Jianlin Liu. 2018. "Role of Carbon Interstitials in Transition Metal Substrates on Controllable Synthesis of High-Quality Large-Area Two-Dimensional Hexagonal Boron Nitride Layers." *Nano Letters* 18 (6): 3352–61. doi:10.1021/acs.nanolett.7b05179.

Tian, Meng, Jiawen Wu, Ruihan Li, Youlin Chen, and Donghui Long. 2019. "Fabricating a High-Energy-Density Supercapacitor with Asymmetric Aqueous Redox Additive Electrolytes and Free-Standing Activated-Carbon-Felt Electrodes." *Chemical Engineering Journal* 363 (May): 183–91. doi:10.1016/j. cej.2019.01.070.

Tie, D., Shifei Huang, Jing Wang, Jianmin Ma, Jiujun Zhang, Yufeng Zhao. 2018. "Hybrid Energy Storage Devices: Advanced Electrode Materials and Matching Principles." *Energy Storage Materials*. doi:10.1016/j.ensm.2018.12.018.

Torregrossa, D., M. Bahramipanah, E. Namor, R. Cherkaoui, and M. Paolone. 2014. "Improvement of Dynamic Modeling of Supercapacitor by Residual Charge Effect Estimation." *IEEE Transactions on Industrial Electronics* 61 (3): 1345–54. doi:10.1109/TIE.2013.2259780.

Trulsson, Martin, Jenny Algotsson, Jan Forsman, and Clifford E. Woodward. 2010. "Differential Capacitance of Room Temperature Ionic Liquids: The Role of Dispersion Forces." *The Journal of Physical Chemistry Letters* 1 (8): 1191–95. doi:10.1021/jz900412t.

Vangari, M., Tonya Pryor, and Li Jiang. 2013. "Supercapacitors: Review of Materials and Fabrication Methods." *Journal of Energy Engineering* 139 (2): 72–79. doi:10.1061/(asce)ey.1943-7897.0000102.

Varghese, Julian, Hainan Wang, and Laurent Pilon. 2011. "Simulating Electric Double Layer Capacitance of Mesoporous Electrodes with Cylindrical Pores." *Journal of the Electrochemical Society* 158 (10): A1106–14. doi:10.1149/1.3622342.

Vatamanu, Jenel, Oleg Borodin, and Grant D. Smith. 2010. "Molecular Insights into the Potential and Temperature Dependences of the Differential Capacitance of a Room-Temperature Ionic Liquid at Graphite Electrodes." *Journal of the American Chemical Society* 132 (42): 14825–33. doi:10.1021/ja104273r.

Vatamanu, Jenel, Oleg Borodin, and Grant D. Smith. 2012. "Molecular Dynamics Simulation Studies of the Structure of a Mixed Carbonate/LiPF6 Electrolyte Near Graphite Surface as a Function of Electrode Potential." *The Journal of Physical Chemistry C* 116 (1): 1114–21. doi:10.1021/jp2101539.

Vazquez, S., S. M. Lukic, E. Galvan, L. G. Franquelo, and J. M. Carrasco. 2010. "Energy Storage Systems for Transport and Grid Applications." *IEEE Transactions on Industrial Electronics* 57 (12): 3881–95. doi:10.1109/TIE.2010.2076414.

Verbrugge, Mark W., and Ping Liu. 2005. "Microstructural Analysis and Mathematical Modeling of Electric Double-Layer Supercapacitors." *Journal of the Electrochemical Society* 152 (5): D79–87. doi:10.1149/1.1878052.

Wander, Matthew C. F., and Kevin L. Shuford. 2010. "Molecular Dynamics Study of Interfacial Confinement Effects of Aqueous NaCl Brines in Nanoporous Carbon." *The Journal of Physical Chemistry C* 114 (48): 20539–46. doi:10.1021/jp104972e.

Wang, B., J. A. Maciá-Agulló, D. G. Prendiville, X. Zheng, D. Liu, Y. Zhang, S. W. Boettcher, X. Ji, and G. D. Stucky. 2014. "A Hybrid Redox-Supercapacitor System with Anionic Catholyte and Cationic Anolyte." *Journal of the Electrochemical Society* 161 (6): A1090–93. doi:10.1149/2.058406jes.

Wang, Baojin, Shengbo Eben Li, Huei Peng, and Zhiyuan Liu. 2015. "Fractional-Order Modeling and Parameter Identification for Lithium-Ion Batteries." *Journal of Power Sources* 293 (October): 151–61. doi:10.1016/j.jpowsour.2015.05.059.

Wang, Faxing, Shiying Xiao, Yuyang Hou, Chenglin Hu, Lili Liu, and Yuping Wu. 2013. "Electrode Materials for Aqueous Asymmetric Supercapacitors." *RSC Advances* 3 (32): 13059–84. doi:10.1039/C3RA23466E.

Wang, Guoping, Lei Zhang, and Jiujun Zhang. 2012. "A Review of Electrode Materials for Electrochemical Supercapacitors." *Chemical Society Reviews* 41 (2): 797–828. doi:10.1039/c1cs15060j.

Wang, Hainan, and Laurent Pilon. 2011. "Accurate Simulations of Electric Double Layer Capacitance of Ultramicroelectrodes." *The Journal of Physical Chemistry C* 115 (33): 16711–19. doi:10.1021/jp204498e.

Wang, Hainan, and Laurent Pilon. 2013. "Mesoscale Modeling of Electric Double Layer Capacitors with Three-Dimensional Ordered Structures." *Journal of Power Sources* 221 (January): 252–60. doi:10.1016/j.jpowsour.2012.08.002.

Wang, Kai, Li Zhang, Bingcheng Ji, and Jinlei Yuan. 2013. "The Thermal Analysis on the Stackable Supercapacitor." *Energy* 59 (September): 440–44. doi:10.1016/j.energy.2013.07.064.

Wang, Yong-Gang, Zi-Dong Wang, and Yong-Yao Xia. 2005. "An Asymmetric Supercapacitor Using RuO_2/TiO_2 Nanotube Composite and Activated Carbon Electrodes." *Electrochimica Acta* 50 (28): 5641–46. doi:10.1016/j.electacta.2005.03.042.

Weddell, A. S., G. V. Merrett, T. J. Kazmierski, and B. M. Al-Hashimi. 2011. "Accurate Supercapacitor Modeling for Energy Harvesting Wireless Sensor Nodes." *IEEE Transactions on Circuits and Systems II: Express Briefs* 58 (12): 911–15. doi:10.1109/TCSII.2011.2172712.

Woo, Sang-Wook, Kaoru Dokko, Hiroyuki Nakano, and Kiyoshi Kanamura. 2008. "Preparation of Three Dimensionally Ordered Macroporous Carbon with Mesoporous Walls for Electric Double-Layer Capacitors." *Journal of Materials Chemistry* 18 (14): 1674–80. doi:10.1039/B717996K.

Wu, Jianzhong. 2006. "Density Functional Theory for Chemical Engineering: From Capillarity to Soft Materials." *AIChE Journal* 52 (3): 1169–93. doi:10.1002/aic.10713.

Wu, Jianzhong, Tao Jiang, De-en Jiang, Zhehui Jin, and Douglas Henderson. 2011. "A Classical Density Functional Theory for Interfacial Layering of Ionic Liquids." *Soft Matter* 7 (23): 11222–31. doi:10.1039/C1SM06089A.

Wu, Jianzhong, and Zhidong Li. 2007. "Density-Functional Theory for Complex Fluids." *Annual Review of Physical Chemistry* 58 (1): 85–112. doi:10.1146/annurev.physchem.58.032806.104650.

Wu, Nae-Lih. 2002. "Nanocrystalline Oxide Supercapacitors." *Materials Chemistry and Physics*, ICMAT 2001 Symposium C (Novel and Advanced Ceramics), 75 (1–3): 6–11. doi:10.1016/S0254-0584(02)00022-6.

Wu, Peng, and Rui Qiao. 2011. "Physical Origins of Apparently Enhanced Viscosity of Interfacial Fluids in Electrokinetic Transport." *Physics of Fluids (1994-Present)* 23 (7): 072005. doi:10.1063/1.3614534.

Xing, L., J. Vatamanu, O. Borodin, and D. Bedrov. 2013. "On the Atomistic Nature of Capacitance Enhancement Generated by Ionic Liquid Electrolyte Confined in Subnanometer Pores." *Journal of Physical Chemistry Letters* 4 (1): 132–40. doi:10.1021/jz301782f.

Xing, Lidan, Jenel Vatamanu, Oleg Borodin, Grant D. Smith, and Dmitry Bedrov. 2012. "Electrode/Electrolyte Interface in Sulfolane-Based Electrolytes for Li Ion Batteries: A Molecular Dynamics Simulation Study." *The Journal of Physical Chemistry C* 116 (45): 23871–81. doi:10.1021/jp3054179.

Xu, Bin, Feng Wu, Yuefeng Su, Gaoping Cao, Shi Chen, Zhiming Zhou, and Yusheng Yang. 2008. "Competitive Effect of KOH Activation on the Electrochemical Performances of Carbon Nanotubes for EDLC: Balance between Porosity and Conductivity." *Electrochimica Acta* 53 (26): 7730–35. doi:10.1016/j.electacta.2008.05.033.

Xu, Jun, Chunting Chris Mi, Binggang Cao, and Junyi Cao. 2013. "A New Method to Estimate the State of Charge of Lithium-Ion Batteries Based on the Battery Impedance Model." *Journal of Power Sources* 233 (July): 277–84. doi:10.1016/j.jpowsour.2013.01.094.

Yang, Hengzhao, and Ying Zhang. 2011. "Self-Discharge Analysis and Characterization of Supercapacitors for Environmentally Powered Wireless Sensor Network Applications." *Journal of Power Sources* 196 (20): 8866–73. doi:10.1016/j.jpowsour.2011.06.042.

Yoshida, Akihiko, Ichiro Tanahashi, and Atsushi Nishino. 1990. "Effect of Concentration of Surface Acidic Functional Groups on Electric Double-Layer Properties of Activated Carbon Fibers." *Carbon* 28 (5): 611–15. doi:10.1016/0008-6223(90)90062-4.

Yu, Xuewen, Jing Wang, Chenglin Wang, and Zhiqiang Shi. 2015. "A Novel Electrolyte Used in High Working Voltage Application for Electrical Double-Layer Capacitor Using Spiro-(1,1′)-Bipyrrolidinium Tetrafluoroborate in Mixtures Solvents." *Electrochimica Acta* 182 (November): 1166–74. doi:10.1016/j. electacta.2015.09.013.

Zaccagnini, P., M. Serrapede, A. Lamberti, S. Bianco, P. Rivolo, E. Tresso, C. F. Pirri, G. Barbero, and A. L. Alexe-Ionescu. 2019. "Modeling of Electrochemical Capacitors under Dynamical Cycling." *Electrochimica Acta* 296 (February): 709–18. doi:10.1016/j.electacta.2018.11.053.

Zhang, Chuanxiang, Donghui Long, Baolin Xing, Wenming Qiao, Rui Zhang, Liang Zhan, Xiaoyi Liang, and Licheng Ling. 2008. "The Superior Electrochemical Performance of Oxygen-Rich Activated Carbons Prepared from Bituminous Coal." *Electrochemistry Communications* 10 (11): 1809–11. doi:10.1016/j. elecom.2008.09.019.

Zhang, Hao, Gaoping Cao, and Yusheng Yang. 2009. "Carbon Nanotube Arrays and Their Composites for Electrochemical Capacitors and Lithium-Ion Batteries." *Energy & Environmental Science* 2 (9): 932–43. doi:10.1039/B906812K.

Zhang, Lei, Xiaosong Hu, Zhenpo Wang, Fengchun Sun, and David G. Dorrell. 2018. "A Review of Supercapacitor Modeling, Estimation, and Applications: A Control/Management Perspective." *Renewable and Sustainable Energy Reviews* 81 (January): 1868–78. doi:10.1016/j.rser.2017.05.283.

Zhang, Lei, Zhenpo Wang, Xiaosong Hu, Fengchun Sun, and David G. Dorrell. 2015. "A Comparative Study of Equivalent Circuit Models of Ultracapacitors for Electric Vehicles." *Journal of Power Sources* 274 (January): 899–906. doi:10.1016/j.jpowsour.2014.10.170.

Zhang, Lei, Zhenpo Wang, Fengchun Sun, and David G. Dorrell. 2014. "Online Parameter Identification of Ultracapacitor Models Using the Extended Kalman Filter." *Energies* 7 (5): 3204–17. doi:10.3390/ en7053204.

Zhang, Li Li, and X. S. Zhao. 2009. "Carbon-Based Materials as Supercapacitor Electrodes." *Chemical Society Reviews* 38 (9): 2520–31. doi:10.1039/B813846J.

Zhang, Qing, Chuan Cai, Jinwen Qin, and Bingqing Wei. 2014. "Tunable Self-Discharge Process of Carbon Nanotube Based Supercapacitors." *Nano Energy* 4 (March): 14–22. doi:10.1016/j.nanoen.2013.12.005.

Zhang, Qing, Jiepeng Rong, Dongsheng Ma, and Bingqing Wei. 2011. "The Governing Self-Discharge Processes in Activated Carbon Fabric-Based Supercapacitors with Different Organic Electrolytes." *Energy & Environmental Science* 4 (6): 2152–59. doi:10.1039/C0EE00773K.

Zhang, Ying, and Hengzhao Yang. 2011. "Modeling and Characterization of Supercapacitors for Wireless Sensor Network Applications." *Journal of Power Sources* 196 (8): 4128–35. doi:10.1016/j. jpowsour.2010.11.152.

Zhang, Yong, Hui Feng, Xingbing Wu, Lizhen Wang, Aiqin Zhang, Tongchi Xia, Huichao Dong, Xiaofeng Li, and Linsen Zhang. 2009. "Progress of Electrochemical Capacitor Electrode Materials: A Review." *International Journal of Hydrogen Energy* 34 (11): 4889–99. doi:10.1016/j.ijhydene.2009.04.005.

Zheng, J. P., J. Huang, and T. R. Jow. 1997. "The Limitations of Energy Density for Electrochemical Capacitors." *Journal of the Electrochemical Society* 144 (6): 2026–31. doi:10.1149/1.1837738.

Zheng, Jim P. 2003. "The Limitations of Energy Density of Battery/Double-Layer Capacitor Asymmetric Cells." *Journal of the Electrochemical Society* 150 (4): A484–92. doi:10.1149/1.1559067.

Zheng, Jim P. 2005. "Theoretical Energy Density for Electrochemical Capacitors with Intercalation Electrodes." *Journal of the Electrochemical Society* 152 (9): A1864–69. doi:10.1149/1.1997152.

Zubieta, L., and Richard Bonert. 2000. "Characterization of Double-Layer Capacitors for Power Electronics Applications." *IEEE Transactions on Industry Applications* 36 (1): 199–205. doi:10.1109/28.821816.

9

Cerium Oxide: Synthesis, Structural, Morphology, and Applications in Electrochemical Energy Devices

Ugochi K. Chime, M. Maaza, and Fabian I. Ezema

CONTENTS

9.1 Introduction 181
9.2 Crystal Structure of Cerium Oxide 182
9.3 Morphology and Synthesis of Cerium Oxide 183
    9.3.1 1-Dimensional Nanoceria 183
    9.3.2 2-Dimensional Nanoceria 185
    9.3.3 3-Dimensional Nanoceria 186
9.4 Applications of Cerium Oxide 188
    9.4.1 In Supercapacitors... 188
    9.4.2 In Fuel Cells.. 191
    9.4.3 In Photocatalytic Water Splitting 192
9.5 Conclusions 194
References 194

9.1 Introduction

In the field of nanotechnology and study of nanomaterials, cerium oxide has been receiving increased attention in recent years. With cerium considered as the most reactive element in the lanthanide series, the oxides of cerium (cerium dioxide and cerium sesquioxide) are considered as some of the most stable oxides in the field of nanotechnology, with cerium dioxide (universally used as cerium oxide) being highly stable. Cerium oxide (also called ceria) is used in various industrial and commercial processes (Younis, Chu, and Li 2016).

Cerium is a shiny, soft, rare earth metal that belongs to the lanthanide series and exists in two oxidation states (+3 and +4) (Rajeshkumar and Naik 2017). According to a report, cerium accounts for 0.0046% of the earth crust's weight, which makes it the most abundant rare earth metal (Constantin, Popescu, and Olteanu 2010). Cerium oxidizes easily in air, forming ceria or cerium oxide. Ceria usually presents as an odourless, yellow to pale, white powder depending on the size and morphology of the particles. It has a density of 7.65 g/cm^3 and a molar mass of 172.11 g. It is insoluble in water and moderately soluble in mineral acids. Its hygroscopic nature makes it useful in absorbing moisture and carbon dioxide from the atmosphere. Ceria has the ability to experience rapid redox cycles, which can be credited to its oxygen storage capacity (Maciel et al. 2012). As very good ion conductors, they are very useful in solid oxide fuel cells and as electrodes for gas sensors (Dahle and Arai 2015).

9.2 Crystal Structure of Cerium Oxide

When crystallized, CeO_2 exist in the fluorite structure with space group Fm-3m. A cell arrangement is comprised of a face-centred cubic structure of cerium cations with oxygen anions occupying the octahedral interstitial sites (Dahle and Arai 2015). In its structure, each Ce^{3+} cation is surrounded by eight neighbouring O^{2-} anions, while each O^{2-} anion is surrounded by four nearest-neighbour Ce^{4+} cations. Its importance is generally due to its ability to switch from its tetravalent state to trivalent state and vice versa with very little potential barrier, while maintaining its cubic fluorite structure. In a reducing condition, Ce^{3+} ions are introduced to the crystal lattice, thereby reducing the O^{2-} coordination number to seven. Since Ce^{3+} ions have a greater ionic radius than Ce^{4+} ions, and coupled with the creation of oxygen vacancies, distortion occurs in the local lattice symmetry (Deshpande et al. 2005). An illustration is shown in Figure 9.1.

X-ray diffraction (XRD) studies of CeO_2 as shown in Figure 9.2a present prominent lattice planes at (111), (220), and (311), with (111) being the most stable plane (Sharma 2013). These planes have been further verified using selected area electron diffraction (SAED) from transmission electron microscopy (TEM) studies (Figure 9.2b). Studies have shown that the lattice expansion of cerium oxide is dependent on the particle size, as evident in Figure 9.2c. From calculations, it was observed that a decrease in particle size led to an increase in the lattice strain. This is because in larger particles, the loss of oxygen

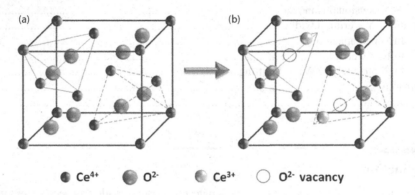

FIGURE 9.1 (a) Intact crystal structure of cerium oxide and (b) distorted crystal structure of nanoceria as a result of replacement of Ce^{4+} ions with Ce^{3+} ions and formation of oxygen vacancies. (Reproduced with permission from Deshpande, S. et al., *Appl. Phys. Lett.*, 87, 133113, 2005. Copyright 2005, AIP Publishing.)

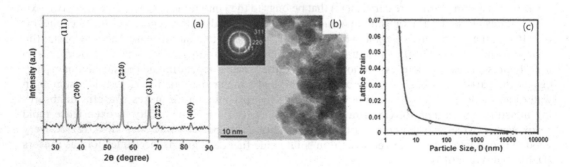

FIGURE 9.2 (a) Typical XRD pattern for cerium oxide. (Reprinted with permission from Sharma, A., *Mat. Sci. Appl.*, 4, 5. Copyright 2013, Scientific Research.); (b) HRTEM of cerium oxide with SAED inset; and (c) lattice strain as a function of particle size of cerium oxide. (Reprinted with permission from Deshpande, S. et al., *Appl. Phys. Lett.*, 87, 133113, 2005. Copyright 2005, AIP Publishing.)

atoms does not cause as much strain as in smaller particles. In fact, for nanoparticles less than 3 nm in size, the loss of a single oxygen atom leads to a very large lattice strain (Deshpande et al. 2005).

The type of precipitating agent and calcination temperature used in the fabrication of CeO_2 play a role in the particle size and crystallinity of the nanoparticle. In work done by Djuričić and Pickering, the precipitations of ceria precursors using different precipitating agents (ammonium hydroxide, hydrogen peroxide, and both) were studied. It was observed that the XRD studies showed that only powders precipitated by ammonia produced well defined peaks, while the others showed broader and less intense peaks. This peak broadening was attributed to the smaller crystallite sizes obtained from the other powders (Djuričić and Pickering 1999). In another study, XRD plots have shown that the calcination of a polyvinyl pyrrolidone/cerium nitrate ($PVP/Ce(NO_3)_3$) amorphous composite leads to the formation of crystalline CeO_2. Also, an increase in calcination temperature from 300°C to 800°C produced sharper peaks, which in turn implied cerium oxide has better crystallization (Cui et al. 2008). This temperature effect has also been observed in the synthesis of CeO_2 using a cathodic electrogeneration of base method. Increasing the bath temperature from 27°C to 80°C led to a nearly linear increase in crystallite size from 6 to 16 nm. However, an increase in the deposition current density led to a decrease in grain size (Zhou and Switzer 1996).

9.3 Morphology and Synthesis of Cerium Oxide

Cerium oxide can be produced using various methods. The composition of the precursor, solvent, surfactant, and synthesis technique has a huge role to play towards the morphology of CeO_2. In this section, we will discuss the synthesis of different dimensions of nanoceria that have been produced using different techniques.

9.3.1 1-Dimensional Nanoceria

1-Dimensional nanoceria (1D CeO_2) materials in the form of nanowires, nanorods, and nanotubes have been synthesized and studied extensively due to their unique properties. In research by Sun et al., 1-dimensional nanorods were prepared using a solvothermal method (see Figure 9.3). It was discovered

FIGURE 9.3 Ceria nanorods obtained using solvothermal synthesis technique. (Reprinted from Sun, C. et al., *Nanotechnology*, 16, 1454–1463 (2005). © IOP Publishing. Reproduced with permission. All rights reserved.)

that the creation of these nanorods was highly dependent on the solvent, precursor, and surfactant. Ethylenediamine solvents played a role in the nanorods formation, as its absence led to the synthesis of larger particles. An optimal concentration of ethylenediamine 0.84 mol l-1 was used in the formation of the ceria nanorods. A further increase in concentration led to a shortening in the length of the nanorods. Cerium (III) chloride precursors create better nanorods than cerium acetate and cerium nitrate. This is because of the presence of chloride anions, which have different kinetics in the hydrolysis that support the nucleation process of nanorod formation (Sun et al. 2005; Ji et al. 2012). Stable nanorods can be produced using a hydrothermal method. In hydrothermal synthesis, nanorods can be formed in the temperature range of 140°C–240°C. However, higher temperatures produce shorter nanorods due to the fast nucleation process (Ji et al. 2012).

The formation of ceria nanorods and nanowires has been reported using two main steps, precipitation and aging. It was reported that after ageing the precipitated cerium nitrate in ammonium hydroxide for 45 days, high definition nanowires and nanorods of diameter between 5 and 30 nanometres were formed (Han, Wu, and Zhu 2005). A surfactant-free approach has been taken towards the formation of ceria nanowires using only cerium nitrate and hydrogen peroxide as starting materials in a hydrothermal synthesis. An increase in the hydrothermal synthesis time from 0 to 20 minutes showed a change in the morphology of ceria from nanoparticles to an aggregation of nanostructured skeletons, and then to nanowires (Tang et al. 2005).

Attempts to produce ordered 1D nanoceria using a template-assisted method have been reported. In work done by La et al., anodic alumina templates were used in the formation of ordered ceria nanowires. Ce^{3+} cations in cerium nitrate solution and $C_2O_4^{2-}$ anions from oxalic acid solution were conveyed into the hexagonal nanochannels of the anodic membranes. Annealing the template at 700°C for 10 hours consequently resulted in the formation of the nanowires in the template. The anodic alumina template can be dissolved in NaOH solution, thereby producing ordered ceria nanowires (see Figure 9.4) (La et al. 2004; Wu et al. 2004). Purified carbon nanotube (CNT) removable templates have also been used in the production of ceria nanotubes for use as a reactive catalyst in carbon monoxide oxidation. This was done by dispersing purified CNT in a $Ce(NO_3)_3 \cdot 6H_2O$ pyridine solution and placing it in an autoclave, which was then heated in a 180°C oven for 24 hours. After cooling naturally, the CNT templates were removed by calcination at different temperatures (400°C–600°C) (Zhang et al. 2009).

FIGURE 9.4 Ceria nanowire arrays partially removed from anodic alumina membranes. (Reprinted from *Mater. Sci. Eng. A*, 368, La, R.-J. et al., Template synthesis of CeO_2 ordered nanowire arrays, 145–148, Copyright 2004, with permission from Elsevier.)

9.3.2 2-Dimensional Nanoceria

The synthesis of 2D ceria nanoplates has been reported. In an experiment performed by Wang et al. (2011), ceria nanoplates were produced through the thermal decomposition of cerium acetate using oleic acid and oleylamine as stabilizers at 320°C–330°C. Using sodium diphosphate as a mineralizer, square nanoplates were produced, while elongated nanoplates were formed using a sodium oleate mineralizer (Wang et al. 2011). The synthesis of ultra-thin ceria nanoribbons has been reported using an aqueous chemical method. Cerium nitrate solution was vigorously stirred in an ammonia solution, forming colloidal CeO_2 particles. The precipitated solution was boiled for about 5 seconds and immediately dipped in ice and allowed to age for 30 days. The precipitate was then centrifuged and washed in water and ethanol to separate the CeO_2 from the solution. The nanoribbons were formed after drying the precipitate overnight at 60°C in air ambience (Sreeremya et al. 2012).

Hexadecyltrimethylammonium bromide (CTAB)-assisted hydrothermal synthesis of ceria nanoplates has been reported. By varying the $CTAB/Ce^{3+}$ molar concentration, reaction time, and temperature, the shapes and sizes of the nanoplates can be controlled. For example, increasing the reaction temperature from 100°C to 160°C at a constant 1:3 $CTAB/Ce^{3+}$ molar ratio and 24 hours reaction time changed the shape of the nanoplates from cubic to hexagonal structure, as in Figure 9.5. Increasing the reaction time from 24 to 48 hours at 100°C also led to a change from a cubic shape to a hexagonal shape and an increase in the average size of these nanoplates. A change in the $CTAB/Ce^{3+}$ molar ratio from 1:3 to 2:3 at a constant 100°C and 24 hours reaction temperature and time, respectively, resulted in a complete shift from a nanoplates to nanorods formation (Pan, Zhang, and Shi 2008).

Using a simple solution combustion process, cerium oxide nanoflakes have been synthesized. Ceric ammonium nitrate solution was mixed evenly with dextrose solution and filtered. The filtrate placed in a ceramic dish was preheated at 400°C in a muffled furnace. The samples kindled and formed spongy materials, which by further annealing at 700°C produced the resultant yellow, silky films, as shown in Figure 9.6 (Umar et al. 2015). Taking a surfactant-free approach, ceria nanoflakes could also be produced by mixing cerium (III) oxide salt solution with dense nitric acid. Then, an ammonium hydroxide solution was dripped into the prepared solution at a rate of 1 ml/min while measuring the pH of the solution. After agitation for 30 mins and drying for 2 hours at 120°C, cerium oxide precipitates were formed. The resultant precipitates were then calcined for 2 hours at different temperatures (200°C–800°C). Increasing the calcination temperature resulted in the formation of larger, better defined hexagonal-shaped nanoflakes (Yu et al. 2007).

FIGURE 9.5 (a) Cubic-shaped and (b) hexagonal- and cubic-shaped ceria nanoplates. (Reprinted from *J. Solid State Chem.*, 181, Pan, C. et al., CTAB assisted hydrothermal synthesis, controlled conversion and CO oxidation properties of CeO_2 nanoplates, nanotubes, and nanorods, 1298–1306, Copyright 2008, with permission from Elsevier.)

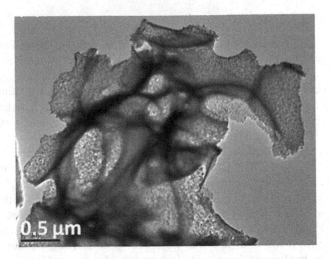

FIGURE 9.6   Cerium oxide nanoflakes obtained from simple combustion process. (Reprinted from *J. Colloid Interf. Sci.*, 454, Umar, A., Growth and properties of well-crystalline cerium oxide (CeO_2) nanoflakes for environmental and sensor applications, 61–68, Copyright 2015, with permission from Elsevier.)

9.3.3 3-Dimensional Nanoceria

Cerium oxide nanocrystals can be prepared using the hydrothermal approach. In research by Kaneko et al., cerium nitrate solution was mixed in NaOH solution and stirred for about 6 hours. After centrifuging and washing in distilled water, decanoic acid was added to the precursor for surface modification and induction of anisotropic growth of the crystals (Kaneko et al. 2007). Ultra-fine ceria nanoparticles have been synthesized using a composite-hydroxide-mediated approach. Cerium nitrate was added to molten-mixed hydroxides of sodium and potassium (NaOH:KOH = 51.5:48.5) of different temperatures and allowed to react. The mixture was then placed in an oven at different temperatures. It was discovered that an increase in temperature increased the particle size of the nanocrystals. High-resolution transmission electron microscopy (HRTEM) images in Figure 9.7b revealed ordered lattice lines over a large surface area, which indicates the face-to-face contact of the ultra-fine particles and coherent interfacial matching (Hu et al. 2006).

Cubic-shaped nanoceria can be synthesized using the typical hydrothermal method. In this case, the synthesis which involves a mixture of cerium nitrate with sodium hydroxide was synthesized for 12 hours at 180°C, as opposed to 100°C for nanorods (Xue et al. 2018). In research by Huang et al., CeO_2 nanocubes (see Figure 9.8) were formed from the hydrolysis of cerium nitrate solution in acetate-acetic acid. When coated with chitosan, these nanocubes can be very effective in accelerating cutaneous wound healing and preventing inflammation (Huang et al. 2018).

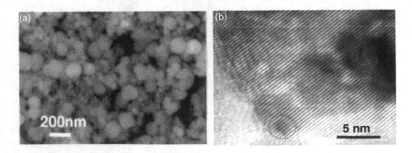

FIGURE 9.7   (a) Ultra-fine nanoceria and (b) HTREM image showing fine lattice lines (Reprinted from Hu, C. et al., *Nanotechnology*, 17, 5983–5987 (2006). © IOP Publishing. Reproduced with permission. All rights reserved.)

FIGURE 9.8 Ceria nanocubes obtained from hydrothermal treatment of cerium nitrate in acetate-acetic acid. (Huang, X., et al., *Inorg. Chem. Front.*, 2018. Reproduced by permission of the Royal Society of Chemistry.)

The formation of ceria hollow spheres has been reported using silica spheres as templates. In this procedure, cerium nitrate was dissolved in water containing silica sol before hydrothermal treatment at 200°C for 24 hours, air drying, and cooling at room temperature. The resultant precipitates were then treated with sodium hydroxide to remove the silica sol, hence, leaving only the ceria hollow spheres. These hollow spheres exhibit properties different from normal particles, which make them very useful as capsule agents in drug delivery (Guo, Jian, and Du 2009). Figure 9.9 shows an example of ceria hollow spheres obtained by this process.

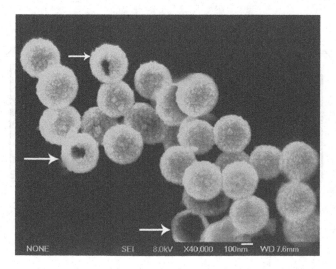

FIGURE 9.9 CeO$_2$ hollow spheres obtained from silica sphere template. (Reprinted from *Scripta Materialia*, 61, Guo, Z. et al., A simple method to controlled synthesis of CeO$_2$ hollow microspheres, 48–51, Copyright 2009, with permission from Elsevier.)

9.4 Applications of Cerium Oxide

9.4.1 In Supercapacitors

Supercapacitors have a unique advantage over batteries, their high-power density. They can be classified into two main types: the electrochemical double-layer capacitors and pseudocapacitors. The electrochemical double-layer capacitor is comprised of two conducting plates as electrodes with charge accumulators stationed between the electrodes and the electrolyte. When the charging process is on, the electrons move from the cathode to the anode through an external connection. In the electrolyte, the positively charged ions move to the cathode, while the negatively charged ions move to the anode. During discharge, the electrons and ions move in the opposite direction. In this type of supercapacitor, there are no redox reactions because charges are not transferred between the electrode and the electrolyte across the interface. This makes electrochemical double-layer capacitors stable and have a longer lifespan. In pseudocapacitors, however, Faradaic redox reactions occur. The application of potential in a pseudocapacitor results in fast and reversible reactions on the electrodes, which generate charges and result in charge transfer across layers. Pseudocapacitors usually produce higher specific capacitance and energy density because the reactions not only occur at the electrode surface, but also within the electrodes. However, pseudocapacitors tend to have lower stability during cycling due to the redox reactions involved (Shi et al. 2014; Okonkwo, Collins, and Okonkwo 2017). Cerium oxide-based supercapacitors display pseudocapacitive behaviour due to their impressive redox characteristics.

Despite its good redox properties, little work has been done to explore the pseudocapacitive properties of cerium oxide due to its relatively low theoretical capacitance of 560 Fg^{-1} (Maiti, Pramanik, and Mahanty 2014). The capacitance values obtained from experiments depend on the type of electrolytes used, the mode of fabrication of the CeO_2 electrodes, the presence of redox enhancers, or the use of CeO_2 as a composite with another material. In research done by Maheswari and Muralidharan, the electrochemical behaviour of four different aqueous electrolytes for ceria nanoparticles (NaCl, KCl, Na_2SO_4, and K_2SO_4) was studied. It was observed that a NaCl electrolyte produced the highest specific capacitance of 523 Fg^{-1} and the lowest charge transfer resistance of 0.84 Ω, followed closely by Na_2SO_4 at 502 Fg^{-1} and 0.9 Ω, respectively. This is as a result of the lower ionic radius of Na^+, which enables better diffusion of ions, which in turn increases capacitance (Maheswari and Muralidharan 2015). Figure 9.10 shows a schematic representation of the electrochemical process of CeO_2 electrodes. The hexagonal structure of the CeO_2 reduces the diffusion length of the cations and, hence, improves the rate of diffusion of Na^+ ions in the material.

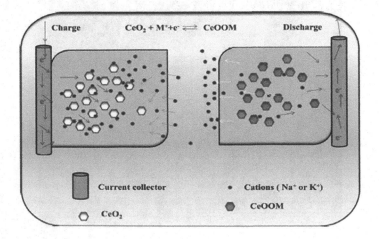

FIGURE 9.10 Schematic representation of the electrochemical process using CeO_2 electrodes. (Reprinted with permission from Maheswari, N., and G. Muralidharan, *Energy Fuels*, 29, 8246–8253, 2015. Copyright 2015 American Chemical Society.)

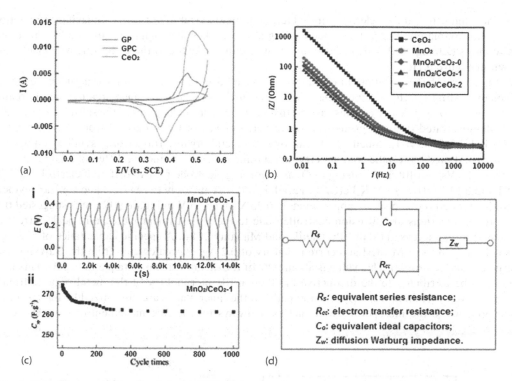

FIGURE 9.11 (a) The cyclic voltammetry curves of pristine CeO_2, 3D graphene (GP), and CeO_2-deposited 3D graphene (GPC) measured at 5 mVs^{-1} scan rate area. (Wang et al., *Dalton Trans.*, 40, 6388–6391, 2011. Reproduced by permission of The Royal Society of Chemistry.); (b) the Bode curves of CeO_2, MnO and MnO/CeO_2 prepared using different methods; (c) (i) charge-discharge curves of MnO/CeO_2 and (ii) Specific capacitance vs cycle time; and (d) equivalent circuit of CeO_2 and the composite MnO/CeO_2. (Reprinted from *Chem. Eng. J.*, 286, Zhang, et al., Hierarchical porous MnO2/CeO_2 with high performance for supercapacitor electrodes, 139–149, Copyright 2016, with permission from Elsevier.)

A way of improving the capacitance of CeO_2 is by forming composites with another material with known good conductivity, such as graphene and manganese oxide. Wang et al. (2011) synthesized a cerium oxide/graphene nanocomposite by using 3D graphene material as a base and depositing CeO_2 nanoparticles onto it. The cyclic voltammetry curves of pristine CeO_2, 3D graphene, and CeO_2-deposited 3D graphene measured at a 5 mVs^{-1} scan rate area, as shown in Figure 9.11a. It can be observed that the redox reaction of the CeO_2/graphene composite is significantly larger than those of the pure CeO_2 and 3D graphene. This indicates that although the pure CeO_2 nanoparticles have poor conductivity, depositing a thin layer of CeO_2 on graphene produces a much better conductivity, which results in a much larger peak current. It was observed that the overall capacitance and power density of the composite material were much higher than that of the sum of the individual materials. This is because there is a great synergy between graphene and CeO_2. Graphene is very conductive, but hydrophobic, while CeO_2 has poor conductivity, but is hydrophilic. Hence, a combination of both produces a better suited material for electrochemical activities (Wang et al. 2011). Similarly, in work by H. Zhang et al., the Bode curves of CeO_2, MnO, and MnO/CeO_2 prepared using different methods were compared (Figure 9.11b). It can be seen that the absolute impedance of the composites are smallest. Between 10^{-2} and 10^0 Hz, the Bode plot is a straight line with a slope of approximately –1. This suggests that the effect of resistances is negligible and that |Z| is equal to the capacitive reactance X_c. Since $X_c = 1/2\pi f C_{sp}$, it implies that as the impedance is reduced, the specific capacitance is increased. The charge-discharge curves of MnO/CeO_2 in Figure 9.11c(i) have good symmetry and are easily reproduced. The specific capacitance C_{sp} decreased only by 1.3% after

20 cycles and after 1000 cycles (Figure 9.11c[ii]), remained stabilized with a capacitance retention rate of approximately 93.9%. This good cycling ability makes the composite suitable for capacitive applications (Zhang et al. 2016). The equivalent circuit of CeO_2 and the composite MnO/CeO_2 are shown in Figure 9.11d.

CeO_2 derived from metal organic frameworks have been reported to produce pseudocapacitance values exceeding the theoretical value. This is because of the larger surface area the metal organic frameworks produce, which enables easy diffusion of ions within the pores of the substructures. Maiti et al. demonstrated the high pseudocapacitive performance of CeO_2 derived from [Ce(1,3,5-BTC) $(H_2O)_6$] (1,3,5-BTC = 1,3,5-benzene-tricarboxylate) metal organic frameworks synthesized using a solvothermal method. Cyclic voltammetry measurements were obtained at different scan rates ranging from 2 mVs^{-1} to 100 mVs^{-1} using CeO_2 as a working electrode for two different electrolytes, 3 M KOH and 3 M KOH + 0.1 M $K_4Fe(CN)_6$. Faradaic redox switches between Ce^{3+} and Ce^{4+} are evident from the defined redox peaks at approximately 0.23 V and 0.35 V in Figure 9.12a. It is suggested that these redox reactions are diffusion controlled due to the almost linear increase in peak current as the scan rate is increased (Maiti, Pramanik, and Mahanty 2014). A specific capacitance of 502 Fg^{-1} was achieved in the 3 M KOH at a current density of 0.2 Ag^{-1} (as calculated from the Galvanostatic charge discharge profile). This value is about 92% of the theoretical capacitance for CeO_2. This high value can be attributed to the distinctive brick-upon-tile morphology of the metal organic frameworks-derived cerium oxide. Ion diffusion paths to the inner planes are minimized by ion buffering reservoirs, while the mesoporous walls act as pathways for the free movement of ions. However,

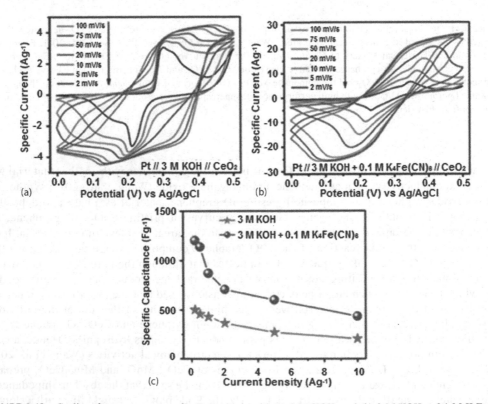

FIGURE 9.12 Cyclic voltammetry curves for CeO_2 anode with (a) 3 M KOH, and (b) 3 M KOH + 0.1 M $K_4Fe(CN)_6$ electrolytes, and (c) specific capacitance vs current density comparing the two electrolytes. (Maiti et al., *Chem. Commun.*, 50, 11717–11720, 2014. Copyright 2014, Reproduced by permission of The Royal Society of Chemistry.)

being materials of poor conductivity, a redox active electrolyte additive such as $K_4Fe(CN)_6$ in KOH electrolyte solution is usually used as a catalyst to improve redox reaction. The introduction of 0.1 M $K_4Fe(CN)_6$ to the 3 M KOH electrolyte resulted in an extraordinarily high capacitance of 1204 Fg^{-1} at 0.2 Ag^{-1} (Figure 9.12c). This high capacitance is due to $Fe(CN)_6^{4-}/Fe(CN)_6^{3-}$ in the electrolyte matching the redox potentials of 0.20/0.37 V, thereby complementing the Ce^{4+}/Ce^{3+} redox reactions (Maiti, Pramanik, and Mahanty 2014).

9.4.2 In Fuel Cells

Research in solid oxides fuel cells (SOFCs) have been on the increase in recent years, as they have the potential to provide clean, sustainable, and reliable electricity. Studies have shown that cerium oxide-based ion conductors have the tendency to provide huge resistance to carbon depositions, leading to a continuous flow of dry hydrocarbon fuels to anodes, which in turn provide continuous electricity (Younis, Chu, and Li 2016; Marina and Mogensen 1999; Gorte and Vohs 2009). A study showed that using SOFCs together with copper and ceria composites yielded a prominent level of electrochemical oxidation of hydrocarbons, such as methane. The anodes used in these studies were made up of a combination of $Cu/CeO_2/YSZ$, with the CeO_2 used to kick-start the catalytic activity for the oxidation reactions in the setup. The anodes were highly stable in the redox reactions and were able to tolerate high concentrations of sulphur without any negative impacts on performance (Park, Vohs, and Gorte 2000; He, Gorte, and Vohs 2005). Reports have shown that ceria-based ceramics exhibit mixed ionic and electronic conductivity. Another reason why the oxides of cerium are preferred in fuel cells is the tendency of the oxides to have well defined spatial attributes (He, Gorte, and Vohs 2005; Chueh et al. 2012). Further research showed that ceria-metal structures were composed of defined interferences in electrocatalytic processes. The research concluded that the design of the ceria nanostructures led to good electrochemical processes, which in turn led to reliable performance (Chueh et al. 2012). Samarium-doped cerium oxide (SDC) has been reported as effective as a buffer layer between a $La_{0.6}Sr_{0.4}Co_{0.2}Fe_{0.8}O_{3-\delta}$ (LSCF) cathode and YSZ electrolyte. This buffer layer prevents the easy reaction between cathode and electrolyte, which improves the stability of the fuel cell. Figure 9.13a shows a comparison of the cell voltages of the SOFCs using blank LSCF and SDC-coated LSCF cathodes. It can be seen that the cell voltage is higher for the LSCF that was infiltrated with SDC. Also, there was a reduction in degradation rate from 5% to 2%. This implies that the presence of SDC not only improved the cell performance, but also produced a more stable cell (Nie et al. 2010).

Cerium oxide is also important in direct alcohol fuel cells (DAFCs). A study had been carried out to observe the ethanol electrooxidation characteristics of electrocatalysts for DAFCs, by using impregnation-hydrothermal methods. The study involved doping three transition metals (Fe, Ni, and Pt) using cerium oxide to prepare nano-composite electrocatalysts. The combinations of the metals and oxides in the experiments were $Fe/PtCeO_x$, $Pt/PdCeO_x$ with PtRu/C. The results showed that mixing commercial PtRu/C with these electrocatalysts greatly improved fuel cell performances. High power densities and performances were also obtained for $Ni/PtCeO_x$ + PtRu/C, $Pt/FeCeO_x$ + PtRu/C, and $Pt/PdCeO_x$ + PtRu/C combinations (Figure 9.13b) (Tapan, Cacan, and Varışlı 2014).

Another study showed the importance of cerium oxide in a membrane electron assembly (MEA). During the process, cerium oxide was used to support the MEA setup by pressing CeO_2-coated electrodes and a perfluorosulfonic acid ionomer membrane together. This process is essential in the creation of polymer electrolyte fuel cells. Findings from the research suggest that by combining the setup with an accelerated stress test, the cerium oxide-supported MEA showed a six times longer lifetime and 40 times lower fluoride emission rate when compared to a MEA setup without cerium oxide (Figure 9.13c and d). The thickness and tensile strength of the membrane in the MEA supported by cerium oxide were also retained. The cerium used in the setup supported the MEA membrane and prevented chemical and mechanical membrane degradation (Lim et al. 2015).

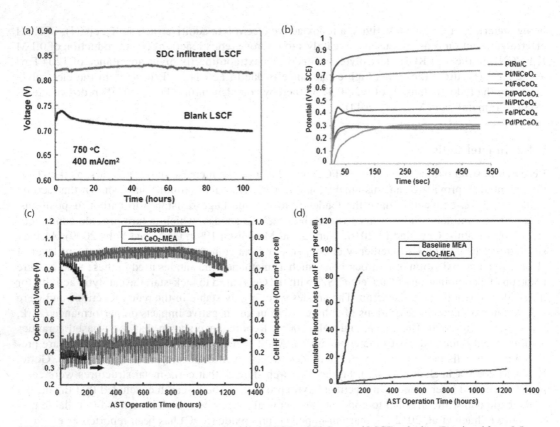

FIGURE 9.13   Cell voltages of the (a) SOFC using blank LSCF and SDC-coated LSCF cathodes. (Reprinted from *J. Power Sources*, 195, Nie, L. et al., $La_{0.6}Sr_{0.4}Co_{0.2}Fe_{0.8}O_{3-\delta}$ cathodes infiltrated with samarium-doped cerium oxide for solid oxide fuel cells, 4704–4708, Copyright 2010, with permission from Elsevier.); (b) DAFC using CeO_x doped with transition metals (Fe, Ni, and Pt). (Reprinted with permission from Tapan, N.A., et al., *Int. J. Electrochem. Sci.*, 9, 4440–4464. Copyright 2014, ESG.) and; (c) MEA CeO_2; and (d) cumulative fluoride loss for MEA CeO_2. (Reprinted with permission from Lim, C. et al., *ECS Electrochem. Lett.*, 4, F29–F31. Copyright 2015, ECS.)

As can be seen from the various researches shown here, the importance of cerium oxide in fuel cells is clear. The use of cerium oxide in proton exchange membrane FCs, DAFCs, AFCs, phosphoric AFCs, molten carbonate FCs, and SOFCs greatly improve their efficiencies and performances (Mori et al. 2012).

9.4.3 In Photocatalytic Water Splitting

Cerium has long been considered as one of the best elements used in water splitting and one of the key components in the process of hydrogen gas production for fuel (Naghavi et al. 2017). Some researchers have even gone a step further to name it as the best element used in the water splitting process. Interest in alternative sources of energy to replace gasoline has caused researchers to work towards a 'hydrogen economy' concept, with the hope of commercializing hydrogen as fuel for transportation. Water splitting is an effective way of generating clean hydrogen. The process involves the use of heat generated from solar radiation to heat cerium oxide to temperatures between 1000°C and 1500°C to enable the separation of the hydrogen from oxygen.

Most of the hydrogen produced around the globe is sourced from natural gas steam reformation, coal gasification, oil reforming, and from the process of water electrolysis. To generate sustainable hydrogen for the future, there is a need to adopt renewable energy approaches (Walter 2017; Bright and Wu 2017; Binotti et al. 2017). It has been argued that photo-electrochemical water splitting is an efficient and sustainable way of producing pure hydrogen, especially when cerium oxide is used (Walter 2017; Bright and Wu 2017). The advantages of using cerium oxide in water splitting are zero corrosions, low recombination after reactions, fast kinetics, solid heat recovery, and its ability to maintain a solid state throughout reactions (Binotti et al. 2017).

Using a temperature programmed desorption (TPD) model, catalytic studies were done to understand how effectively CeO_x can be oxidized by water. The first cycle of TPD on c-Ce_2O_3 interacting with H_2O yielded a high concentration of hydrogen (up to 8×1014 cm^{-2}), which can be collected in bulk in c-Ce_2O_3 and subsequently released at a high temperature. Figure 9.14a shows the H_2 recombination yield obtained using TPD for c-Ce_2O_3, $CeO_{1.67}$, and i-Ce_7O_{12}. It can be seen that $CeO_{1.67}$ and i-Ce_7O_{12} produced much less H_2 than c-Ce_2O_3. This is attributed to the lower concentration of O vacancies on the c-Ce_2O_3 surface (Dvořák et al. 2018). Another researcher compared the performance of CeO_2 nanoparticles produced using a biopolymer templated synthesis method (CeO_2 (A)), commercial CeO_2 nanoparticles (CeO_2 (B)), and Au-doped CeO_2 (A) and (B) at different concentrations of Au. It was observed that the specially prepared CeO_2 showed a higher oxygen evolution rate compared to the commercial nanoparticles. However, the evolution of oxygen was only possible in the ultra-violet (UV) region of the light spectrum. Doping CeO_2 with Au leads to impressive oxygen evolution in the visible light region (Figure 9.14b and c) (Primo et al. 2011). Similar results have been obtained by Clavijo-Chaparro et al., where doping nanorods and nanocubes of CeO_2 with Cu led to an increased hydrogen evolution (Figure 9.14d) (Clavijo-Chaparro et al. 2016).

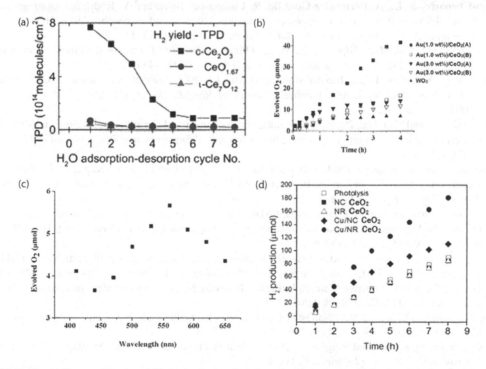

FIGURE 9.14   (a) H_2 recombination yield obtained using TPD for c-Ce_2O_3, $CeO_{1.67}$, and i-Ce_7O_{12}. (Reprinted with permission from Dvořák, F., et al., *ACS Catal.*, 8, 4354–4363, 2018. Copyright 2018 American Chemical Society.); (b) O_2 evolution using Au-doped CuO_2 at 400 nm; (c) O_2 evolution for Au(1.0 wt%)CeO_2 (A) (CeO_2 nanoparticles produced using a biopolymer templated synthesis method as stated in the article). (Reprinted with permission from Primo, A. et al., *J. Am. Chem. Soc.*, 133, 6930–6933, 2011. Copyright 2011 American Chemical Society.) (d) H_2 production using CeO_2 and Cu-doped CeO_2. (Reprinted from *J. Mol. Catal. Chem.*, 423, Clavijo-Chaparro, S.L. et al., Water Splitting Behavior of Copper-Cerium Oxide Nanorods and Nanocubes Using Hydrazine as a Scavenging Agent, 143–150, Copyright 2016, with permission from Elsevier.)

9.5 Conclusions

Cerium oxide is an important semiconductor due to its ability to switch between its trivalent and tetravalent states. CeO_2 can take different nanostructures, such as nanoflakes, nanorods, nanocubes, etc., and these different morphologies are highly dependent on the fabrication technique. Though on its own, cerium oxide has a low specific capacitance, research has shown that doping ceria with certain metals and semiconductors increases its performance tremendously. Very few studies with dopants to study their electrochemical properties have been done. Hence, it is recommended more interest should be explored in this subject.

REFERENCES

Binotti, Marco, Gioele Di Marcoberardino, Mauro Biassoni, and Giampaolo Manzolini. 2017. "Solar Hydrogen Production with Cerium Oxides Thermochemical Cycle." *In AIP Conference Proceedings*, 1850:100002. AIP Publishing.

Bright, Joeseph, and Nianqiang Wu. 2017. "Photoelectrochemical Water Splitting of Metal Oxide Photoanode Enhanced with a Cerium (III/IV) Redox Mediator." *In Meeting Abstracts*, 1899–1899. The Electrochemical Society.

Chueh, William C., Yong Hao, WooChul Jung, and Sossina M. Haile. 2012. "High Electrochemical Activity of the Oxide Phase in Model Ceria–Pt and Ceria–Ni Composite Anodes." *Nature Materials* 11 (2): 155.

Clavijo-Chaparro, S. L., A. Hernández-Gordillo, R. Camposeco-Solis, and V. Rodríguez-González. 2016. "Water Splitting Behavior of Copper-Cerium Oxide Nanorods and Nanocubes Using Hydrazine as a Scavenging Agent." *Journal of Molecular Catalysis A: Chemical* 423: 143–50.

Constantin, Virgil, Ana-Maria Popescu, and Mircea Olteanu. 2010. "Electrochemical Studies on Cerium (III) in Molten Fluoride Mixtures." *Journal of Rare Earths* 28 (3): 428–434.

Cui, Qizheng, Xiangting Dong, Jinxian Wang, and Mei Li. 2008. "Direct Fabrication of Cerium Oxide Hollow Nanofibers by Electrospinning." *Journal of Rare Earths* 26 (5): 664–69. doi.org/10.1016/S1002-0721(08)60158-1.

Dahle, Jessica T., and Yuji Arai. 2015. "Environmental Geochemistry of Cerium: Applications and Toxicology of Cerium Oxide Nanoparticles." *International Journal of Environmental Research and Public Health* 12 (2): 1253–1278.

Deshpande, Sameer, Swanand Patil, Satyanarayana V.N.T. Kuchibhatla, and Sudipta Seal. 2005. "Size Dependency Variation in Lattice Parameter and Valency States in Nanocrystalline Cerium Oxide." *Applied Physics Letters* 87 (13): 133113.

Djuričić, Boro, and Stephen Pickering. 1999. "Nanostructured Cerium Oxide: Preparation and Properties of Weakly-Agglomerated Powders." *Journal of the European Ceramic Society* 19 (11): 1925–1934. doi:10.1016/S0955-2219(99)00006-0.

Dvořák, Filip, Lucie Szabová, Viktor Johánek, Matteo Farnesi Camellone, Vitalii Stetsovych, Mykhailo Vorokhta, Andrii Tovt, Tomáš Skála, Iva Matolínová, and Yoshitaka Tateyama. 2018. "Bulk Hydroxylation and Effective Water Splitting by Highly Reduced Cerium Oxide: The Role of O Vacancy Coordination." *ACS Catalysis* 8: 4354–4363.

Gorte, R. J., and J. M. Vohs. 2009. "Nanostructured Anodes for Solid Oxide Fuel Cells." *Current Opinion in Colloid & Interface Science* 14 (4): 236–244.

Guo, Zhiyan, Fangfang Jian, and Fanglin Du. 2009. "A Simple Method to Controlled Synthesis of CeO_2 Hollow Microspheres." *Scripta Materialia* 61 (1): 48–51.

Han, Wei-Qiang, Lijun Wu, and Yimei Zhu. 2005. "Formation and Oxidation State of CeO_{2-x} Nanotubes." *Journal of the American Chemical Society* 127 (37): 12814–12815.

He, Hongpeng, Raymond J. Gorte, and John M. Vohs. 2005. "Highly Sulfur Tolerant Cu-Ceria Anodes for SOFCs." *Electrochemical and Solid-State Letters* 8 (6): A279–A280.

Hu, Chenguo, Zuwei Zhang, Hong Liu, Puxian Gao, and Zhong Lin Wang. 2006. "Direct Synthesis and Structure Characterization of Ultrafine CeO_2 Nanoparticles." *Nanotechnology* 17 (24): 5983.

Huang, Xue, Lin-Dong Li, Guang-Ming Lyu, Bai-Yu Shen, Yan-Fei Han, Jing-Lin Shi, Jia-Li Teng, Li Feng, Shao-Yan Si, and Ji-Hua Wu. 2018. "Chitosan-Coated Cerium Oxide Nanocubes Accelerate Cutaneous Wound Healing by Curtailing Persistent Inflammation." *Inorganic Chemistry Frontiers* 5: 386–393.

Ji, Zhaoxia, Xiang Wang, Haiyuan Zhang, Sijie Lin, Huan Meng, Bingbing Sun, Saji George, Tian Xia, André E. Nel, and Jeffrey I. Zink. 2012. "Designed Synthesis of CeO_2 Nanorods and Nanowires for Studying Toxicological Effects of High Aspect Ratio Nanomaterials." *ACS Nano* 6 (6): 5366–5380.

Kaneko, Kenji, Koji Inoke, Bert Freitag, Ana B. Hungria, Paul A. Midgley, Thomas W. Hansen, Jing Zhang, Satoshi Ohara, and Tadafumi Adschiri. 2007. "Structural and Morphological Characterization of Cerium Oxide Nanocrystals Prepared by Hydrothermal Synthesis." *Nano Letters* 7 (2): 421–425.

La, Ren-Jiang, Zhong-Ai Hu, Hu-Lin Li, Xiou-Li Shang, and Yu-Ying Yang. 2004. "Template Synthesis of CeO_2 Ordered Nanowire Arrays." *Materials Science and Engineering: A* 368 (1–2): 145–148.

Lim, C., A. Sadeghi Alavijeh, M. Lauritzen, J. Kolodziej, S. Knights, and E. Kjeang. 2015. "Fuel Cell Durability Enhancement with Cerium Oxide under Combined Chemical and Mechanical Membrane Degradation." *ECS Electrochemistry Letters* 4 (4): F29–F31.

Maciel, Cristhiane Guimarães, Tatiana de Freitas Silva, Marcelo Iuki Hirooka, Mohamed Naceur Belgacem, and Jose Mansur Assaf. 2012. "Effect of Nature of Ceria Support in CuO/CeO_2 Catalyst for PROX-CO Reaction." *Fuel* 97: 245–252.

Maheswari, Nallappan, and Gopalan Muralidharan. 2015. "Supercapacitor Behavior of Cerium Oxide Nanoparticles in Neutral Aqueous Electrolytes." *Energy & Fuels* 29 (12): 8246–8253.

Maiti, Sandipan, Atin Pramanik, and Sourindra Mahanty. 2014. "Extraordinarily High Pseudocapacitance of Metal Organic Framework Derived Nanostructured Cerium Oxide." *Chemical Communications* 50 (79): 11717–11720.

Marina, Olga A., and Mogens Mogensen. 1999. "High-Temperature Conversion of Methane on a Composite Gadolinia-Doped Ceria–Gold Electrode." *Applied Catalysis A: General* 189 (1): 117–126.

Mori, Toshiyuki, Ding Rong Ou, Jin Zou, and John Drennan. 2012. "Present Status and Future Prospect of Design of Pt–Cerium Oxide Electrodes for Fuel Cell Applications." Progress in Natural Science: *Materials International* 22 (6): 561–571.

Naghavi, S. Shahab, Antoine A. Emery, Heine A. Hansen, Fei Zhou, Vidvuds Ozolins, and Chris Wolverton. 2017. "Giant Onsite Electronic Entropy Enhances the Performance of Ceria for Water Splitting." *Nature Communications* 8 (1): 285. doi:10.1038/s41467-017-00381-2.

Nie, Lifang, Mingfei Liu, Yujun Zhang, and Meilin Liu. 2010. "$La_{0.6}Sr_{0.4}Co_{0.2}Fe_{0.8}O_{3-\delta}$ Cathodes Infiltrated with Samarium-Doped Cerium Oxide for Solid Oxide Fuel Cells." *Journal of Power Sources* 195 (15): 4704–4708.

Okonkwo, P.C., E. Collins, and E. Okonkwo. 2017. "18 – Application of Biopolymer Composites in Super Capacitor." In *Biopolymer Composites in Electronics*, edited by K.K. Sadasivuni, D. Ponnamma, J. Kim, J.-J. Cabibihan, and M.A. AlMaadeed, 487–503. Elsevier. doi:10.1016/B978-0-12-809261-3.00018-8.

Pan, Chengsi, Dengsong Zhang, and Liyi Shi. 2008. "CTAB Assisted Hydrothermal Synthesis, Controlled Conversion and CO Oxidation Properties of CeO_2 Nanoplates, Nanotubes, and Nanorods." *Journal of Solid State Chemistry* 181 (6): 1298–1306.

Park, Seungdoo, John M. Vohs, and Raymond J. Gorte. 2000. "Direct Oxidation of Hydrocarbons in a Solid-Oxide Fuel Cell." *Nature* 404 (6775): 265.

Primo, Ana, Tiziana Marino, Avelino Corma, Raffaele Molinari, and Hermenegildo Garcia. 2011. "Efficient Visible-Light Photocatalytic Water Splitting by Minute Amounts of Gold Supported on Nanoparticulate CeO_2 Obtained by a Biopolymer Templating Method." *Journal of the American Chemical Society* 133 (18): 6930–6933.

Rajeshkumar, S., and Poonam Naik. 2017. "Synthesis and Biomedical Applications of Cerium Oxide Nanoparticles – A Review." *Biotechnology Reports* 17: 1–5.

Sharma, Ashutosh. 2013. "Effect of Synthesis Routes on Microstructure of Nanocrystalline Cerium Oxide Powder." *Materials Sciences and Applications* 4 (9): 5. //www.scirp.org/journal/PaperInformation. aspx?PaperID=36550.

Shi, Fan, Lu Li, Xiu-li Wang, Chang-dong Gu, and Jiang-ping Tu. 2014. "Metal Oxide/Hydroxide-Based Materials for Supercapacitors." *RSC Advances* 4 (79): 41910–41921. doi:10.1039/C4RA06136E.

Sreeremya, Thadathil S., Asha Krishnan, Srividhya J. Iyengar, and Swapankumar Ghosh. 2012. "Ultra-Thin Cerium Oxide Nanostructures through a Facile Aqueous Synthetic Strategy." *Ceramics International* 38 (4): 3023–3028.

Sun, Chunwen, Hong Li, Huairuo Zhang, Zhaoxiang Wang, and Liquan Chen. 2005. "Controlled Synthesis of CeO_2 Nanorods by a Solvothermal Method." *Nanotechnology* 16 (9): 1454.

Tang, Bo, Linhai Zhuo, Jiechao Ge, Guangli Wang, Zhiqiang Shi, and Jinye Niu. 2005. "A Surfactant-Free Route to Single-Crystalline CeO$_2$ Nanowires." *Chemical Communications*, 28: 3565–3567.

Tapan, Niyazi Alper, Umut B. Cacan, and Dilek Varışlı. 2014. "Ceria Based Nano-Composite Synthesis for Direct Alcohol Fuel Cells." *International Journal of Electrochemical Science* 9: 4440–4464.

Umar, Ahmad, R. Kumar, M. S. Akhtar, G. Kumar, and S. H. Kim. 2015. "Growth and Properties of Well-Crystalline Cerium Oxide (CeO$_2$) Nanoflakes for Environmental and Sensor Applications." *Journal of Colloid and Interface Science* 454: 61–68.

Walter. 2017. "Cerium's Properties Enhance Water Splitting." Research & Development. October 24. https://www.rdmag.com/article/2017/10/ceriums-properties-enhance-water-splitting.

Wang, Dianyuan, Yijin Kang, Vicky Doan-Nguyen, Jun Chen, Rainer Küngas, Noah L. Wieder, Kevin Bakhmutsky, Raymond J. Gorte, and Christopher B. Murray. 2011. "Synthesis and Oxygen Storage Capacity of Two-Dimensional Ceria Nanocrystals." *Angewandte Chemie International Edition* 50 (19): 4378–4381.

Wang, Yi, Chun Xian Guo, Jiehua Liu, Tao Chen, Hongbin Yang, and Chang Ming Li. 2011. "CeO$_2$ Nanoparticles/Graphene Nanocomposite-Based High Performance Supercapacitor." *Dalton Transactions* 40 (24): 6388–6391.

Wu, G. S., T. Xie, X. Y. Yuan, B. C. Cheng, and L. D. Zhang. 2004. "An Improved Sol–Gel Template Synthetic Route to Large-Scale CeO$_2$ Nanowires." *Materials Research Bulletin* 39 (7–8): 1023–1028.

Xue, Yingfei, Sricharani Rao Balmuri, Akhil Patel, Vinayak Sant, and Shilpa Sant. 2018. "Synthesis, Physico-Chemical Characterization, and Antioxidant Effect of PEGylated Cerium Oxide Nanoparticles." *Drug Delivery and Translational Research* 8 (2): 357–367.

Younis, Adnan, Dewei Chu, and Sean Li. 2016. "Cerium Oxide Nanostructures and Their Applications." In Muhammad Akhyar Farrukh (Ed.) *Functionalized Nanomaterials*. London, InTech, pp. 53–68.

Yu, X., P. Ye, L. Yang, S. Yang, P. Zhou, and W. Gao. 2007. "Preparation of Hexagonal Cerium Oxide Nanoflakes by a Surfactant-Free Route and Its Optical Property." *Journal of Materials Research* 22 (11): 3006–13.

Zhang, Dengsong, Chengsi Pan, Liyi Shi, Lei Huang, Jianhui Fang, and Hongxia Fu. 2009. "A Highly Reactive Catalyst for CO Oxidation: CeO$_2$ Nanotubes Synthesized Using Carbon Nanotubes as Removable Templates." *Microporous and Mesoporous Materials* 117 (1–2): 193–200.

Zhang, Huaihao, Jiangna Gu, Jie Tong, Yongfeng Hu, Bing Guan, Bin Hu, Jing Zhao, and Chengyin Wang. 2016. "Hierarchical Porous MnO$_2$/CeO$_2$ with High Performance for Supercapacitor Electrodes." *Chemical Engineering Journal* 286: 139–149.

Zhou, Yanchun, and Jay A. Switzer. 1996. "Growth of Cerium(IV) Oxide Films by the Electrochemical Generation of Base Method." *Journal of Alloys and Compounds* 237 (1): 1–5. doi:10.1016/0925-8388(95)02048-9.

10

Multifunctional Energy Storage: Piezoelectric Self-charging Cell

Blessing N. Ezealigo, M. Maaza, and Fabian I. Ezema

CONTENTS

10.1 Introduction 197
10.2 Nanogenerator Energy Harvesters 199
    10.2.1 Piezoelectric Nanogenerator 199
    10.2.2 Triboelectric Nanogenerator 199
    10.2.3 Pyroelectric Nanogenerator 200
    10.2.4 Components of a Piezo-Self-charging Power Cell 202
10.3 Piezo-Charging of Power Cell: Working Principle of a Non-integrated System 202
    10.3.1 Piezo-Self-charging Cell (Integrated System): Working Principle 203
        10.3.1.1 Mechanism of a Piezo-Self-charging Power Cell 203
10.4 The Piezoelectricity and Charge Distribution in Polarized PVDF Films 204
    10.4.1 Mechanism of Energy Storage 205
    10.4.2 Self-charging Performance of SCPC 206
10.5 Applications of Self-charging Power Cells 208
10.6 Advantages of SCPC 208
10.7 Challenges 208
References 209

10.1 Introduction

The growing threat of pollution and global warming resulting from the overdependence on fossil fuels has currently made the quest for renewable alternative energy resources very important. There is a rise in the development of technological devices, such as mobile electronic gadgets, health care equipment, and environmental surveillance with low power consumption. These mobile devices are mostly powered by rechargeable batteries, therefore increasing the number of batteries in circulation. In order to successfully replace batteries or make batteries self-sufficient (that is without the need of an external charging source), there is a global effort targeted towards the development of energy harvesting systems, such as solar energy, wind energy, thermal energy, mechanical energy, and biofuels (Hu et al. 2007, 110–14; Pop-Vadean et al. 2017; Digital Transformation Monitor 2017). Energy production from the ambient environment is still less than 5% of the total energy production as of 2017 (Figure 10.1). The aim of self-powered systems is to develop reliable independent power sources which can function over a wide range of environmental conditions for a long time duration (Wang and Wu 2012, 11700).

The advancement in technology has led to the need for multifunctional devices, hence, reducing the bulkiness of existing electronic gadgets. Multifunctional devices have improved performance and efficiency by the reduction of the redundancy between subsystem components and functions (Christodoulou and Venables 2003, 39). The increase in portable electronic devices has led to advances in devices consisting of both energy harvesting and storage, hence, being self-powered. The combination of a piezoelectric

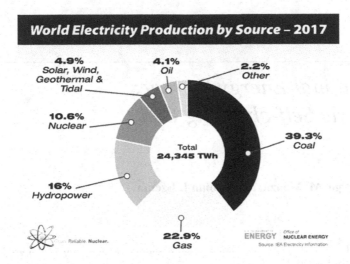

FIGURE 10.1   World electrical energy production.

energy harvester with an electrochemical energy storage function in one device, to offer a multifunctional piezoelectric self-powered cell, is advantageous in that the device does not rely on a conventional charging power source, and its independent operation significantly reduces its weight and volume.

Multifunctional supercapacitors such as electrochromic (Tian et al. 2014, 2150; Yun et al. 2019, 3141; Zhu et al. 2015, 21321; Yuksel et al. 2015, A2805; Yun et al. 2017; Zhou et al. 2018, 290), piezoelectric (Panchal et al. 2016, 1960; Ramadoss et al. 2015a, 4337; Wang et al. 2019, 868), thermal self-charging (Kim et al. 2016; Lim et al. 2014; Wang et al. 2015, 5784), and photo self-charging (Zhang et al. 2014, 466; Chen et al. 2014, 1897) supercapacitors are some of the recent advances in supercapacitor devices where energy harvesters are incorporated into the supercapacitor system to enhance their application in modern devices (Wang et al. 2017, 6816). Also, in recent studies, multifunctional structural supercapacitors have been developed where structural enhancement and electrochemical storage is performed by a single material (Christodoulou et al. 2003, 39–45; Qian et al. 2013, 6113–22).

Energy harvesters derive energy from external sources, e.g., solar, heat, wind, and motion, which are stored for use in small independent devices. These energy harvesters facilitate the development or fabrication of devices without the need of batteries for sensing, wearable, and biomedical applications (Yang et al. 2013, 1744–49; Saravanakumar et al. 2013, 16646–56; Hwang et al. 2014, 4880–87). Energy harvesting from the environment has had a great impact in the energy sector, but it still pose some limitations, as it is dependent on natural factors which cannot be controlled by man. Hence, the need for other controllable energy harvesters is essential and has led to energy harvesting through piezoelectric, triboelectric, and pyroelectric/thermoelectric effect.

It is very convenient and profitable to autonomously power devices with a small energy-harvesting source without batteries that need regular replacement. The environment provides countless ambient energy sources, such as piezoelectric energy, thermal energy, vibration energy, and photovoltaic energy, but at low power, which is below the required power for most devices. Therefore, a battery or a supercapacitor acts as a power buffer to store enough energy to deliver the power bursts needed. These energy-storage devices are charged at low power and deliver the burst power when required (Mars 2012, 40–42).

Piezoelectric energy harvesters convert mechanical strain into electrical energy. They have attracted great attention in the last 15 years (Wang and Song 2006, 242; Li et al. 2015, 8926; Ottman et al. 2002, 669; Erturk and Inman 2011; Shu and Lien 2006, 1499; Kim, Kim and Kim 2011, 1129; Lefeurve et al. 2005, 865; Kang et al. 2016; Priya et al. 2017; Yoon et al. 2018; Han et al. 2019, 26; Du et al. 2019, 1; Chamanian et al. 2018, 2739; Yu et al. 2019, 3479; Li et al. 2016, 6988; Abbasipour et al. 2019, 279) due to the simple configuration and high conversion efficiency compared with electrostatic and electromagnetic harvesters (Roundy et al. 2003, 7663).

A supercapacitor is a high capacity capacitor for energy storage application. It bridges the gap between electrolytic capacitors and batteries. It has a high-power density compared with batteries and consists of electrodes, separators, and electrolytes. Novel materials have been explored for the different components of a supercapacitor. However, the performance of the electrode material of a supercapacitor is strongly affected by the porosity, surface area, mechanical stability, electrical conductivity, and electrochemical stability (Purkait et al. 2018).

10.2 Nanogenerator Energy Harvesters

A nanogenerator converts mechanical or thermal energy produced by a small physical change to electricity. Nanogenerators can be classified into three categories.

10.2.1 Piezoelectric Nanogenerator

This type of nanogenerator converts externally applied mechanical (kinetic) energy to electrical energy with the aid of nanostructured piezoelectric materials. Piezoelectric energy harvesting was first introduced in 2006 (Wang and Song 2006, 242–46), since then until now, a lot of breakthroughs in research and innovation have been recorded. Piezoelectric materials work in a dual mode, such that when they are mechanically stressed, they generate voltage, also when an external voltage is applied to them their physical shape/size changes.

Mechanical energy scavenging from our immediate surroundings is an important renewable source of power for various applications. The renewable power sources range from large-scale power generators (convert mechanical actuation found in nature, e.g., waterfalls and wind, to electricity) (Lu, McElroy, and Kiviluoma 2009, 10933–77; Scruggs and Jacob 2009, 1176–78) to small-scale energy harvesters, which harness energy from mechanical vibration sources in automobiles and human body movements (Paradiso and Starner 2005, 18–25; Donelan et al. 2008, 807–809; Chang et al. 2010, 727–30).

When mechanical stress is applied to a piezoelectric material, the crystal structure of the material is distorted and leads to the movement of electric charges. The polarization charge density arising from the electrical moment is proportionate to the applied mechanical stress, represented by Equation (10.1) (Lee et al. 2016, 7985):

$$\rho = dX, \tag{10.1}$$

where ρ is the polarization charge density, d is the piezoelectric coefficient, and X is the applied stress. Then, the charge density results in an electric field with potential as follows, Equation (10.2) (Lee et al. 2016, 7985):

$$\nabla E = \frac{\rho}{\varepsilon}, \tag{10.2}$$

where ∇E is the divergence of the electric field, ρ is the charge density, and ε is the permittivity. Table 10.1 shows some piezoelectric materials reportedly used to fabricate piezoelectric nanogenerators. These materials have been subjected to various forms of external mechanical energy, such as vibrations, pressing, bending, stretching, and muscle movements (Cha et al. 2011, 5145; Lee et al. 2015, 3204; Kwon et al. 2012, 8972; Lee et al. 2014, 767; Wang et al. 2007, 2967) to generate electrical energy.

10.2.2 Triboelectric Nanogenerator

This type of nanogenerator converts mechanical energy to electricity by the combination of contact electrification (triboelectric effect) and electrostatic induction (Fan, Tian, and Wang 2012, 328).

TABLE 10.1

Comparison of the Reported Piezoelectric Materials

Material	Structure	Type	References
ZnO	Wurtzite	Semiconductor	Lu et al. (2009, 1223–27); Zhu et al. (2010, 3131–55)
InN	Wurtzite		Huang et al. (2010a, 4008–13)
GaN	Wurtzite		Huang et al. (2010b, 4766–71)
CdS	Wurtzite		Lin et al. (2014)
BaTiO$_3$	Perovskite	Insulator	Wang et al. (2007, 2966–69); Yan et al. (2016, 15700–9)
PbZrTiO$_3$	Perovskite		Kwon et al. (2012, 8970–75)
KNbO$_3$	Perovskite		Joung et al. (2014, 18547–53)
NaNbO$_3$	Perovskite		Jung et al. (2011, 10041–46)
PVDF/P(VDF-TrFE)	Polymer		Chang et al. (2010, 726–31) Cha et al. (2011, 5142–47); Pi et al. (2014, 33–41)

An electrostatically charged material creates a potential that drives induced electrons to flow through the electrodes by periodic contact and separation of the two materials. The generated electric potential V can be estimated from Equation (10.3):

$$V = -\frac{\rho d}{\varepsilon_o},\qquad(10.3)$$

where ρ is the triboelectric charge density, ε_o is the permittivity of free air, and d is the interlayer distance. The generated current I in the connected load is given by Equation (10.4):

$$I = C\frac{\partial V}{\partial t} + V\frac{\partial C}{\partial t},\qquad(10.4)$$

where C is the capacitance and V is the voltage across the electrodes. The first term is the change in potential due to the triboelectric charges. The second term is the change in the capacitance due to the mechanical deformation (Lee at al. 2016, 7986).

10.2.3 Pyroelectric Nanogenerator

A pyroelectric nanogenerator generates electricity by harvesting thermal energy from time-dependent fluctuation or sources, it was first discovered in 2012 (Yang et al. 2012, 2833–38). The mechanism of harvesting energy in a pyroelectric nanogenerator is different from the conventional thermoelectric effect (creates permanent voltage) that depends on the Seebeck effect in which charge carriers are driven by the temperature gradient between the ends of the device, but the pyroelectric effect (temporary voltage) operates on temperature change over a period of time.

The working mechanism of a pyroelectric nanogenerator can be subdivided into two:

1. The primary pyroelectric effect results from a charge generated due to a change in polarization with temperature.
2. The secondary pyroelectric effect is the contribution from the piezoelectrically induced charge by the thermal expansion of a pyroelectric material with temperature change. The total pyroelectric effect is the addition of the primary and secondary pyroelectric effects. The pyroelectric coefficient, p, is defined by Equation (10.5):

$$p = \frac{d\rho}{dT},\qquad(10.5)$$

where p is the pyroelectric coefficient, ρ is the spontaneous polarization, and T is the temperature. The electric current generated by the pyroelectric effect is given by Equation (10.6):

$$I = \frac{dQ}{dT} = \mu e A \frac{dT}{dt}, \qquad (10.6)$$

where Q is the induced charge, μ is the absorption coefficient of radiation, A is the surface area, and dT/dt is the rate of temperature change. Hence, when pyroelectric materials are heated ($dT/dt > 0$) or cooled ($dT/dt < 0$), the total polarization arising from the dipole moment is diminished or improved, which causes a current to flow in the circuit, as shown in Figure 10.2 (Lee at al. 2016, 7986–87; Yang et al. 2012, 5357–62). The degree of dipole oscillation is determined by the statistical thermal variations. The higher the thermal fluctuations, the smaller the polarization.

Figure 10.3 shows the classification of materials and the relationship between piezoelectrics and pyroelectrics. Since dielectric materials are insulators that become polarized by the application of an electrical field, this property further enhances the energy storage potential of piezo-supercapacitors.

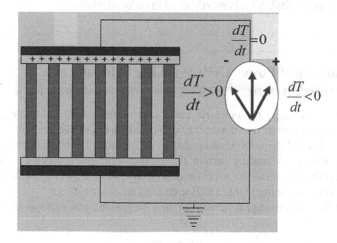

FIGURE 10.2 Schematic of a pyroelectric nanogenerator at various conditions: negative dipole at room temperature $\frac{dT}{dt} = 0$, heated $\frac{dT}{dt} > 0$, and cooled $\frac{dT}{dt} < 0$. (From Yang, Y. et al.: Flexible Pyroelectric Nanogenerators Using a Composite Structure of Lead-Free KNbO$_3$ Nanowires. *Advanced Materials.* 2012. 24. 5357–5362. Copyright Wiley-VCH Verlag GmbH & Co. KGaA. Adapted with permission.)

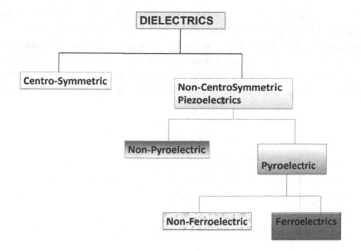

FIGURE 10.3 Classification of materials.

10.2.4 Components of a Piezo-Self-charging Power Cell

The self-charging power cell (SCPC) consists of:

- Electrodes: The positive and negative electrodes: On these electrodes are deposited the active materials (e.g., MnO_2, $NiOH$, $LiCoO_2$, etc.). Easily compressible materials, such as aluminium foil (Ramadoss et al. 2015a, 4337–45) and functionalized carbon cloth (Song et al. 2015, 14963–70) are used as electrode material so as to enable compressional forces to act on the device
- Separator: The conventional separator is replaced with a piezoelectric material, which acts as both a separator and a potential generator, to produce the piezo-potential required for charging the power cell. For instance, in a self-charging battery power cell, the polyethylene separator is replaced with polarized polyvinylidene difluoride or polyvinylidene fluoride (PVDF) film in a self-charging battery (Xue et al. 2012a, 5048–5054)
- Electrolyte: The electrolyte serves the purpose of conducting current between the electrodes and prevents the passage of electrons to the load. In a SCPC, a gel electrolyte or a solid-piezo electrolyte (He et al. 2007, 594) is mostly used to reduce leakage.

10.3 Piezo-Charging of Power Cell: Working Principle of a Non-integrated System

The traditionally non-integrated piezoelectric energy harvester consists of three parts: Piezoelectric nanogenerator, harvesting circuit, and the storage device, as shown in Figure 10.4.

Vibrational or mechanical dynamic loading causes deformation in the structure of the piezoelectric element, and then generates alternating current electrical energy, this generated electrical energy is fed into the circuitry, and then converted to direct current electrical energy through a full bridge rectifier, the direct current is used to charge the storage device (battery or capacitor).

The sinusoidal current and voltage generated in the piezoelectric material can be given by:

$$i(t) = I \sin wt, \tag{10.7}$$

$$V = \frac{I}{wC}, \tag{10.8}$$

where I is the peak amplitude of the current, w is the frequency of the source, and C is the internal capacitance. This technique has been reportedly used to charge a 40 mA Ni rechargeable metal hydride battery from a completely discharged state to full capacity in less than an hour (Sodano et al. 2005, 67–76), supercapacitors (Yuan et al. 2012, 5018–22), and electrolytic capacitors (Umeda et al. 1997, 3146–51). Guan et al. (2008, 67180) investigated the charge/discharge efficiencies for the supercapacitor, lithium, and Ni metal hydride rechargeable batteries and reported it to be 95%, 92%, and 65%, respectively.

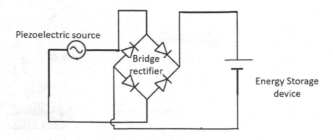

FIGURE 10.4 Traditional piezoelectric energy harvesting circuit.

The efficiency is affected by the internal series resistance and the leakage resistance of energy storage cells. They also reported that the leakage loss of the Ni metal hydride rechargeable battery is much larger than those from the lithium rechargeable battery and supercapacitor, hence, making lithium-ion batteries and supercapacitors a better choice for piezoelectric energy harvesting systems.

In order to optimize the charging process of the non-integrated system, Ottman, Hofmann and Lesieutre (2003, 696–703) proposed a two-stage harvesting circuit to harvest power from various sources, where power from the piezoelectric element is first rectified and stored in a temporary storage device whose voltage is kept at the optimal rectifier voltage and afterwards transferred to the energy storage device through a converter. However, Guan et al. (2007, 498–505) evaluated the efficiency of the one-stage and two-stage energy harvesting charging process and reported that the one-stage energy harvesting scheme can achieve a higher efficiency than the two-stage scheme in a range of energy storage voltages.

10.3.1 Piezo-Self-charging Cell (Integrated System): Working Principle

In an integrated system, the external circuitry is not required, as the piezo-self-charging system directly transfers the generated potentials to the storage device to charge it.

The piezoelectric nanogenerator produces piezoelectric potential (piezo-potential) from the externally applied force (strain), which drives the movement of electrons in the device (Yang et al. 2009, 34–39). The electricity generated from this process is stored by the supercapacitor or battery, thus hybridizing the system called a SCPC.

The fabrication of a self-charging device is based on the combination of the piezoelectric and electrochemical properties of materials, Figure 10.5a. The SCPC is made up of the electrodes (positive and negative), electrolyte, and separator. The conventional separator is replaced with a piezoelectric material (usually PVDF-based). The piezoelectric material film creates a piezo-potential in the presence of applied stress (Figure 10.5b), which converts applied stress (mechanical energy) to electrical energy and also drives the electrode active ions (Xue et al. 2012a, 5049).

The piezoelectric performance of PVDF-based separator film without an electrolyte is shown in Figure 10.5b, which is a typical output produced by a piezoelectric nanogenerator. When the PVDF film is subjected to a compressive strain, a piezo-potential is produced within the film that causes a temporary flow of free electrons to the external load. The applied stress drives the electrons to flow back and forth in the external circuit, resulting in an alternating output. The difference in the magnitude of the output voltage depends on the straining rate of the PVDF-based film (Zhang et al. 2014).

10.3.1.1 Mechanism of a Piezo-Self-charging Power Cell

The charging of a piezoelectric supercapacitor relies on the piezoelectric potential-driven electrochemical oxidation and reduction reaction. The device at the onset is in a discharge state (chemical non-equilibrium). The application of compressive stress causes the piezoelectric separator layer (i.e., PVDF-based film) to

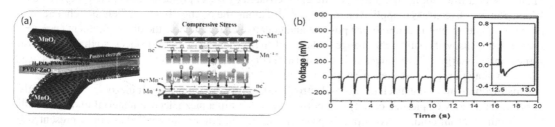

FIGURE 10.5 (a) The schematic of a self-charging supercapacitor power cell. (Reprinted with permission from Ramadoss, A. et al., *ACS Nano*, 9, 4337, 2015a. Copyright 2015, American Chemical Society.) (b) The piezoelectric output of a polarized PVDF film. (Reprinted with permission from Xue et al., *Nano Lett.*, 12, 5048–5054, 2012b. Copyright 2012, American Chemical Society.)

generate piezo-potential with polarity towards the opposite electrode. Under the influence of the piezo-potential, the electrode-active ions also migrate, leading to the charging process. When chemical equilibrium is attained, the charging process ends, hence, a cycle of self-charging is completed (Xue et al. 2012a, 5050). As the applied stress is released, the piezo-potential vanishes, breaking the electrostatic equilibrium, as the electrode-active ions return to the cathode. The self-charging process can also be described by the Nernst equation (Bard and Faulkner 2001, 418–20) by the relationship between electrode potentials and H^+ concentration.

For a reduction reaction as:

$$M^+ + ze^- \Leftrightarrow M,$$ (10.9)

where M^+ is the oxidized electrode-active material, z is the number of electrons, e^- is an electron, and M is the reduced electrode-active material.

The change in the Gibbs free energy, ΔG, can be related to the change in the standard state Gibbs free energy, ΔG^θ, by:

$$\Delta G = \Delta G^\theta + RTInQ.$$ (10.10)

$$\Delta G = -zF\varphi.$$ (10.11)

The electrochemical potential φ of the electrochemical reaction is defined as the decrease in the Gibbs free energy per Coulomb of charge:

$$\varphi_{cell} = \varphi_{cell}^\theta - \frac{RT}{zF} InQ,$$ (10.12)

where Q is the reaction quotient of the cell reaction, R is the universal gas constant, z is the number of electrons transferred in the cell reaction, and T is the temperature in Kelvin. φ_{cell} is the cell potential, φ_{cell}^θ is the *standard* cell potential, and F is the Faraday constant.

The potential of a charge ion, z, is given by the concentration of the ion inside and outside the cell:

$$\varphi = \frac{RT}{zF} In \frac{[ions]_{outside}}{[ions]_{inside}}.$$ (10.13)

10.4 The Piezoelectricity and Charge Distribution in Polarized PVDF Films

The piezoelectricity of the PVDF-based films arises from the spontaneous electric polarization in its polar β phase, where the centre of positive charges and negative charges of all molecules do not match. When the PVDF-based film is polarized, the electric dipole moments align in the direction of the electric field, such that the surface of the PVDF-based film forms bonded charges. For a strain-free PVDF film in electric equilibrium, the bond charges are compensated by the opposite space charges (Figure 10.6a). When compressive stress is applied on the PVDF film along the poling direction, there is a change in the distance (d) between the centre of the positive and negative charge, hence, the dipole moment ($p = q \cdot d$) is changed. This will result in the change of the bond charge density on both surfaces, so that the opposite signed space charge will no longer be in equilibrium (Figure 10.6b) to balance the bonded charges, leading to a voltage drop across the surfaces (Xue et al. 2012b, 7).

This mechanism of the mechanical-to-electrochemical process applies to all piezoelectric materials.

From the theory of piezoelectricity, the current generated in a piezoelectric material under the influence of an external compressive strain is proportional to the rate of the applied strain, represented by Equation (10.14) (Chag et al. 2010, 727–28).

$$\frac{dQ}{dt} = d_{33}EA\frac{d\varepsilon}{dt},$$ (10.14)

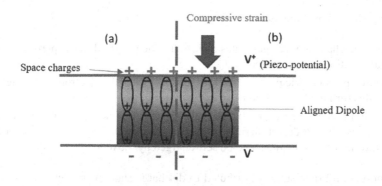

FIGURE 10.6 The generation of electricity and distribution of charge in a polarized PVDF film: (a) without compressive stress (piezo-potential = 0). (b) Under compressive strain (piezo-potential ≠ 0). (Adapted with permission from Xue, X. et al., *Nano Lett.*, 12, 5048–5054, 2012b, Copyright 2012, American Chemical Society.)

where Q is the generated charge, t is time duration, $d_{33}(\frac{\partial D_3}{\partial T_3}$; D is the electric displacement, T is the stress) is the piezoelectric constant, E is Young's modulus, A is the cross-sectional area, and ε is the applied strain. The strain rate during the relaxation process is relatively less than that from the compressive impact. The generated electric field in the relaxation process is in the opposite direction and has a smaller amplitude compared with that generated in the compressive process (Song et al. 2015, 14970).

PVDF films possess good piezoelectric and mechanical properties, with chemical stability and weathering characteristics (Bard and Faulkner 2001; Kawai 1969, 975–76; Holmes-Siedie, Wilson, and Verrall 1984, 910–18; Lovinger 1982) and are used for sensing and actuation applications (Chen et al. 2007; Lee, Elliot, and Gardonio 2003, 541–48). Nevertheless, unpolarized PVDF possesses α, β, and γ crystalline phases and requires mechanical stretching and electrically poling to obtain the β phase essential for generating piezoelectricity (Calvert 1975, 694; Davis 1987).

10.4.1 Mechanism of Energy Storage

The electrode-active material undergoes an electrochemical reaction with the electrolyte, evenly distributed on the surfaces of the electrode and separator. At the onset, there is equilibrium between the electrode material and the electrolyte, hence, no electrochemical reaction. Once a compressive force is applied to the device, polarization is initiated in the PVDF film by the piezoelectric effect. The polarized ions produce a potential difference across the separator. The positive and negative ions generated by the piezo-potential drive the ions in the electrolyte to the electrodes. A cationic (H^+) movement is also initiated in the electrolyte by the piezoelectric field. This cationic movement screens the piezo-potential across the separator, hence, creating an electrochemical inequality between the electrolyte and the electrodes. To regain chemical equilibrium, oxidation and reduction reactions take place at the surface of the electrode material as follows:

$$X + H^+ + e^- \leftrightarrow XOOH, \tag{10.15}$$

where X is the active material deposited on the electrode.

An ideal supercapacitor system uses an electrochemical workstation to initiate/study the charging and discharging process (electron transfer), but in an SCPC charging process, the electron transfer mechanism is inherent in the system. To estimate the electron flow, a multimetre is used to measure the stored voltage resulting from compression and release of the mechanical load. When a continuous compression force is applied, the produced piezoelectric potential charges the energy storage device, hence, the applied mechanical energy is converted to electrochemical energy.

The SCPC is governed by three main processes:

1. The applied mechanical force generates a piezoelectric potential in the piezoelectric material, which is also acting as a separator.
2. The generated piezoelectric potential, in turn, causes ions in the electrolytes to migrate towards the electrodes through the separator.
3. Oxidation and reduction reactions occur at the positive and negative electrodes, respectively, through the insertion or desertion of H^+ ions. This change in the concentration of the H^+ ions in the electrodes causes the device to be self-charged by the redox potential.

The chemical non-equilibrium state is attributed to discharge state of the device, and the equilibrium state represents the charged state of the device.

10.4.2 Self-charging Performance of SCPC

The self-charging capability depends on the applied compressive force. For the self-charging performed under varying compressive forces (Figure 10.7), it was reported that with an increase in the compressive force, the self-charging capability improved as a result of increased piezo-potential arising from a higher compressive force (Ramadoss et al. 2015a, 4339–40; Xue et al. 2012a, 5050–54), which is consistent with Equation (10.1) because the polarization charge density arising from the electrical dipole moment is proportional to the applied mechanical stress.

When a continuous mechanical deformation (compressive force) is applied to the device for a period of time, the voltage in the device increases by some mV, which shows the device is charging (Figure 10.8a). When the compressive force is removed, the stored energy is sustained for few seconds and begins to discharge. The intermittent bending and releasing state that charges at a particular frequency (rate) creates a sinusoidal (alternating) power output (Figure 10.8b). This process can be repeated intermittently, hence, showing the ability of the device to charge itself under mechanical deformation.

Since the self-charging process depends on the generated piezoelectric potential in the separator resulting from the mechanical stress or strain, an increase in strain is advantageous for the production of more piezo-potential. The cycling performance of the SCPC tested under compressive force confirmed the consistency of the charging process.

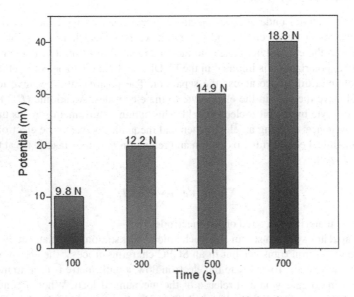

FIGURE 10.7 The transfer of electrons across the electrodes during the self-charging process. (Adapted with permission from Ramadoss, A. et al., *ACS Nano*, 9, 1–13, 2015b. Copyright 2015, American Chemical Society.)

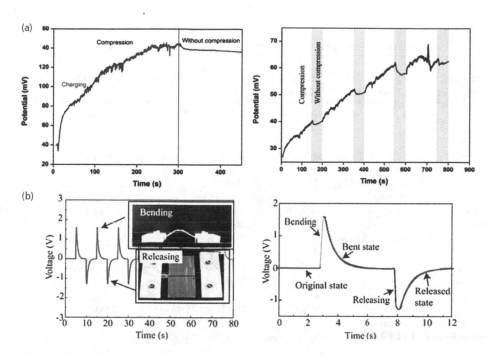

FIGURE 10.8 Self charging performance: (a) By human palm contact (Reprinted with permission from Ramadoss, A. et al., *ACS Nano* 9, 1–13, 2015b. Copyright 2015, American Chemical Society.) (b) Open circuit voltage and short circuit current during bending (From Lu, X. et al., *Scientific Rep.*, 7, 2907, 2017.)

The piezo-potential and the charging duration of self-charging supercapacitor power cells (SCSPCs) reported so far in literature are 110 mV in 300 s (Ramadoss et al. 2015a, 4337), 100 mV in 40 s (Song et al. 2012, 14963), 500 mV in 20 s (Gilsheteyn et al. 2018), and 708 mV in 100 s (Pazhamalai et al. 2018, 1800055).

The SCPC can be self charged independently under the repeated deformation with/without an external circuitry, and the charges are transferred from one electrode to the other inside the device so that the energy is successfully stored after harvesting. The electrons move inside the system during the self-charging process via the electrolyte – PVDF-based separator – between the two electrodes (Sloop, Kerr, and Kinoshita 2003, 330–37). The summary of reported works on piezoelectric self-charging supercapacitor power cells is presented in Table 10.2 and the piezoelectric self-charging battery power cell in Table 10.3.

TABLE 10.2

Summary of Piezoelectric Self-charging Supercapacitor Power Cell Reported in Literature

Material	Electrolyte	Capacitance	Energy Density	Power Density	References
Functionalized carbon cloth (FCC)	PVA/H$_2$SO$_4$ gel	357.6 F m^{-2}	400 mW m^{-2}	49.67 mWh m^{-2}	Song et al. (2015)
MnO$_2$	PVA/H$_3$PO$_4$	455 mFg^{-1}	58 mWh Kg^{-1}	9.9 kW Kg^{-1}	Ramadoss et al. (2015)
MoSe$_2$	(PVDF-co-HEP/ TEABF$_4$)-(PVDF/ NaNbO$_3$) piezopolymer	18.93 mF cm^{-2}	37.90 mJ cm^{-2}	268.91 μW cm^{-2}	Pazhamalai et al. (2018)
SWCNT with (PVDF-TrFE) Metalized film/PET electrodes	PVA/H$_2$SO$_4$	50 Fg^{-1}	182.25 Wg^{-1}	5.06 Wh g^{-1}	Gilshteyn et al. (2018)
NiCoOH@Cu (positive electrode) rGO@Cu (negative electrode)	PVA-KOH gel	165.1 Fg^{-1} (330 mF cm^{-2})	51.6 Whkg^{-1}	750 Wkg^{-1}	Maitra et al. (2017, 633–45)

TABLE 10.3

Summary of Piezoelectric Self-charging Battery Power Cell Reported in Literature

Piezo-Material/ Anode Material	Electrolyte	Compression Force/ Frequency	Charging Capacity (mV)	Discharge Current (μA)	Stored Capacity/ Energy	References
PZT-PVDF nanocomposite film/MWCNTs	LiPF₆	10 N/1.5 Hz	210–297.6 in 240 s	1 μA in 37 s	0.01 μAh	Zhang et al. (2014)
PVDF/TiO₂ NTs	1 M LiPF₆ in 1:1 ethylene carbonate: dimethyl carbonate	45 N/2.3 Hz	327–395 in 240 s	1 μA in 130 s	0.036 μAh	Xue et al. (2012a)
PVDF/flexible graphene	1 M LiPF₆ in 1:1 ethylene carbonate: dimethyl carbonate	34 N/1 Hz	500–832 in 500 s	1 μA in 957 s	0.266 μAh	Xue et al. (2014)
PVDF/CuO piezo-anode	LiPF₆	18 N/1 Hz	50–144 in 240 s	1 μA in 89 s	0.0247 μAh/6.12 μJ	Xue et al. (2013, 2615–20)
PVDF/graphite, conductive carbon, and binder (carboxymethyl cellulose)	Solid piezo-electrolyte of PVDF-LiPF₆	30 N/1 Hz	25–473 in 240 s	5 μA in 85 s	0.118 μAh	He et al. (2017)
PVDF/LiCoO₂	Carbonate-based electrolyte containing 1 M LiPF₆ salt	282 mJ (2.82 N)/1 Hz	1200–1400 in ~ 200 s	0.01 mA	0.4 μAh	Kim et al. (2015, 77–86)

10.5 Applications of Self-charging Power Cells

The self-charging powered cells can be utilized in devices or systems, such as wireless sensors, medical equipment, environmental monitoring, military devices, and portable electronics (Xu et al. 2010, 366–73). The development of SCPCs from materials which are ecologically friendly, biocompatible, and decomposable is essential.

10.6 Advantages of SCPC

1. Self-charging power cells hybridize an energy harvester unit with an energy storage unit, without energy being lost to external circuitry, therefore, decreasing energy conversion loss (Lee et al. 2016).
2. The lead-free nanogenerator utilized in an SCPC solves the problem of using toxic piezoelectric materials (Lin et al. 2012).
3. Ubiquitous sources from moving parts having diverse frequency and power range.

10.7 Challenges

1. There exists a problem of scalability, as this research is still in the early stage.
2. The piezoelectric nanogenerator has inadequate power to charge the high capacity storage unit (Luo et al. 2015). The triboelectric nanogenerator may possibly resolve this problem since it has a better mechanism to efficiently harvest mechanical energy.

3. Low voltage response, though there has been an improvement in this aspect in recent studies, but not yet commensurate for commercial applications.

4. A high voltage source having very low leakage current is required to power analogue circuits and other electronic devices. Attempts to solve this problem have been through the choice of electrodes, electrolyte, and the integration of a tandem device of sandwiched piezo-films with several supercapacitors connected in parallel (Davies 1987; Niu et al. 2015).

5. The design/fabrication flow for the development of self-charging powered cells should be susceptible to scale up and compatible with the various fabrication technologies. To date, virtually all prototypes are at a level inappropriate for large-scale production. Advanced fabrication techniques were rarely employed, which severely restricted the wider range of application of the self-charging powered cells.

6. Low efficiency and discontinuous output (Wang and Wu 2012, 11721).

REFERENCES

Abbasipour, Mina, Khajavi Ramin, Yousefi, Ali Akbar, Yazdanshena Mohammad, Razaghian Farhad, and Akbarzadeh Abdolhamid. 2018. "Improving Piezoelectric and Pyroelectric Properties of Electrospun PVDF Nanofibers Using Nanofillers for Energy Harvesting Application." *Polymers for Advanced Technologies* 30 (2): 279–91. https://doi.org/10.1002/pat.4463.

Bard, J. Allen, and Faulkner R. Larry. 2001. *Electrochemical Methods: Fundamentals and Applications*, 2nd ed., 412–20. New York: John Wiley & Sons.

Calvert, P. 1975. "Piezoelectric Polyvinylidene Fluoride." *Nature* 56: 694.

Cha, Seung Nam, Kim Seong Min, Kim HyunJin, Ku JiYeon, Sohn Jung Inn, Park Young Jun, Song Byong Gwon, et al. 2011. "Porous PVDF as Effective Sonic Wave Driven Nanogenerators." *Nano Letters* 11: 5142–47.

Chamanian, Salar, Hasan Ulus, Aziz Koyuncuoğlu, Ali Muhtaroğlu, and Haluk Külah. 2019. "An Adaptable Interface Circuit with Multistage Energy Extraction for Low-Power Piezoelectric Energy Harvesting MEMS." 34 (3): 2739–47. https://doi.org/10.1109/TPEL.2018.2841510.

Chang, Chieh, Tran H. Van, Wang Junbo, Fuh Yiin-Kuen, Lin Liwei. 2010."Direct-Write Piezoelectric Polymeric Nanogenerator with High Energy Conversion." *Nano Letters* 10: 726–31.

Chen, Xuli, Sun Hao Yang, Zhibin Guan, Guozhen Zhang Zhitao, Qiu Longbin, and Huisheng Peng. 2014. "A Novel 'Energy Fiber' by Coaxially Integrating Dye-Sensitized Solar Cell and Electrochemical Capacitor." *Journal of Materials Chemistry A* 2: 1897–1902.

Chen, Zheng, Yantao Shen, Ning Xi, and Xiaobo Tan. 2007. "Integrated Sensing for Ionic Polymer–Metal Composite Actuators Using PVDF Thin Films." *Smart Materials and Structure* 16: S262–71. https://doi.org/10.1088/0964-1726/16/2/S10.

Christodoulou, Leo, and John D. Venables. 2003. "Multifunctional Material Systems: The First Generation." *The Journal of the Minerals, Metals & Materials Society* 55: 39–45.

Davis, G. T. 1987. "Piezoelectric Polymer Transducers." *Advances in Dental Research* 1: 45–49.

Digital Transformation Monitor. "Energy Harvesting to Power the Rise of the Internet of Things Energy Harvesting to Power the Rise of the Internet of Things." July (2017). https://ec.europa.eu/growth/tools-databases/dem/monitor/sites/default/files/DTM_Energy%20harvesting%20v1_0.pdf.

Donelan, Max, Li Qingguo, V. Naing, Joaquin Andres Hoffer, Weber J. Douglas, and A. D. Kuo. 2008. "Biomechanical Energy Harvesting: Generating Electricity During Walking with Minimal User Effort." *Science* 319 (5864): 807–10.

Du, Sijun, Yu Jia, Chun Zhao, G. A. J. Amaratunga, and A. A. Seshia. 2019. "A Fully Integrated Split-Electrode SSHC Rectifier for Piezoelectric Energy Harvesting." *IEEE Journal of Solid-State Circuits* PP(99): 1–11. https://doi.org/10.1109/JSSC.2019.2893525.

Erturk, Alper, and Daniel J. Inman. 2011. *Piezoelectric Energy Harvesting*. Hoboken, NJ: John Wiley & Sons.

Fan, Feng-Ru, Tian Zhong-Qun, and Wang Zhong Lin. 2012. "Flexible Triboelectric Generator." *Nano Energy* 1 (2): 328–34.

Fu, See Lijun, Qingming Zhou, and Yuping Wu. 2017. "As Featured in: Latest Advances in Supercapacitors: From New Electrode Materials to Novel Device Designs." *Chemical Society Reviews* 46: 6816–54. https://doi.org/10.1039/c7cs00205j.

Gilshteyn, P. Evgenia, Amanbaev Daler, Silibin V. Maxim, Sysa Artem, Kondrashov A. Vladislav, Anisimov S. Anton, Kallio Tanja, and Nasibulin G. Albert. 2018. "Flexible Self-powered Piezo-Supercapacitor System for Wearable Electronics." *Nanotechnology* 29: 325501. https://doi.org/10.1088/1361-6528/aac658.

Guan, M. J., and Liao Wei-Hsin. 2007. "On the Efficiencies of Piezoelectric Energy Harvesting Circuits towards Storage Device Voltages." *Smart Materials and Structures* 16 (2): 498–505.

Guan, M. J., and Liao Wei-Hsin. 2008. "Characteristics of Energy Storage Devices in Piezoelectric Energy Harvesting Systems." *Journal Intelligent Materials System and Structures* 19: 671–80.

Han, Mengdi, Heling Wang, Yang Yiyuan, Liang Cunman, Bai Wubin, Yan Zheng, et al. 2019. "Three-Dimensional Piezoelectric Polymer Microsystems for Vibrational Energy Harvesting, Robotic Interfaces and Biomedical Implants." *Nature Electronics* 2: 26–35. https://doi.org/10.1038/s41928-018-0189-7.

He, Haoxuan, Fu Yongming, Zhao Tianming, Gao Xuchao, Xing Lili, Zhang Yan, and Xue Xinyu. 2017. "All-Solid-State Flexible Self-charging Power Cell Basing on Piezo-Electrolyte for Harvesting/Storing Body-Motion Energy and Powering Wearable Electronics." *Nano Energy* 39: 590–600.

Holmes-Siedle, A. G., P. D. Wilson, and A. P. Verrall. 1984. "PVDF: An Electronically-Active Polymer for Industry." *Materials and Design* 4: 910–18.

Hu, Youfan, Long Lin, Yan Zhang, and Zhong Lin Wang. 2012. "Replacing a Battery by a Nanogenerator with 20V Output." *Advanced Materials* 24 (1): 110–14. https://doi.org/10.1002/adma.201103727.

Huang, Chi-Te, Song Jinhui, Tsai Chung-Min, Lee Wei-Fan, Lien Der-Hsien, Gao Zhiyuan, Hao Yue, Chen Lih-Juann, and Wang Zhong Lin. 2010a. "Single-InN-Nanowire Nanogenerator with up to 1 V Output Voltage." *Advanced Materials* 22: 4008–13.

Huang, Chi-Te, Song Jinhui, Tsai Chung-Min, Lee Wei-Fan, Lien Der-Hsien, Gao Zhiyuan, Hao Yue, Chen Lih-Juann, and Wang Zhong Lin. 2010b. "GaN Nanowire Arrays for High-Output Nanogenerators." *Journal of the American Chemical Society* 132 (13): 4766–71.

Hwang, Geon-tae, Hyewon Park, Jeong-ho Lee, Sekwon Oh, Kwi-il Park, Myunghwan Byun, Hyelim Park, et al. 2014. "Self-Powered Cardiac Pacemaker Enabled by Flexible Single Crystalline PMN-PT Piezoelectric Energy Harvester." *Advanced Materials* 26: 4880–87. https://doi.org/10.1002/adma.201400562.

Joung, Mi-ri, Haibo Xu, In-tae Seo, Dae-hyeon Kim, Joon Hur, Sahn Nahm, Chong-yun Kang, Seok-jin Yoon, and Hyun-min Park. 2014. "Piezoelectric Nanogenerators Synthesized Using KNbO3 Nanowires with Various Crystal Structures." *Journal of Materials Chemistry A: Materials for Energy and Sustainability* 2: 18547–53. https://doi.org/10.1039/C4TA03551H.

Jung, Jong Hoon, Lee Minbaek, Hong Jung-Il, Ding Yong, Chen Chih-Yen, Chou Li-Jen, and Wang Zhong Lin. 2011. "Lead-Free NaNbO₃ Nanowires for a High Output Piezoelectric Nanogenerator." *ACS Nano* 5: 10041–46.

Kang, Min-Gyu, Jung Woo-Suk, Kang Chong-Yun, and Yoon Seok-Jin. 2016. "Recent Progress on PZT Based Piezoelectric Energy Technologies." *Actuators* 5 (1): 5. https://doi.org/10.3390/act5010005.

Kawai, Heiji. 1969. "The Piezoelectricity of Poly(Vinylidene Fluoride)." *Japanese Journal of Applied Physics* 8 (7): 975–76.

Kim, Heung Soo, Kim Joo-Hyong, and Kim Jaehwan. 2011. "A Review of Piezoelectric Energy Harvesting Based on Vibration." *International Journal of Precision Engineering and Manufacturing* 12 (6): 1129–41.

Kim, Sang-Woo, and Jung Ho. 2016. "All-in-One Energy Harvesting and Storage Devices." *Journal of Materials Chemistry A* 4: 7983–99. https://doi.org/10.1039/c6ta01229a.

Kim, Suk Lae, Henry Taisun Lin, and Choongho Yu. 2016. "Thermally Chargeable Solid-State Supercapacitor," *Advanced Energy Materials* 6 (18): 1600546. https://doi.org/10.1002/aenm.201600546.

Kim, Young-Soo, Xie Yannan, Wen Xiaonan, Wang, Sihong, Kim Sang Jae, Song Hyun-Kon, and Wang Zhong Lin. 2015. "Highly Porous Piezoelectric PVDF Membrane as Affective Lithium Ion Transfer Channels for Enhanced Self-charging Power Cell." *Nano Energy* 14: 77–86.

Kwon, Junggou, Seung Wanchul, Sharma K. Bhupendra, Kim Sang-Woo, and Ahn Jong-Hyun. 2012. "A High Performance PZT Ribbon-Based Nanogenerator Using Graphene Transparent Electrodes." *Energy and Environmental Science* 5: 8970–75.

Lee, Ju-Hyuck, Keun Young Lee, Manoj Kumar Gupta, Tae Yun Kim, Dae-yeong Lee, Junho Oh, Changkook Ryu, et al. 2014. "Highly Stretchable Piezoelectric-Pyroelectric Hybrid Nanogenerator." *Advanced Materials* 26 (5): 765–69. https://doi.org/10.1002/adma.201303570.

Lee, Ju-Hyuck, Yoon Hong-Joon, Kim Tae Yun, Gupta Manoj Kumar, Lee Jeong Hwan, Seung Wanchul, Ryu Hanjun, and Kim Sang-Woo. 2015. "Micropatterned P(VDF-TrFE) Film Based Piezoelectric Nanogenerators for Highly Sensitive Self-Powered Pressure Sensing System." *Advanced Functional Materials* 25: 3203–209.

Lee, Ju-Hyuck, Jeonghun Kim, Tae Yun Kim, Md Shahriar Al Hossain, Sang-Woo Kim, and Jung Ho Kim. 2016. "All-in-One Energy Harvesting and Storage Devices." *Journal of Materials Chemistry A* 4: 7983–99.

Lee, Young-Sup, Elliott J. Stephen, and Paolo Gardonio. 2003. "Matched Piezoelectric Double Sensor/Actuator Pairs for Beam Motion Control." *Smart Materials and Structure* 12 (4): 541–48.

Lefeuvre, Elie, Badel Adrien, Richard Claude, and Guyomar Daniel. 2005. "Piezoelectric Energy Harvesting Device Optimization by Synchronous Electric Charge Extraction." *Journal of Intelligent Material Systems and Structures* 16 (10): 865–76.

Li, Baozhang, Chengyi Xu, Feifei Zhang, Jianming Zheng, and Chunye Xu. 2015. "Self-Polarized Piezoelectric Thin Films: Preparation, Formation Mechanism and Application." *Journal of Materials Chemistry C* 3: 8926–31. https://doi.org/10.1039/C5TC01869B.

Li, Baozhang, Zhang Feifei, Guan Shian, Zheng Jianming, and Xu Chunye. 2016. "Wearable Piezoelectric Device Assembled by One-Step Continuous Electrospinning." *Journal of Materials Chemistry C* 4: 6988–95.

Lim, Hyuck, and Yu Qiao. 2014. "Performance of Thermally-Chargeable Supercapacitors in Different Solvents." *Physical Chemistry Chemical Physics* 16: 12728–30. https://doi.org/10.1039/c4cp01610f.

Lin, Yi-Feng, Jinhui Song, Yong Ding, Shih-Yuan Lu, Zhong Lin Wang, Yi-Feng Lin, Jinhui Song, Yong Ding, Shih-Yuan Lu, and Zhong Lin Wang. 2014. "Piezoelectric Nanogenerator Using CdS Nanowires Piezoelectric Nanogenerator Using CdS Nanowires." *Applied Physics Letters* 92 (2): 022105. https://doi.org/10.1063/1.2831901.

Lin, Zong-Hong, Yang Ya, Wu Jyh Ming, Liu Ying, Zhang Fang, and Wang Zhong Lin. 2012. "BaTiO₃ Nanotubes-Based Flexible and Transparent Nanogenerators." *Journal of Physical Chemistry Letters* 3 (23): 3599–604.

Lovinger, J. Andrew. 1982. "Poly(Vinylidene Fluoride)." *Developments in Crystalline Polymers* 1: 195–273.

Lu, Ming-Pei, Jinhui Song, Ming-Yen Lu, Min-Teng Chen, Yifan Gao, Lih-Juann Chen, and Zhong Lin Wang. 2009. "Piezoelectric Nanogenerator Using P-Type ZnO Nanowire Arrays 2009." *Nano Letters* 9 (3): 1223–27.

Lu, Xi, McElroy B. Michael, and Juha Kiviluoma. 2009. "Global Potential for Wind Generated Electricity," *Proceedings of the National Academy of Sciences* 106 (27): 10933–38. doi:10.1073/pnas.0904101106.

Lu, Xin, Qu Hang, and Skorobogatiy Maksim. 2017. "Piezoelectric Microstructured Fibers via Drawing of Multimaterial Preforms." *Scientific Reports* 7 (1): 2907.

Luo, Jianjun, Fan Feng Ru, Jiang Tao, Wang Zhiwei, Tang Wei, Zhang Cuiping, Liu Mengmeng, Cao Guozhong, and Wang Zhong Lin. 2015. "Integration of Micro-supercapacitors with Triboelectric Nanogenerators for a Flexible Self-charging Power Unit." *Nano Research* 8 (12): 3934–43.

Maitra, Anirban, Karan Sumanta Kumar, Paria Sarbaranjan, Das Amit Kumar, Bera Ranadip, Halder Lopamudra, Si Suman Kumar, and Khatua Bhanu Bhusan. 2017. "Fast Charging Self-powered Wearable and Flexible Asymmetric Supercapacitor Power Cell with Fish Swim Bladder as an Efficient Natural Bio-piezoelectric Separator." *Nano Energy* 40: 633–45.

Mars, Pierre. 2012. "Coupling a Supercapacitor with a Small Energy Harvesting Source." *EDN* 2012: 39–42.

Niu, Simiao, Wang Xiaofeng, Yi Fang, Zhou Yu Sheng, and Wang Zhong Lin. 2015. "A Universal Self-charging System Driven by Random Biomechanical Energy for Sustainable Operation of Mobile Electronics." *Nature Communications* 6: 8975.

Ottman, K. Geffrey, Heath F. Hofmann, Archin C. Bhatt, and George A. Lesieutre. 2002. "Adaptive Piezoelectric Energy Harvesting Circuit for Wireless Remote Power Supply," *IEEE Transactions on Power Electronics* 17 (5): 669–76.

Ottman, K. Geffrey, Hofmann Heath, and Lesieutre A. George. 2003. "Optimized Piezoelectric Energy Circuit Using Step-Down Converter in Discontinuous Conduction Mode." *IEEE Transactions on Power Electronics* 18 (2): 696–703.

Panchal, Vasu, Avi Gupta, and Mili Gandhi. 2016. "Development of Piezoelectric Nano- Generator with Super-Capacitor." *International Journal of Advanced Engineering, Management and Science* 2 (12): 1960–1962.

Paradiso, Joseph, and Starner Thad. 2005. "Energy Scavenging for Mobile and Wireless Electronics." *IEEE Pervasive Computing* 4 (1): 18–27. doi:10.1109/MPRV.2005.9.

Pazhamalai, Parthiban, Krishnamoorthy Karthikeyan, Mariappan Vimal Kumar, Sahoo Surjit, Manoharan Sindhuja, and Kim Sang-Jae. 2018. "A High Efficacy Self-Charging MoSe$_2$ Solid-State Supercapacitor Using Electrospun Nanofibrous Piezoelectric Separator with Ionogel Electrolyte." *Advanced Materials Interfaces* 5: 1800055-64. https://doi.org/10.1002/admi.201800055.

Pi, Zhaoyang, Zhang Jingwei, Wen Chenyu, Zhang, Zhi-bin, and Wu Dongping. 2014. "Flexible Piezoelectric Nanogenerator Made of Poly(vinylidenefluoride-co-trifluoroethylene) (PVDF-TrFE) Thin Film." *Nano Energy* 7: 33–41.

Pop-Vadean, Adina, Petrica Paul Pop, Latinovic Tihomir, Cristian Barz, and C. Lung. 2017. "Harvesting Energy an Sustainable Power Source, Replace Batteries for Powering WSN and Devices on the IoT." *IOP Conference Series: Materials Science and Engineering* 200: 012043. https://doi.org/10.1088/1757-899X/200/1/012043.

Priya, Shashank, Hyun-cheol Song, Yuan Zhou, Ronnie Varghese, and Anuj Chopra. 2017. "A Review on Piezoelectric Energy Harvesting: Materials, Methods, and Circuits." *Energy Harvesting and Systems* 4 (1): 3–39. https://doi.org/10.1515/ehs-2016-0028.

Purkait, Taniya, Guneet Singh, Dinesh Kumar, Mandeep Singh, and Ramendra Sundar Dey. 2018. "High-Performance Flexible Supercapacitors Based on Electrochemically Tailored Three-Dimensional Reduced Graphene Oxide Networks." *Scientific Reports* 8 (1): 640. https://doi.org/10.1038/s41598-017-18593-3.

Qian, Hui, Kucernak R. Anthony, Greenhalgh S. Emile, Bismarck Alexander, and Shaffer S. P. Milo. 2013. "Multifunctional Structural Supercapacitor Composites Based on Carbon Aerogel Modified High Performance Carbon Fiber Fabric." *ACS Applied Materials and Interfaces* 5: 6113–22.

Ramadoss, Ananthakumar, Balasubramaniam Saravanakumar, Woo Lee, Young-Soo Kim, Sang Jae Kim, and Zhong Lin Wang. 2015a. "Piezoelectric-Driven Self-Charging Supercapacitor Power Cell." *ACS Nano* 9 (4): 4337–45. https://doi.org/10.1021/acsnano.5b00759.

Ramadoss, Ananthakumar, Balasubramaniam Saravanakumar, Woo Lee, Young-Soo Kim, Sang Jae Kim, and Zhong Lin Wang. 2015b. "Piezoelectric-Driven Self-Charging Supercapacitor Power Cell." *ACS Nano* 9 (4): Supporting Information, 1–13.

Roundy, Shad, Wright Paul Kenneth, and Rabaey Jan. 2003. *Energy Scavenging for Wireless Sensor Networks-with Special Focus on Vibrations*, 51–85. Boston, MA: Kluwer Academic Publishers.

Saravanakumar, Balasubramanim, Rajneesh Mohan, Kaliannan Thiyagarajan, and Sang-Jae Kim. 2013. "Fabrication of a ZnO Nanogenerator for Eco-friendly Biomechanical Energy Harvesting." *RSC Advances* 3: 16646–56.

Scruggs, Jeff, and Jacob Paul. 2009. "Harvesting Ocean Wave Energy." *Science* 323 (5918): 1176–78.

Shu, Y. C., and I. C. Lien. 2006. "Analysis of Power Output for Piezoelectric." *Smart Materials and Structures* 15 (6): 1499–512. https://doi.org/10.1088/0964-1726/15/6/001.

Sloop, Steven, Kerr John, and Kinoshita Kim. 2003. "The Role of Li-ion Battery Electrolyte Reactivity in Performance Decline and Self-discharge." *Journal of Power Sources* 2003119: 330–37.

Sodano, A. Henry, Daniel J. Inman, and Gyuhae Park. 2005. "Generation and Storage of Electricity from Power Harvesting Devices." *Journal of Intelligent Material Systems and Structures* 16 (1): 67–75. https://doi.org/10.1177/1045389X05047210.

Song, Ruobing, Jin Huanyu, Li Xing, Fei Linfeng, Zhao Yuda, Huang Haitao, Chan Helen Lai-Wa, Wang Yu, and Chai Yang. 2015. "A Rectification-Free Piezo-Supercapacitor with a Polyvinylidene Fluoride Separator and Functionalized Carbon Cloth Electrodes." *Journal of Materials Chemistry A* 3: 14963–70.

Tian, Yuyu, Cong Shan, Su Wenming, Chen Hongyuan, Li Qingwen, Geng Fengxia, and Zhao Zhigang. 2014. "Synergy of W$_{18}$O$_{49}$ and Polyaniline for Smart Supercapacitor Electrode Integrated with Energy Level Indicating Functionality." *Nano Letters* 14 (1): 2150–56.

Umeda, Mikio, Nakamura Kentaro, and Ueha, Sadayuki. 1997. "Energy Storage Characteristics of a Piezo-Generator Using Impact Induced Vibration." *Japanese Journal of Applied Physics* Part 1, 35 (5B): 3146–51.

Wang, Jianjian, Feng Shien-Ping, Yang Yuan, Hau Nga Yu, Munro Mary, Ferreira-Yang Emerald, and Gang Chen. 2015. "'Thermal Charging' Phenomenon in Electrical Double Layer Capacitors." *Nano Letters* 15 (9): 5784–90. https://doi.org/10.1021/acs.nanolett.5b01761.

Wang, Ning, Wei Dou, Chao Jiang, Saifei Hao, Dan Zhou, Xiaomin Huang, and Xia Cao. 2018. "Tactile Sensor from Self-chargeable Piezoelectric Supercapacitor." *Nano Energy* 56: 868–74.

Wang, Zhaoyu, Hu Jie, Suryavanshi P. Abhijit, Yum Kyungsuk, and Yu Min-Feng. 2007. "Voltage Generation from Individual BaTiO$_3$ Nanowires under Periodic Tensile Mechanical Load." *Nano Letters* 7: 2966–69.

Wang, Zhong Li, and Jinhui Song. 2006. "Piezoelectric Nanogenerators Based on Zinc Oxide Nanowire Arrays." *Science* 312 (5771): 242–46.

Wang, Zhong Lin, and Wenzhuo Wu. 2012. "Nanotechnology-Enabled Energy Harvesting for Self- Powered Micro-/Nanosystems." *Angewandte Chemie* 51: 11700–721. https://doi.org/10.1002/anie.201201656.

Xu, Sheng Yu, Qin Yong Hua, Xu Chen Yan, Wei Yaguang, Yang Rusen, and Wang Zhong Lin. 2010. "Self-powered Nanowire Devices." *Nature Nanotechnology* 5: 366–73.

Xue, Xinyu, Deng Ping, He Bin, Nie Yuxin, Xing Lili, and Zhang Yan. 2013. "CuO/PVDF Nanocomposite Anode for a Piezo-Driven Self-charging Lithium Battery." *Energy and Environmental Sciences* 6: 2615–20.

Xue, Xinyu, Deng Ping, He Bin, Nie Yuxin, Xing Lili, Zhang Yan, and Wang Zhong Lin. 2014. "Flexible Self-Charging Power Cell for One-Step Energy Conversion and Storage." *Advanced Energy Materials* 4: 1301329. https://doi.org/10.1002/aenm.201301329.

Xue, Xinyu, Sihong Wang, Wenxi Guo, Yan Zhang, and Zhong Lin Wang. 2012a. "Hybridizing Energy Conversion and Storage in a Mechanical-to-Electrochemical Process for Self-Charging Power Cell." *Nano Letters* 12: 5048–54.

Xue, Xinyu, Sihong Wang, Wenxi Guo, Yan Zhang, and Zhong Lin Wang. 2012b. "Hybridizing Energy Conversion and Storage in a Mechanical-to-Electrochemical Process for Self-Charging Power Cell." *Nano Letters* supporting information, 1–15.

Yan, Jing, and Jeong Young Gyu. 2016. "High Performance Flexible Piezoelectric Nanogenerators based on BaTiO$_3$ Nanofibers in Different Alignment Modes." *ACS Applied Materials Interfaces* 8 (24): 15700–709.

Yang, Rusen, Yong Qin, Liming Dai, and Zhong Lin Wang. 2009. "Power Generation with Laterally Packaged Piezoelectric Fine Wires." *Nature Nanotechnology* 4: 34–39.

Yang, Y., Hulin Zhay, Jun Chen, Sangmin Lee. 2013. "Simultaneously Harvesting Mechanical and Chemical Energies by a Hybrid Cell for Self-powered Biosensors and Personal Electronic." *Energy and Environmental Science* 6: 1744–49.

Yang, Y., Jung Jong Hoon, Yun Byung Kil, Zhang Fang, Pradel C. Ken, Guo Wenxi, and Wang Zhong Lin. 2012. "Flexible Pyroelectric Nanogenerators Using a Composite Structure of Lead-Free KNbO$_3$ Nanowires." *Advanced Materials* 24: 5357–62.

Yang, Ya, Wenxi Guo, Ken C. Pradel, Guang Zhu, Yusheng Zhou, Yan Zhang, Youfan Hu, Long Lin, and Zhong Lin Wang. 2012. "Pyroelectric Nanogenerators for Harvesting Thermoelectric Energy." *Nano Letters* 12 (6): 2833–38.

Yoon, Sanghyun, Kim Jinhwan, Cho Kyung Ho, Ko Younh Ho, Lee Sang Kwon, Koh Jun-Hyuk. 2018. "Piezoelectric Energy Generators Based on Spring and Inertial Mass." *Materials* 11 (11): 2163.

Yu, Xiaole, Yudong Hou, Haiyan Zhao, Jing Fu, Mupeng Zheng, and Mankang Zhu. 2019. "The Role of Secondary Phase in Enhancing Transduction Coefficient of Piezoelectric Energy Harvesting Composites." *Journal of Materials Chemistry C* 7: 3479–85.

Yuan, Longyan, Xu Xiao, Tianpeng Ding, Junwen Zhong, Xianghui Zhang, Yue Shen, Bin Hu, Yunhui Huang, Jun Zhou, and Zhong Lin Wang. 2012. "Paper-Based Supercapacitors for Self-Powered Nanosystems." *Angewandte Chemie* 124: 5018–22.

Yuksel, Recep, Can Cevher, Ali Cirpan, and Levent Toppare. 2015. "All-Organic Electrochromic Supercapacitor Electrodes." *Journal of Electrochemical Society* 162 (14): 2805–10.

Yun, Tae Gwang, Donghyuk Kim, Yong Ho Kim, Minkyu Park, Seungmin Hyun, and Seung Min Han. 2017. "Photoresponsive Smart Coloration Electrochromic Supercapacitor." *Advanced Materials* 29 (32): 1606728:1–10.

Yun, Tae Gwang, Park Minkyu, Kim Dong-Ha, Kim Donghyuk, Cheong Jun Young, Bae Jin Gook, Han Seung Min, and Kim Il-Doo. 2019. "All-Transparent Stretchable Electrochromic Supercapacitor Wearable Patch Device." *ACS Nano* 13 (3): 3141–50.

Zhang, Yan, Zhang Yujing, Xue Xinyu, Cui Chunxiao, He Bin, Nie Yuxin, Deng Ping, and Wang Zhong Lin. 2014. "PVDF–PZT Nanocomposite Film Based Self-charging Power Cell." *Nanotechnology* 25: 105401.

Zhang, Zhitao, Xuli Chen, Peining Chen, Guozhen Guan, Longbin Qiu, and Huijuan Lin. 2013. "Integrated Polymer Solar Cell and Electrochemical Supercapacitor in a Flexible and Stable Fiber Format." *Advanced Materials* 26: 466–70.

Zhou, Kailing, Hao Wang, Jinting Jiu, Jingbing Liu, Hui Yan, and Katsuaki Suganuma. 2018. "Polyaniline Films with Modified Nanostructure for Bifunctional Flexible Multicolor Electrochromic and Supercapacitor Applications." *Chemical Engineering Journal.* https://doi.org/10.1016/j.cej.2018.03.175.

Zhu, Guang, Yang Rusen, Wang Sihong, and Wang Zhong Lin. 2010. "Flexible High-Output Nanogenerator Based on Lateral ZnO Nanowire Array." *Nano Letters* 10 (8): 3151–55.

Zhu, Minshen, Huang Yang, Huang Yan, Meng Wenjun, Gong Qingchao, Li Guangming, and Zhi Chunyi. 2015. "An Electrochromic Supercapacitor and Its Hybrid Derivatives: Quantifiably Determining Their Electrical Energy Storage by an Optical Measurement." *Journal of Materials Chemistry A* 3 (1): 21321–27.

11

The Contributions of Electrolytes in Achieving the Performance Index of Next-Generation Electrochemical Capacitors (ECs)

Innocent S. Ike, Iakovos J. Sigalas, Sunny E. Iyuke, and Egwu E. Kalu

CONTENTS

11.1   Introduction..215
11.2   Electrolyte's Materials and Compositions for Ultracapacitors...............................218
        11.2.1   Aqueous Electrolytes...219
        11.2.2   Strong Acid Electrolytes...219
        11.2.3   Strong Alkaline Electrolytes..219
        11.2.4   Neutral Electrolytes..219
11.3   Organic Electrolytes..220
        11.3.1   Organic Solvents..221
        11.3.2   Conducting Salts for Electrolytes...222
        11.3.3   Ionic Liquid-Based Electrolytes..222
11.4   Solid- or Quasi-Solid-State Electrolytes...223
11.5   Redox-Active Electrolytes...224
        11.5.1   Aqueous Electrolytes with Redox-Active Species.................................224
        11.5.2   Non-aqueous Electrolytes with Redox-Active Species.............................224
        11.5.3   Solid Electrolytes with Redox-Active Species....................................224
11.6   Separators...225
11.7   Contribution of Electrolytes in Achieving the Next-Generation
        Electrochemical Capacitors..225
11.8   Progress on Electrolytes of Electrochemical Capacitors/Ultracapacitors.......................227
        11.8.1   Factors Determining the Cell Voltage of Electrochemical Capacitors................228
11.9   Challenges of Ultracapacitors Electrolytes...230
11.10 Summary..231
Acknowledgements...233
References...233

11.1 Introduction

The electrolyte, which is usually electrolyte salt and solvent and is a main component of electrochemical capacitors (ECs) also known as ultracapacitors, provides the ionic conductivity that makes charge exchange on each electrode easy. The electrolyte inside the ultracapacitor executes a key role in the formation of electric double layers (EDLs), reversible oxidation and reduction reaction (charge storage mechanism) in pseudocapacitors, and influences or contributes to the device performance. Type and size of ions, concentration of ions and solvent, interactions between ion and solvent, interactions among electrolyte and electrode, as well as potential range are the characteristics of electrolytes that influence EDL capacitance, pseudocapacitance, energy, power densities, and the ultracapacitors cycle life (Chmiola et al. 2008).

Various types of electrolytes were produced and presented in literature and are basically categorized into either liquid electrolytes or solid/quasi-solid-state electrolytes. The liquid electrolytes are again classified as aqueous, organic, and ionic liquid (IL) electrolytes, while the solid or quasi-solid-state electrolytes are generally classified as organic and inorganic electrolytes (Huang, Sumpter, and Meunier 2008). An ideal electrolyte with all the characteristics mentioned above has not emerged because each type has some merits and demerits associated with it. For instance, ultracapacitors with aqueous electrolytes exhibit high conductivity and capacitance, but have a low potential window because of small decay in the voltage of aqueous electrolytes. Organic electrolytes and ionic liquids have relatively high potential windows and can work at high voltages, but have low ionic conductivity, while the solid-state electrolytes do not have the challenges of leakage of potential relative to the liquid electrolytes, but their conductivity is also low (Huang, Sumpter, and Meunier 2008).

Since the energy density is strongly dependent on the square of device voltage, increasing the voltage of ultracapacitors is a more successful approach to enhancing ECs energy density compared to enhancing the capacitance of electrodes. Moreover, electrolytes/solutions are vital in structuring other essential properties needed in ECs practical application like power density, internal resistance, operating temperature window, cycling life, rate of self-discharge, toxicity, etc. (Largeot et al. 2008). Thus, more attention needs to be channelled towards the design and formulation of next-generation electrolytes/solutions with greater potential windows, rather than concentrating on developing new electrode materials. The ultracapacitor working voltage range is mainly decided by the electrolyte's electrolyte stable potential window (ESPW) if the electrodes are firm and steady within the working voltage range.

In order to develop the next-generation ultracapacitors, the global research interests are currently focused on enhancing the energy density through optimization of the electrode's pore size distribution and utilization of non-aqueous electrolytes. In the breakthrough work of the Gogotsi group (Chmiola et al. 2008), carbide-derived electrodes having unimodal micropores (Huang, Sumpter, and Meunier 2008) less than 1 nm were synthesized. They discovered that the materials show an outstanding growth in capacitance in acetonitrile organic electrolytes, when compared to similar electrodes of pore sizes more than 2 nm, where the capacitance increased little with pore size increase. Their findings were contrary to the idea that pores whose sizes are less than those of solvated electrolyte ions do not participate in the charge/energy storage phenomenon. This outstanding capacitance growth in subnanometer pores is attributed to desolvation of ions entering the subnanometer pores (Chmiola et al. 2006, 2008), as confirmed by an experimental study using an ionic liquid electrolyte that does not have a solvation shell around the electrolyte ions (Largeot et al. 2008). The quantity of energy stored in an EC, E is directly related to the capacitance and square of applied voltage:

$$E = \frac{1}{2}CV^2. \qquad (11.1)$$

Generally, power P is the rate of releasing energy and is fundamentally given by:

$$P = IV = \frac{V^2}{4 \times ESR}, \qquad (11.2)$$

where I is the applied current, V is the applied voltage, and ESR is the equivalent series resistance of the device's components to current flow (Ban et al. 2013; Burke 2000; Chu and Braatz 2002).

Higher energy densities are attainable by increasing either the electrode's capacitance via the use of electrode substances having high capacitance or the electrolyte's voltage window by utilizing non-aqueous electrolytes with a greater electrochemical window stability. It is more effective to improve the energy density by elevating the device voltage window going by the fact that energy density is proportional to voltage squared. Also, there are minimal or no challenges related to growth in a resistor-capacitor (RC)-time constant for growing capacitance to attain higher energy density (Pell and Conway 2001a).

Whereas renewed efforts were focused on the production of enhanced electrodes in a bid to enhance energy densities, theoretical expressions showed that ion concentration and decomposition voltage of electrolytes usually restricted the ultracapacitors' energy densities (Zheng 2003, 2005; Zheng, Huang, and Jow 1997). Also, researches presented that an ultracapacitors' power densities could be restricted by electrolytes (Pell and Conway 2001b; Conway and Pell 2003). Therefore, research outcomes emphasize that optimizing electrolytes is as essential as optimizing the electrode in order to obtain energy and power densities close to the theoretical values of ECs. The type of electrolyte employed greatly determines the actual electrical resistance of the devices. The key factors which influence an electrolyte's electrical conductivity are: (1) the electrolyte concentration and its ability to separate into ions, which are the free charge carriers, and (2) the mobility of ions in electrolytes (Pilon, Wang, and d'Entremont 2015).

Theoretically, ultracapacitors give the chances of minimum or no degradation, very high cycle efficiency, and successful safety for limitless charge-discharge cycles, but their applicability are realistically restricted by their cell voltage and low energy density (Simon and Gogotsi 2008). The charge-storage features of ultracapacitors have fortunately been successfully enhanced via an enlarged surface area, electrode pore size, and size distribution optimization by the development of a ranking porous formation without compromising the electrical conductivity (Wang et al. 2014; Zhu et al. 2011). The electrolytes' composition has equally been examined to improve the density of ion adsorption onto the electrode surface (He et al. 2016; Lian et al. 2016). Alternatively, increasing the operating voltage is more functional for enhancing energy density because it is dependent on the square of the operating voltage according to equation (11.1), and this is effectively achievable by choosing an acceptable electrolyte that has a wider potential window. Thus, organic liquid and ionic liquid electrolytes are the potential electrolytes having operating potential ranges usually up to 2.5–3.0 V and >4.0 V, respectively. Researchers have developed different new organic electrolytes with very few toxicities and wide working potential windows capable of replacing commercial organic electrolytes, like acetonitrile (ACN) (Chiba, Ueda, Yamaguchi, Oki, Saiki, et al. 2011; Chiba, Ueda, Yamaguchi, Oki, Shimodate, et al. 2011; Zhong et al. 2015). Even upon a great improvement in energy density by using non-aqueous electrolytes to increase device voltage, the ultracapacitors usage is still restricted, although further and enhanced functions have been shown via the integration of lithium ion batteries (LIBs) and ECs (Aida et al. 2007; Cao and Zheng 2012). Since the highest operating voltage of an ordinary ultracapacitor using an ACN electrolyte is restricted to 2.7–2.85 V in organic electrolytes (Balducci et al. 2007; Ishimoto et al. 2009), the mismatched applied voltages between LIBs and ECs cause the parallel/series concern. To date, it is a serious problem to design and produce the next-generation ultracapacitors with a long life to meet the immediate demand for clean energy usage (Lu et al. 2019; Naoi et al. 2012). Electrolytes and electrode materials of ultracapacitors are greatly destroyed when the electrolytes' voltage restrictions are exceeded, and reasonable side reactions like the evolution of gas and the formation of passive film occurs on the surface of electrodes (Ishimoto et al. 2009).

Thermal stability and the working temperature window of ultracapacitors are hugely influenced by the electrolyte's viscosity, boiling, and freezing points. ECs aging and failure are equally associated with the electrolyte's electrochemical decomposition. Fortunately, different types of new ultracapacitors like flexible or solid-state ECs and micro-ECs, which depend greatly on novel electrolytes, such as solid-state electrolytes have been developed (Brandt et al. 2013). Several types of electrolytes like aqueous electrolytes, organic electrolytes, ionic liquid electrolytes, redox-type electrolytes, and solid or semi-solid electrolytes were investigated with huge achievements in the past decades (Fic et al. 2012). Unfortunately, concerns in the use of ILs as EC electrolytes include their high viscosity and working temperature. High electrolyte viscosity endangers low ionic mobility and conductivity. Low ionic conductivity is expensive for EC performance, as shown in Figure 11.1, thereby hampering the commercial potential of ECs with ILs. Also, the investigation for solid or semi-solid electrolytes resulted in the conception of flexible or solid-state ultracapacitors that have been shown to possess no self-discharge challenges like devices using liquid electrolytes (Choudhury, Sampath, and Shukla 2008; Gao and Lian 2014). Redox-type electrolytes were suggested as ultracapacitor electrolytes in recent years due to the extra pseudocapacitance that emanates from the electrolytes' redox reaction at the interface of electrode/electrolyte (Lota and Frackowiak 2009).

Globally, a perfect electrolyte is required to have the following: (1) wide potential range; (2) high ionic conductivity; (3) stability both chemically and electrochemically; (4) device components to show elevated chemical and electrochemical inertness; (5) wide working temperature window; (6) suitability for electrodes materials; (7) non-volatility and flammability; (8) environmental friendliness; and (9) low

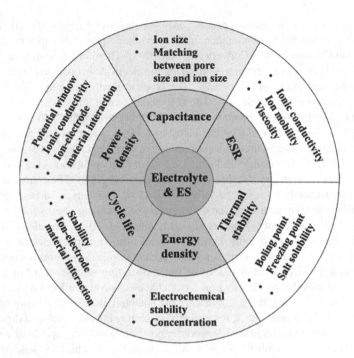

FIGURE 11.1 The electrolyte effects on EC performance. (Zhong, C. et al., *Chem. Soc. Rev.*, 44, 7484–7539, 2015. Reproduced by permission of The Royal Society of Chemistry.)

cost. In practical terms, electrolytes can hardly satisfy all the demands, and each has its merits and demerits. This has prompted a large amount of studies to enhance the entire performance of electrolytes and their related ultracapacitors. Outstanding reviews regarding electrode materials abound (Augustyn, Simon, and Dunn 2014; Yan et al. 2014; G. Wang, Zhang, and Zhang 2012; Zhang and Zhao 2009; Zhai et al. 2011; Lee et al. 2011; Béguin et al. 2014), while reviews with particular attention on ultracapacitor electrolytes are limited (Brandt et al. 2013; Choudhury, Sampath, and Shukla 2008; Gao and Lian 2014; Béguin et al. 2014).

Although many outstanding overviews of solid polymer electrolytes for ultracapacitors were formerly presented (Choudhury, Sampath, and Shukla 2008; Gao and Lian 2014), it is widely accepted that an overview or review that captures the most recent improvements in the area and provides an understanding of formulation and design of electrolytes is inevitable. This will assist in the comprehension of factors influencing performance in electrolyte design and optimization. Furthermore, knowledge of interactions among EC's electrolytes and current collectors, binders, and separators are equally crucial in improving device performance.

In a bid to improve the energy and power densities of ultracapacitors, raising both applied voltage, V, and capacitance, C, and reducing both cell weight, W, and ESR are inevitable. This is because energy and power densities are strongly related to the square of the working voltage, and the voltage growth will certainly contribute greatly in the enhancement of the device's energy and power densities, more than improving the capacitance or decreasing ESR. Generally, the highest working voltage of ultracapacitors firmly relies upon the ESPW or the electrolyte potential range.

11.2 Electrolyte's Materials and Compositions for Ultracapacitors

The electrolyte is the main component of ECs that provides the ionic conductivity and thereby promotes the charge storage process on device electrodes. Meanwhile, the majority of the existing ultracapacitors use organic electrolytes that have cell voltages between the range of 2.5–2.8 V (Conway 1999;

Burke and Miller 2011). An electrolyte plays cardinal functions in deciding the ECs' performance, in particular the solid-state ECs, like ion supplement, electron conduction, and adhesion of electrode particles. The fundamental demands for an appropriate ultracapacitor's electrolyte are: wide decomposition potential range, very stable electrochemically, high concentration of ions, low resistivity, low toxicity, etc.

11.2.1 Aqueous Electrolytes

Generally, aqueous electrolytes are categorized as acid, alkaline, and neutral solutions, with the representative and most commonly used given as H_2SO_4, KOH, and Na_2SO_4, respectively. The major demerit of aqueous electrolytes is the relatively low potential range, that are limited by water decomposition (Burke 2000). Aqueous electrolytes have safety concerns because gas evolution potentially causes a rupture of the ultracapacitor cell and reduces the performance. The voltages of both acid and alkaline electrolytes are restricted to around 1.3 V despite the type of electrode material employed. The maximum cell voltage for the neutral electrolytes is 2.2 V (Fic et al. 2012). Apart from the working temperature window, ultracapacitors using aqueous electrolytes are equally limited to the abovementioned water freezing point and under or below the boiling point.

11.2.2 Strong Acid Electrolytes

The greater ionic conductivity of H_2SO_4 compared with those of neutral electrolytes results in the equivalent series resistance of ultracapacitors using an acid (H_2SO_4) electrolyte smaller than that of devices using neutral electrolytes (Zhang et al. 2012; Wu et al. 2013; Jiménez-Cordero et al. 2014). Numerous previous researchers have equally shown that an electrode's specific capacitance is related with the electrolyte conductivity, that is, the specific capacitance grows along with growth in the electrolytes' conductivity (Torchała, Kierzek, and Machnikowski 2012). The mobility of ions which is strongly connected to the electrolyte conductivity gives the reason for this relationship. Other types of acid electrolytes, such as perchloric acid, hexafluorosilicic acid, and tetrafluoroboric acid, can be employed for ultracapacitors, although few of them, because of safety issues, have been examined for ECs application (Perret et al. 2011). Also, self-discharge challenges in cells using concentrated electrolytes is an issue of serious concern, especially when metal ion contaminants and oxygen are present (Andreas, Lussier, and Oickle 2009; Oickle and Andreas 2011).

11.2.3 Strong Alkaline Electrolytes

Apart from the moves to employ strong acid electrolytes, reasonable studies dedicated to enhancing ultracapacitors' energy densities via base electrolytes have been done by raising the capacitance and/or enlarging the working voltage range through the design of asymmetric ultracapacitors (Patil et al. 2008). Corrosion of electrodes is one challenge associated with the use of concentrated electrolytes, and this will obviously attack the electrodes' surface (Patil et al. 2008). Thus, optimization of electrolyte concentration in terms of the entire performance of the ultracapacitor is very essential. Increasing the electrolyte temperature will obviously lead to ESR reduction and a subsequent increase in the given capacitance because of the improved ion diffusion phenomenon in the electrolyte (Su et al. 2014). Similar to the examinations of ECs that use acidic electrolytes, enlarging the working voltage range of ultracapacitors using alkaline electrolytes is the present direction of studies to reasonably enhance the energy density as well as suppress the electrolyte's side reactions (Wang et al. 2013; Hulicova-Jurcakova et al. 2009).

11.2.4 Neutral Electrolytes

The neutral electrolytes are commonly employed in pseudocapacitors and hybrid ultracapacitors, but a few researches have used neutral electrolytes in electric double-layer capacitor (EDLCs). Neutral electrolytes are used very often and were shown to have a high potential as an electrolyte for pseudocapacitive electrodes, such as MnO_2 is Na_2SO_4. Several research works have shown that specific capacitances of

EDLCs employing neutral electrolytes are low compared to those that utilized H_2SO_4 or KOH electrolytes (Zhang et al. 2012; Wu et al. 2013; Jiménez-Cordero et al. 2014; Torchała, Kierzek, and Machnikowski 2012). Ultracapacitors employing neutral electrolytes generally have lower equivalent series resistance than those with H_2SO_4 or the KOH electrolytes because of the lower ionic conductivities of neutral electrolytes (Zhang et al. 2012; Wu et al. 2013; Jiménez-Cordero et al. 2014; Torchała, Kierzek, and Machnikowski 2012). The working voltages of carbon-based capacitors using neutral electrolytes are large compared with those using both acidic and alkaline aqueous electrolytes because of the elevated ESPWs (Fic et al. 2012; Bichat, Raymundo-Piñero, and Béguin 2010; Demarconnay, Raymundo-Piñero, and Béguin 2010). Generally, the influence of a neutral electrolyte on device performance is equally dependent upon the type of electrolyte (Fic et al. 2012; Tsay, Zhang, and Zhang 2012).

Recently, some novel types of electrolytes like Li–SiW, Na–SiW, and K–SiW were studied by Gao et al. (2014) as ultracapacitor aqueous-neutral electrolytes. These types of electrolytes displayed much higher conductivities in comparison with the other types with C anions due to a great number of dissociated cations and a larger anion movement of the Keggin anions. Different factors of the neutral electrolytes that affect the ultracapacitors are pH, charge carrier type and anion species (Kuo and Wu 2006), ion concentrations (B. Xu, Wu, Su, et al. 2008; Xu et al. 2013; Wen et al. 2009), additives (Komaba et al. 2013), and solution temperature. Different alkaline metal or alkaline-earth metal cations of various ionic sizes, size of the hydrated ion, as well as various diffusivities and ionic conductivities greatly affect device specific capacitance and ESR. Li et al. (2012) presented that the values of given capacitance and the energy and power densities of devices using MnO_2 oxide electrodes with mesopores were in an order: $Li_2SO_4 \geq Na_2SO_4 \geq K_2SO_4$. The introduction of additives in neutral electrolytes were studied in order to enhance the ultracapacitor performance (Komaba et al. 2013).

Use of neutral aqueous electrolytes in electrochemical capacitors mitigate the corrosion issues, reduce costs, and equally enhance the working voltage as well as energy density in an environmentally friendly manner. Nonetheless, further enhancement of the ultracapacitor performance using neutral electrolytes is still required so as to raise or improve the energy density and cycle life.

11.3 Organic Electrolytes

Although serious examinations and efforts have been channelled to the aqueous electrolyte-based ultracapacitors, organic electrolyte-based ultracapacitors have taken over the market due to their wider working potential range, which is from 2.5 to 2.8 V. An enhanced working voltage has the capacity to yield a great increase in energy and power densities. Also, the organic electrolyte permits the application of low-cost substances like aluminium for device current collectors and packages. Nevertheless, the use of organic electrolytes in ultracapacitors have some concerns associated with them including: capital intensiveness, low capacitance, low conductivity and flammability, volatility, and toxicity. Also, an organic electrolyte needs rigorous purification and packaging processes to eliminate residual impurities that result in high self-discharge and great performance degradation (Conway 1999). Specific capacitances obtained from carbon-based ECs using organic electrolytes are usually low compared with those from similar devices using aqueous electrolytes (Barranco et al. 2010; McDonough et al. 2012; Yang et al. 2007).

The size of a solvated ion of organic electrolytes is large with small dielectric constants, hence, resulting in smaller electric double layer capacitance. Also, carbon electrodes contribute insignificant pseudocapacitance in organic electrolytes, such as $TEABF_4/ACN$, which is understandable going by the nature of surface functionalities (Barranco et al. 2010; Vaquero et al. 2012). Accessibility into the electrode pores is strongly related to the organic electrolyte's ionic sizes, species type, and the interaction between ion and solvent. It is essential to match electrode pore size with the size of ions to increase the capacitance, since pores of small size enhance specific surface area and limit ionic access. Reasonable efforts have been dedicated to understanding the connection between size of ion and capacitive aptitude of electrodes of various pore size distributions and optimization of pore size and ionic size matching (McDonough et al. 2012; Yang et al. 2007). A basic comprehension of ion dynamics during charging and discharging and estimations of the EDLC using organic electrolytes have been given reasonable consideration (H. Wang et al. 2011, 2013; Levi et al. 2009; Deschamps et al. 2013) to be used as effective guidelines for a process blueprint of ionic forms at the interface (H. Wang et al. 2011; Tanaka et al. 2010; Boukhalfa et al. 2014), from a minute view. The theoretical interface, ion electro-adsorption, and the pore size influence methods through molecular dynamics estimation and density functional theory estimations. Monte Carlo (MC) computations yield reasonable insights into organic solvents' ionic solvation, EDL form and capacitance, and pore size effects and their morphologies (Forse et al. 2013). The organic electrolyte's electrochemical stable potential window is dependent upon such factors as the kind of conducting salts, solvent, and amount of impurities. These are essential features that determine the ECs working cell voltage (Jiang et al. 2012; Yang et al. 2009). Several of the experimental methods regarding interactions between electrolyte and electrode have been performed to get in-depth comprehension of electrolyte behaviour under operating conditions (Ue, Ida, and Mori 1994; Kurzweil and Chwistek 2008). Different instrumental analytical approaches like nuclear magnetic resonance (H. Wang et al. 2011, 2013; Deschamps et al. 2013; Blanc, Leskes, and Grey 2013), quartz-crystal microbalance (Levi et al. 2009), in situ Raman micro spectrometry (Bonhomme, Lassègues, and Servant 2001), and in situ small angle neutron scattering (Boukhalfa et al. 2014) were equally used to analyze performance.

Degradation and failure associated with organic electrolyte-based ultracapacitors are serious concerns that should be handled. The degradation could basically be as a result of or a combination of one of the following: (i) accelerated oxidation of electrode materials when the employed working cell voltage window is wider than certain values (higher than 3 V), which were used to enhance the energy density. This also causes the evolution of gas due to electrolyte decomposition (Naoi et al. 2012; Ishimoto et al. 2009); (ii) intercalation of ions or the reaction of the organic electrolytes leading to ECs performance decay (Ishimoto et al. 2009; Hardwick et al. 2006); and (iii) harsh operating conditions, such as high operating voltage and maximum temperature can lead to rapid decay of the ECs performance (Vaquero et al. 2012). Ishimoto et al. (2009) and Naoi et al. (2012), in an attempt to comprehend the failure pattern, examined the products of gas evolution in the electrode chambers of a H-type cell after float-tests. Thus, understanding of degradation and failure mechanisms is essential for the development of next-generation ECs that have wider voltage windows. In order to enhance the intercalation or de-intercalation of ions, the most frequently employed organic electrolytes for pseudocapacitors have Li ions because of their small ion size. The typical salts employed in the organic electrolytes were $LiClO_4$ and $LiPF_6$ (Cho et al. 2014; Cai et al. 2014; Hanlon et al. 2014; Lim et al. 2014), while the organic solvents were propylene carbonate (PC), acetonitrile (ACN), or a combination of distinct solvents like ethylene carbonate – diethylene carbonate (EC–DEC) (X. Huang et al. 2014; Hanlon et al. 2014; Lim et al. 2014), ethylene carbonate – dimethyl carbonate (EC–DMC) (Cho et al. 2014; Cai et al. 2014; Qu et al. 2014), ethylene carbonate – ethylmethyl carbonate (EC–EMC), EC–DMC–EMC, and EC–DMC–DEC, which are generally utilized in Li-ion batteries.

11.3.1 Organic Solvents

Solvent is an integral part of the electrolyte for the ultracapacitor. It plays a key role in the performance (high or low) of the ultracapacitor. A perfect organic solvent should have good solvation capacity in a given conducting salt, high electrolyte stability potential window, and low viscosity in the working temperature window. The low flash point and toxicity of ACN are the main challenges, which made Japan ban the use of ACN in ultracapacitors because of its high toxicity (Naoi 2010; Simon and Gogotsi 2013; Ue 2007), and great attention has been given to PC-based electrolytes which are less toxic and have a higher flash point, although low energy efficiency and power density are anticipated since they have lower conductivity at low temperatures (Liu, Verbrugge, and Soukiazian 2006). The interest to further enhance the working voltage of organic electrolyte capacitors has led to developing other new sulfonated solvents with higher working voltages including electrolytes containing sulfolane and dimethyl sulphide (Naoi 2010). Chiba et al. (2011) in a bid to reduce the sulfolane-based electrolytes viscosity and melting point examined various kinds of linear sulfones, which have relatively low molecular weights.

Examination of carbonates and sulphites like EC, DEC, and DMC (Laheäär et al. 2009; Jänes et al. 2014; Perricone, Chamas, Cointeaux et al. 2013) and diethyl sulphite and 1,3-propylene sulphite which change the electrolyte viscosity and conductivity employed in binary PC solvents abound in literature (Jänes and Lust 2006). Perricone, Chamas, Leprêtre, et al. (2013) suggested an introduction of organic ester additives having methoxy or fluorinated groups to enhance ECs' safety against the addition of organic esters like ethyl acetate and methyl acetate.

11.3.2 Conducting Salts for Electrolytes

Conducting salts in organic electrolytes give the ionic charge for the working of ultracapacitors. Thus, ion concentration and mobility are very crucial in the determination of the value of electrolytes' ionic conductivity. Apart from many other salts examined for the purpose of enhancing one or all of the following: solubility, conductivity, stability, and temperature performance in literature. TEABF$_4$ is presently the most frequently used salt with wide applications in commercial ECs. Ion concentration in the electrolyte influences both ionic conductivity and the highest energy density of its related ultra-capacitors (Zheng and Jow 1997). Ue, Ida, and Mori (1994) studied the conductivities of certain basic additives which play essential roles in an organic electrolyte's ESPW (Ue 2007; K. Xu, Ding, and Jow 2001) in various solvents (Ue, Ida, and Mori 1994) like PC, g-butyrolactone, N, N-dimethylformamide (DMF), and ACN. Organic salts relying upon the inorganic ions like Li$^+$ (Laheäär et al. 2009; Laheäär, Jänes, and Lust 2011; Q. Li et al. 2011), Na$^+$ (Väli et al. 2014), and Mg^{2+} (Chandrasekaran et al. 2010), have drawn serious attention in addition to organic ions. Numerous researches focused on employing lithium salt in organic electrolytes for ultracapacitors (Laheäär et al. 2009, 2011).

11.3.3 Ionic Liquid-Based Electrolytes

ILs, equally called low or room temperature molten salts, commonly contain salts made up of mainly ions of melting points below 100°C (Rogers and Voth 2007). Ionic liquids commonly have a big asymmetric organic cation and inorganic or an organic anion, and the peculiar integration of the given ions lowers the melting point (Armand et al. 2009). Basically, ILs have numerous potential merits like elevated thermal, chemical, and electrochemical stability, and are non-volatile, and non-flammable (Armand et al. 2009; Brandt et al. 2013). Their visible and chemical features are significantly modifiable due to their unlimited combinations of cations and anions (Freemantle 1998) and are categorized based on their composition as aprotic, protic, and zwitterionic kinds (Armand et al. 2009). ILs having imidazolium can generally give greater ionic conductivity, whereas those that have pyrrolidinium provide broadened electrochemical stable potential windows (Lewandowski and Galinski 2007; Galiński, Lewandowski and Stępniak 2006).

Numerous examinations of ultracapacitors employing ILs electrolytes gave working cell voltage higher than 3 V (Lewandowski and Galinski 2007; Lewandowski et al. 2010) and also showed that solvent-free ILs have handled the safety challenges related to the organic solvents. [EMIM][BF$_4$] and [BMIM][BF$_4$] ILs have high viscosities compared to those of organic electrolytes (Huddleston et al. 2001; Zhou, Matsumoto, and Tatsumi 2004), but their specific capacitance at an elevated scan rate or charging and discharging are quite low compared to those of aqueous and organic electrolytes (Chen et al. 2011; B. Xu, Wu, Chen, et al. 2008). However, challenges like large viscosity, small ionic conductivity, and elevated cost that greatly increase the ESR value are associated with ultracapacitors using ILs electrolytes, especially at room and low temperatures (Lewandowski and Galinski 2007; Lewandowski et al. 2010).

In an attempt to enhance the performance of ECs with IL electrolytes, the optimal choice of the IL composition and cell design has been explored experimentally and theoretically (Feng et al. 2011; Merlet et al. 2012; Fedorov and Kornyshev 2014) to get the basic knowledge of electrochemical characteristics, form, and the capacitive features of EDL at the interface of the IL/electrode. Improvements in this direction have been presented in successful reviews (Fedorov and Kornyshev 2014; Baldelli 2008; Burt, Birkett, and Zhao 2014; Feng et al. 2013), and it was reported that some cations of the ILs have similar structure with those of surfactants (Kunze, Jeong, et al. 2011; Kunze, Paillard, et al. 2011). Studies in this area are basically to understand the EDL form and capacitance in ionic liquids, develop EC's ionic liquids with pseudocapacitive electrodes for enhanced charge storage capability (Kruusma et al. 2014;

Romann et al. 2014), enhance conductivity of ions, viscosity, and electrolyte stability potential window of ILs by tuning the ions, and use combinations of ionic liquids or organic solvents to enhance the entire IL electrolytes' performance.

McEwen, McDevitt, and Koch (1997) in a bid to decrease viscosity and elevate ionic conductivity of ionic liquids via the mixture of ILs and organic solvents discovered that the conductivity of ionic liquids containing carbonates like [EMIM][PF_6] and [EMIM][BF_4] are about 25% more than that of a $TEABF_4$ organic electrolyte. Orita, Kamijima, and Yoshida (2010) examined varieties of alkyl functionalized IL electrolytes containing organic solvents like PC for ultracapacitors. Zhang et al. (2013) discovered that the mixture of DMF and [BMIM][PF_6] ionic liquids enhanced the capacitance and also reduced the ESR of asymmetric EC with AC//MnO_2 electrodes, due to the improved electrolyte permeation and ion movement. The mixture of pyrrolidinium-based ILs and organic solvents to get composite electrolytes for electrochemical capacitors have been developed and studied due to fairly high ESPWs, and the results showed a large increase in conductivity, a decrease in viscosity, and increase in working voltage to 3.2–3.5 V (Krause and Balducci 2011; Pohlmann et al. 2014; Ruiz et al. 2012).

11.4 Solid- or Quasi-Solid-State Electrolytes

The solid-state electrolytes work as ionic conducting media and the electrode separators and have drawn great attention recently. Key merits of solid-state electrolytes are that the ultracapacitors are easy to fabricate and package and liquid does not leak. The main kinds of solid-state electrolytes produced to date contain polymers, with few studies on ceramic materials (Francisco et al. 2012; Ulihin, Mateyshina, and Uvarov 2013). The solid electrolytes containing polymers employed in ultracapacitors are categorized into the solid polymer electrolyte, equally called dry polymer electrolytes, gel polymer electrolyte (GPE), and polyelectrolyte, also known as quasi-solid-state electrolytes because of the existence of liquid in it (Fan et al. 2014; Łatoszyńska et al. 2015). The GPEs depending on their composition can have fairly weak mechanical strength and a low working temperature range if dissolved in water, and this causes internal short circuits, which is a serious safety concern. Also, they have low ionic conductivities and small contact surface areas among electrolytes and electrodes, which increase the equivalent series resistance value and reduce the performance. Numerous reviews published recently focused on the solid-state electrolytes, these noted that it is very challenging for this electrolyte to satisfy all the following conditions: greater ionic conductivity, higher chemical, electrochemical, and thermal stability, and reasonable mechanical robustness and stable dimensionally (Choudhury, Sampath, and Shukla 2008; Gao and Lian 2014; Samui and Sivaraman 2010; Agrawal and Pandey 2008).

Besides water, solvents like PC (Duay et al. 2012), EC and DMF (Huang et al. 2012), or the combinations like PC–EC (Sudhakar, Selvakumar, and Bhat 2013), PC–EC–DMC (Ramasamy, Palma del vel, and Anderson 2014), and PC–EC (Huang et al. 2012) were mainly utilized as GPEs' plasticizers. The ratio of polymer to plasticizer mainly determines the plasticization level, thereby influencing the GPEs' glass-transition temperature (Choudhury, Sampath, and Shukla 2008). Several solid-state electrochemical capacitors have been produced based on poly-vinyl alcohol (PVA)-based hydrogels, namely: flexible ECs, stretchable ECs, flexible micro-ECs, printable micro-ECs on chip micro-ECs, three-dimensional (3D) micro-ECs yarn ECs, wire or fibre-like ECs, transparent ECs, ultrathin ECs, wearable ECs, paper-like ECs, and integrated ECs with other devices (X. Wang et al. 2013; Niu et al. 2013; Gengzhi Sun et al. 2015; S. Liu et al. 2014; Si et al. 2013; Kou et al. 2014; Meng et al. 2014; Gengzhi Sun et al. 2014; Peng et al. 2013; Ren et al. 2013; Yuan et al. 2013; T. Chen et al. 2014). Some works which attracted little attention in comparison with the polymer-based electrolytes were dedicated to the production of inorganic solid-state electrolytes (Francisco et al. 2012; Ulihin et al., Mateyshina, and Uvarov 2013; Q. Zhang et al. 2014), which are robust mechanically and stable thermally. Francisco et al. (2012) presented the application of glass-ceramic electrolytes ($Li_2S–P_2S_5$) having fairly high Li-ion conductivity as the ion conductor, as well as a separator for all solid-state ECs. Ulihin, Mateyshina, and Uvarov (2013) reported a composite solid electrolyte ($LiClO_4–Al_2O_3$) with $LiClO_4–Al_2O_3$ ratio of 4:6 for both symmetric and asymmetric ultracapacitors.

11.5 Redox-Active Electrolytes

A novel approach was recently investigated in order to enhance the ECs' capacitance by encouraging the pseudocapacitivity emanating from redox-active electrolytes. The redox reactions taking place in the electrolytes give additional pseudocapacitance to the ECs, different from the one coming from the pseudocapacitive electrode materials (Lota and Frackowiak 2009; Fic, Frackowiak, and Béguin 2012).

11.5.1 Aqueous Electrolytes with Redox-Active Species

A common example of a redox-active aqueous electrolyte is the iodide/iodine redox pair electrolyte employed in ultracapacitors with carbon electrodes by Lota and Frackowiak (2009). Phosphotungstic acid and Silicotungstic acid, which provide great proton conductivities, as well as many fast electron transfer oxidation and reduction activities (Lian and Li 2009; Suárez-Guevara, Ruiz, and Gomez-Romero 2013; Tian and Lian 2010) were investigated as redox-active electrolytes' mediators. Suárez-Guevara, Ruiz, and Gomez-Romero (2013) discovered that employing the phosphotungstic acid ($H_3PW_{12}O_{40}$, PW_{12}) electrolyte in ECs with activated carbon (AC) electrodes has the capacity to give pseudocapacitance from the electrolyte and elevate the device working voltage to 1.6 V due to PW_{12}'s elevated overpotential to the evolution of H_2. Redox-active electrolytes containing a pair of metallic ions have equally been examined for ultracapacitors (Tanahashi 2005; Mai et al. 2013), but fast self-discharge is associated with this type of electrolytes. Also, reasonable research works have been directed towards the organic redox mediators like hydroquinone, methylene blue, indigo carmine, p-phenylenediamine, m-phenylenediamine, lignosulfonates, sulfonated polyaniline, and humic acids as pseudocapacitive sources (Roldán, Granda, et al. 2011; Roldán et al. 2012; L. Chen, Bai, et al. 2014; Roldán, González, et al. 2011; Wu et al. 2012; Yu, Fan, et al. 2012; Wasiński, Walkowiak, and Lota 2014; Lota and Milczarek 2011; L. Chen, Chen, et al. 2014). Introduction of these mediators can significantly enhance the capacitance of ultracapacitors using carbon electrodes by two- to fourfold. Electrochemical capacitors with a hydroquinone redox-active electrolyte exhibited quick self-discharge, and Chen et al. (L. Chen, Bai, et al. 2014) showed that movement of the redox-active species among the electrodes is responsible for quick self-discharge. Chen et al. (L. Chen, Bai, et al. 2014) presented that using a Nafion ion-exchange membrane separator and choosing electrolytes which are reversibly transformed to an insoluble species while charging and discharging the device successfully suppressed the self-discharge. This was possible because stopping the movement of the redox species among the electrodes will quench the self-discharge.

11.5.2 Non-aqueous Electrolytes with Redox-Active Species

In an attempt to increase the applied voltage and energy density, organic and ionic liquids electrolytes (Ionica-Bousquet et al. 2010; Tooming et al. 2014; Yamazaki et al. 2012) were examined and presented in the literature. The research showed that the improvement of the NaI/I_2 mediator was very noticeable compared with the $K_3Fe(CN)_6/K_4Fe(CN)_6$ mediator in the studies. Also, the addition of a redox mediator into the GPE could greatly enhance ionic conductivity, mostly when the GPE has ACN salts in the solution (J. Zhou et al. 2011; Yin et al. 2011; Ma, Li, et al. 2014).

11.5.3 Solid Electrolytes with Redox-Active Species

The approach of utilizing a redox-mediator has equally been tried in ultracapacitors using a solid- or quasi-solid-state electrolyte and reasonable performance improvements have been noticed. The common salts in solid-state electrolytes examined in this direction were iodides like NaI and KI, $K_3Fe(CN)_6$, Na_2MoO_4, mediators like hydroquinone, p-phenylenediamine, p-benzenediol, and methylene blue, redox salts like KI–$VOSO_4$, etc. (Yuan et al. 2013; J. Zhou et al. 2011; Yin et al. 2011; Ma, Li, et al. 2014; Senthilkumar et al. 2013, 2012; Ma, Feng, et al. 2014). The mediators are basically similar to those employed in liquid electrolytes. Most studies on solid-state electrolytes with a polymer were concentrated on the GPEs hosted in PVA, PEO, (Yin et al. 2011; Ma, Li, et al. 2014; Senthilkumar et al. 2012, 2013), and Nafion (J. Zhou et al. 2011; Mansour, Zhou, and Zhou 2014). Because the active salts are usually in liquid GPEs, their influence and operating mode are identical to those in aqueous or organic electrolytes.

11.6 Separators

The essential requirements a suitable separator has to meet include having a minimal or zero ionic resistance inside the electrolyte, but have strong electronic insulating ability, have great chemical and electrochemical stabilities, and have outstanding mechanical toughness for durability. Separators are usually made up of cellulose, polymer membranes, and glass fibres. Although cellulose separators function well in organic solvents (Bittner et al. 2012), they degrade in an acidic electrolyte. Bittner et al. (2012) discovered that any little quantity of water in a $TEABF_4/ACN$ electrolyte can accelerate ultracapacitors' ageing or degradation if a cellulose separator was employed. The ionic conductivity of viscous electrolytes like ILs in a separator can change internal resistance (ESR), thereby affecting the device performance (Tõnurist et al. 2012). A separator's composition, thickness, porosity, pore size distribution, and surface morphology affect different electrochemical capacitors' performance indexes differently, such as capacitance, ESR, energy, power densities, etc. (Tõnurist et al. 2012). Liu and Pickup (2008) studied four kinds of separators for a RuO_2 electrodes-based EC using 0.5 M H_2SO_4 as an electrolyte. The selection of a suitable separator in ultracapacitors with redox-active electrolytes positively affects the device performance by suppressing the self-discharge processes (L. Chen, Bai, et al. 2014; L. Chen, Chen, et al. 2014). Shulga et al. (2014) presented that novel separator materials like graphene oxide (GO) films (Shulga et al. 2014) and eggshell membranes (H. Yu, Tang, et al. 2012), might be a suitable supercapacitor separator by displaying proton conductivity after absorbing a H_2SO_4 electrolyte.

11.7 Contribution of Electrolytes in Achieving the Next-Generation Electrochemical Capacitors

The electrolyte is one of the key components of ECs, which gives the ionic conductivity and enhanced charge storage process on electrodes. Currently, most of the existing ultracapacitors employ organic electrolytes that have applied voltages ranging from 2.5 V to 2.8 V (Conway 1999; Burke and Miller 2011). An electrolyte plays cardinal functions like ion supplementary conduction of an electric charge, electron conduction, and adhesion of electrode particles in determining the performance of ECs, such as solid-state devices. An appropriate ultracapacitor's electrolyte should basically have: broad decomposition potential window, elevated electrochemical stability, high concentration of ions, low resistivity, low toxicity, etc. The design of novel electrolyte components is taken as one of the significant approaches to achieve the next-generation ultracapacitors with enhanced energy (Balducci 2016), as shown if Figure 11.2.

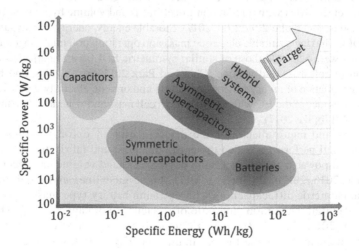

FIGURE 11.2 Ragone plot displaying the specific power against specific energy of different energy storage devices. (Noori, A. et al., *Chem. Soc. Rev.*, 48, 1272–1341, 2019. Reproduced by permission of The Royal Society of Chemistry.)

A lot of works have been presented that suggested that it is more effective to improve energy density by increasing the applied voltage window going by the fact that energy and power densities are proportional to voltage squared according to equations (11.1) and (11.2), and this is successively achievable by selecting suitable electrolytes that have a broader potential window.

Advanced electrolytes for ultracapacitors with high electrochemical stability are important in enhancing a devices suitability for various usages (Heß, Wittscher, and Balducci 2019). Investigation of electrolytes using propylene carbonate and butylene carbonate solvents were carried out. A comparison of various solvent-salt combinations demonstrated that 1.0 M N-butyl-N-methylpyrrolidinium bis(trifluoromethanesulfonyl)imide (Pyr$_{14}$TFSI) in butylene carbonate is superior to electrolytes using ordinary PC as regards voltage window and capacitance. Several groups developed different electrolytes with high ionic conductivity, ESPWs, and electrochemical stability for the purpose of getting high ECs (Yu et al. 2015; Tian et al. 2019; Kumar et al. 2019; Li et al. 2019; Lian et al. 2018; Iwama et al. 2019; Liu et al. 2019; Lee et al. 2019; Wang et al. 2019; Menzel, Frackowiak, and Fic 2019; Zhong et al. 2015; Chen and Lin 2019).

Research has provided expressions and guidelines to determine the practical electrode thickness, optimum charging current, and utilization of materials in devices with a specific practical electrode and electrolyte conductivity for next-generation ECs (Ike, Sigalas, and Iyuke 2017c). Theoretical expressions for different performance parameters of various ultracapacitors have been optimized subject to electrodes and electrolytes using optimal coefficients associated to battery-type material K_{BMopt} and electrolyte material K_{Eopt} (Ike 2017; Ike, Sigalas, and Iyuke 2017b; Choi and Park 2014). The dependence of various ECs' performance on electrode and electrolyte fabricating conditions, like mass ratio of electrode, electrode type, materials reaction, potential window of electrolyte, and capacitance through modelling and simulation have been presented (Ike 2017; Ike, Sigalas, and Iyuke 2017b; Choi and Park 2014). Having a perfect knowledge of the effects of each structural parameter, such as electrode thickness, porosity, separator thickness and porosity, electrode and electrolyte effective conductivities, and operating conditions like charging current densities and times is crucial in the design and fabrication of next-generation ultracapacitors with high performance.

The research performances of asymmetric ECs with suitable electrode mass and operating potential range ratios employing aqueous electrolytes, and those with suitable electrode mass, operating potential range ratios with organic electrolytes of appropriate operating potential range, and specific capacitance were 2.20- and 5.56-fold, respectively, higher than those of symmetric EDLCs and asymmetric ECs employing the same aqueous electrolyte (Choi and Park 2014; Ike, Sigalas, and Iyuke 2017b; Ike 2017). This improvement was associated with a reduction in cell mass and volume. Storable and deliverable energies of the asymmetric ECs with suitable electrode mass and operating potential range ratios employing proper organic electrolytes were also a factor of 12.9 greater than those of symmetric EDLCs employing aqueous electrolytes with a reduction in cell mass and volume by a factor of 1.73 (Choi and Park 2014; Ike, Sigalas, and Iyuke 2017b; Ike 2017). Storable energy, energy density, and power density of asymmetric EDLCs having a suitable electrode mass and operating potential range ratios, with proper organic electrolytes, were 5.56-fold those of similar symmetric EDLCs using aqueous electrolytes with 1.77-fold reduction in cell mass and volume (Choi and Park 2014; Ike, Sigalas, and Iyuke 2017b; Ike 2017). Also, the introduction of an asymmetric electrode and organic electrolyte was very successful in enhancing the performance of the EC with a reduction in cell mass and volume (Choi and Park 2014; Ike, Sigalas, and Iyuke 2017b; Ike 2017).

Useful guidelines and requirements for the determination of optimum ratios, proper organic electrolytes for optimal performance, and the entire blueprint and fabrication of ultracapacitors of enhanced energy and power with a reduction in device mass and volume have been presented (Ike, Sigalas, and Iyuke 2016, 2017b, 2017c; Ike 2017). These guidelines and requirements aid in the development of asymmetric ultracapacitors with optimum battery-type mass ratio, potential range ratio, maximum potential range ratio, and ratio of capacitance of capacitor-type (Ike, Sigalas, and Iyuke 2016, 2017b, 2017c; Ike 2017). Also, EC models that incorporated various self-discharge mechanisms were used to determine the minimum impurity or redox species concentration and optimum total thickness of separators and anodes for self-discharge suppression (Ike 2017; Ike, Sigalas, and Iyuke 2017a, 2018).

11.8 Progress on Electrolytes of Electrochemical Capacitors/Ultracapacitors

Electrolytes have influence on the ultracapacitor's performance like ion supplementary, conduction of electric charge, and electrode particles adhesion in solid-state ECs. Fundamental demands of a suitable electrolyte for ECs are: wide range of decomposition potential, great electrochemical stability, large ionic concentration, low resistivity, low toxicity, etc.

In order to get high working voltage in electrochemical capacitors, replacing aqueous electrolytes with organic electrolytes is essential and successful because increasing the applied voltage affects the energy density in a quadratic manner as compared with enhancing the cell capacitance. Both symmetric and asymmetric ultracapacitors using organic electrolytes give high energy densities compared with similar capacitors using aqueous electrolytes, not minding that electrodes in organic electrolytes basically display a lower specific capacitance because of a low ionic conductivity, high viscosity, and large ionic size, as shown in Figure 11.3. Flammability and toxicity issues of organic electrolytes must be addressed not minding the cost intensiveness of assembling the cell and solvents.

The highest cell voltage of commercial ultracapacitors using PC or ACN organic electrolytes is 2.7 V, though a voltage of 2.85 V were recently reported (Balducci et al. 2007). Considering the likely side reactions occurring at cell voltages higher than 2.7 V, Hahn et al. (2005) and Ishimoto et al. (2009) discovered CO_2, propene, and H_2 as products of decomposition from the cathode and anode when CV cycling is beyond 2.6 V in $TEABF_4$/PC. Recently, many researchers introduced the mixed-solvent electrolytes into electrochemical capacitors in order to obtain a next-generation ultracapacitor with elevated energy, power, and wide working temperature window [Ikc et al. 2016, 2017b, 2017c, Ike 2017]. Some researchers mixed ethyl carbonate (EC), DMC, or both with commercial organic electrolytes because of EC's high dielectric constant and DMC's low viscosity (Laheäär et al. 2009; X. Yu et al. 2015; Tian et al. 2014). A g-butyrolactoneacetonitrile system, which is mixed-solvent electrolytes containing ACN that have high conductivity and low viscosity can increase the working temperature window, reduce the equivalent series resistance, and attain high applied voltage (Ding et al. 2004). Ionic liquids are stable organic salts of ions, which are successful electrolytes for next-generation ECs. ILs have the following outstanding features because of their peculiar physicochemical properties: wide voltage, low flammability, poor fluidity, and wide working temperature, although they have the following issues of concern: high viscosity, low ionic conductivity, and high cost.

At least voltages of 3.5 V and above were achieved in ultracapacitors utilizing ILs electrolytes due to their high chemical and electrochemical stability (Lian et al. 2016; Liu et al. 2010; Iamprasertkun,

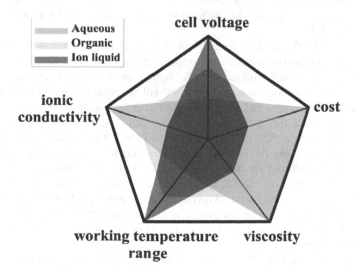

FIGURE 11.3 Comparisons of different desirable features of aqueous, organic, and ionic liquid electrolytes used in ECs showing superiority of electrolytes. (Reprinted from *Carbon*, 145, Liu, C.-F. et al., Carbon materials for high-voltage supercapacitors, 529–548, Copyright 2019, with permission from Elsevier.)

Krittayavathananon, and Sawangphruk 2016; Huang et al. 2015). Qin et al. (C. Lian et al. 2016) discussed the potential windows of general ionic liquids and noticed that graphene electrodes in 1-methyl-1-propylpiperidinium bis(trifluoromethyl sulfonyl)imide electrolyte could work at 4 V. Chang et al. (Huang et al. 2015) presented that various ionic liquids display various decomposition potentials and a larger potential range is achievable by combining butylmethylpyrrolidinium and dicyanamide. The materials of every component of high-voltage ECs have to be stable in the concerned potential range, and as such, the stabilities of other device components like binders and current collectors have to be accounted for (Balducci et al. 2005; Kühnel and Balducci 2014; Varzi, Balducci, and Passerini 2014). Aubert et al. (Dagousset et al. 2017) showed that the mixture of g-butyrolactone and ionic liquids such as TFSI has the capacity to operate at elevated temperatures and not combust. Although Liaw et al. (2012) pointed out that ionic liquid's side products can burn and are prone to fire outbreak. Romann et al. (2017) presented ECs with electrolytes that contain tetracyanoborate, which display a self-stopping process that encourages electrode surface passivation to suppress more reactions. A composite of ionic liquids and organic solvents give a reduced voltage range between 2.5 and 3.5 V and are restricted to operate at fairly low temperatures to avert decomposition of the organic solvents (Liu et al. 2010; Dagousset et al. 2017; Leyva-García et al. 2016; Zhang et al. 2016).

Ionic liquid electrolytes operate comfortably at high temperatures due to improved conductivity, although they have the capacity to function in a broad temperature range as shown by the galvanostatic/charge discharge (GCD) cycling test, floating test, and low and high temperature tests (Dagousset et al. 2017; Lin, Taberna, and Simon 2016; Sun et al. 2012). Because the temperature of the surroundings affects ionic liquid conductivity, device capacitance at reduced temperatures is particularly essential (Lin, Taberna, and Simon 2016). The performance of ultracapacitors with ionic liquid electrolytes operating at elevated temperatures is certainly higher than that of similar devices with non-IL electrolytes operating at low temperatures, although working at elevated temperatures restricts the real use of ECs using ionic liquids.

11.8.1 Factors Determining the Cell Voltage of Electrochemical Capacitors

The main issue restricting the ultracapacitor's applied voltage is the unwanted side reactions taking place at an electrode and electrolyte interface, and numerous researchers have attempted to find ways to increase supercapacitor's cell voltage without or with minimal side reactions in the past decade. Three possible approaches to successfully increase the ultracapacitor's voltage so as to enhance the capacitor's energy are: use of organic solvents or ionic liquids (Chi et al. 2016); electrode surface passivation to suppress the reaction and interaction between solvents, salts, and electrode surface defects (Shen and Hu 2014); and synthesis of electrodes with highly inert surfaces to allow a wider potential range in normal electrolytes (Chi et al. 2016).

Numerous studies in literature have focused on how to modify the electrode surface in a bid to acquire a wide operating potential window and suppress the side reactions. Brandt and Balducci (2012), Pohlmann, Ramirez-Castro, and Balducci (2015), Schroeder et al. (2013) produced a novel solvent and salts based on AC to increase the ultracapacitors' voltage beyond 3.0 V. The main reason several researchers recently affirmed that the specific surface area is now an inferior factor in deciding the ECs performance is because the applied voltage influences capacitor energy in a quadratic manner compared with the capacitance (K. Xu, Ding, and Richard Jow 2001). Brandt and Balducci showed that electrochemical capacitors using 0.7 M TEABF$_4$ in adiponitrile solvent displayed an applied voltage range of up to 3.5 V (Brandt and Balducci 2012; Ruther et al. 2017). Balducci (2016) presumed that acquiring applied voltages of 2.7, 2.9, 3.2, and 3.5 V were realizable in devices using the TEABF$_4$-, TEATFSI-, Pyr$_{14}$BF$_4$-(Pyr$_{14}$: 1-butyl-1-methylpyrrolidinium), and Pyr$_{14}$TFSI, respectively. Apart from changing organic solvents, the use of various salts as the complementary electrolyte also have the potential to affect the working voltage range of organic-based ultracapacitors (Ue, Ida, and Mori 1994; Pohlmann et al. 2014; Yu et al. 2014).

Xu et al. (K. Xu, Ding, and Jow 2001) examined different ions in similar ultracapacitor system, from where they noticed a general pattern that larger cations are highly resistant to the electrochemical reduction, and this agrees with results presented earlier by other groups (Morita, Goto, and Matsuda 1992). Simon et al. (Chmiola et al. 2006; Zhang et al. 2008) discovered that electrode pores of diameters smaller than solvated ions size can increase the device capacitance. A saturation process was discovered for the sub-nanometre porous electrodes in ultracapacitors (Mysyk, Raymundo-Piñero, and Béguin 2009). The size consistency among active material pore sizes and electrolyte ionic sizes conclusively shows massive effect on capacitance and suitable potential range. The approaches to increase both ultracapacitors' capacitance and charge storage in organic electrolytes are: withdrawal of surface functional groups, integrating heteroatoms in electrodes, and addition of hydrophobic groups to the electrode surface. Nitrogen functional groups doped onto electrodes greatly affect electrochemical capacitor function in organic electrolytes (Salinas-Torres et al. 2015; Li et al. 2015). Salina-Torres et al. (2015) demonstrated that the existence of positively charged N species lowers the chances of the nucleophilic effect on electrolyte molecules and thereby enhances the electrolyte stability. Surface tuning such as surface deposits created the electrolyte's disintegration potential window, which was seen to increase the operating potential range of dried ACs due to a passivation influence in the 1 M $TEABF_4$/PC electrolyte after polarization.

Additives in the LIB application also operate effectively in lithium ion capacitors (Michan et al. 2016). The salts in LIBs are targeted at modifying the solid electrolyte interphase's composition film and to quench disintegration of an electrolyte. Some works show that fluoroethylene carbonate has the capacity to decrease the quantity of non-conductive inorganic compounds like HCO_2Li, $Li_2C_2O_4$, and Li_2CO_3. Gustafsson et al. (Xu et al. 2015) disclosed that fluoroethylene carbonate can quench $LiPF_6$ mortification, whereas Boltersdorf et al. (2018) demonstrated that vinylene carbonate and tris (trimethylsilyl) phosphate can enhance both the rate performance and cycle life. Surface tuning is a successful approach to quench electrolyte decomposition in order to obtain a wider potential window. The cardinal idea of hybrid electrochemical capacitor (HEC) use for elevated voltage and elevated energy implementations is to integrate two electrode substances having a battery-kind and capacitor-kind. In order to obtain a high cell voltage, the positive (cathode) and negative (anode) electrodes must have a wide and stable positive working potential window and a wide and stable negative potential window, respectively (Rajkumar et al. 2015). Ultracapacitors that successfully use the concept of the HECs are lithium-ion capacitors, where AC and graphite due to their highly negative lithium intercalation potential work as cathode and anode, respectively. The combination gave an applied voltage of 3.8–4.1 V (Budarin et al. 2000) due to the higher potential limit of the AC anode, which can attain 4.3 V vs Li/Liþ. Numerous groups took lithium ion capacitors as the new generation ultracapacitors because they have the capability to give a higher energy density compared with the commercial organic ECs because of LIBS intercalation/de-intercalation phenomena (Amatucci et al. 2001; Naoi and Simon 2008; F. Zhang et al. 2013). Another prospective HEC system is the $Li_4Ti_5O_{12}$//AC, due to a $Li_4Ti_5O_{12}$ perfect cathode which makes for a high power LIB owing to its 'zero strain' lithium-ion insertion features (He et al. 2012). Studies have shown that main charges balance via a mass loading and right evaluation of charge capacities (Hsu et al. 2014) of the anode and cathode to concurrently enhance the performances of HECs using a one battery-type electrode.

Ruther et al. (2017) presented that $NaPF_6$ in 1,2-dimethoxyethane or monoglyme electrolytes can be employed to suppress the aluminium current collectors' corrosion, enhance ionic conductivity and applied voltage to 3.5 V under certain tests. Krause and Balducci (2011) and Hess and Balducci (2018) equally produced novel non-aqueous electrolytes for enhancing the ultracapacitor's cycling stability. Fang et al. (2005) and Fang and Binder (2006) showed that the existence of hydrophobic groups elevate the inclination of AC to organic electrolytes, thereby improving the capacitance of AC further.

Interestingly, a reasonable enhancement of energy density is achievable by a fairly novel method using lithium-ion intercalation material like $LiFePO_4$/AC, stabilized lithium metal powder//AC, $Li_4Ti_5O_{12}$//metal-organic structure-derived carbon, and fluorine-doped Fe_2O_3//AC as the cathode (Cao and Zheng 2013; Banerjee et al. 2014; Karthikeyan et al. 2014; Shellikeri et al. 2018). The crucial factors in the development of new-generation ultracapacitors with high voltage and outstanding performance are: the design of electrolytes with enhanced features like ionic conductivity, viscosity, and electrochemical/chemical stability. Such modification will permit high applied voltages, low equivalent series resistance, and long cycling. The mass ratio of the anode and cathode were shifted for purposes of balancing the charges so as to obtain the maximum applied voltage and long life; and the asymmetric arrangements were needed to obtain capacitors with elevated voltage, power, and energy densities (Béguin et al. 2014).

11.9 Challenges of Ultracapacitors Electrolytes

Increasing energy and power densities as well as improving the durability of electrochemical capacitors are usually the key problem in developing or achieving the next-generation ultracapacitors. It was discovered that the electrolyte is one of the major components that decide the ECs performance: the ESPW value, ionic conductivity value, chemical and thermal stabilities, and working temperature window, all have great influence on device performance and real application. Much progress has been made to date in the area of an ultracapacitor's electrolyte, but there are still extensive difficulties militating against the achievement of the new-generation electrochemical capacitors:

1. *Insufficient value of ESPWs and its influence on energy density:* The ultracapacitors working voltage which influences energy and power densities is greatly determined by the electrolytes' ESPW. Electrolytes with a very high electrochemical stability potential window permit higher applied voltage, which largely enhances energy density; the elevated voltage usually leads to degradation of other features like ionic conductivities, viscosity, etc. ECs using organic electrolytes have higher applied voltages compared with similar cells with aqueous electrolytes, but they have lower ionic conductivities than one using aqueous electrolytes. ECs with IL electrolytes can enhance the applied voltage to 4 V, but their ionic conductivities are quite low compared with those with organic electrolytes. Also, devices using organic or ionic liquid electrolytes usually exhibit lower capacitance than those using aqueous electrolytes. Thus, it is very difficult to produce ultracapacitors that have enhanced energy density, while maintaining other merits like elevated power density and durability.

2. *High value of equivalent series resistance and its influence on power density:* Organic and ionic liquid electrolytes, due to their high ESPWs, are suitable for enhancing the ECs energy density, but their ESRs are much higher because of lower ionic conductivities and high viscosities (Vaquero et al. 2012). The ESR values of ultracapacitors with IL electrolytes are higher compared with those using organic and non-aqueous electrolytes due to low ionic conductivities and high viscosity. On this understanding, the use of electrolytes with higher electrochemical stability potential windows with ECs usually results in a higher ESR and a reduced power performance, except that the power density degradation can be cushioned by the increase in cell working voltage.

3. *Electrolyte associated impurities and their influence on ESPW and self-discharge:* Associated impurities such as little quantities of water could significantly reduce the electrochemical stability potential window of organic or IL electrolytes. These impurities or remnant gas inside electrolytes trigger self-discharge and its phenomena because they are prone to adulterants and remnant gas. Several researchers have presented that self-discharge processes are triggered by given quantities of metallic contaminant, water, and oxygen existing in electrolytes (Andreas, Lussier, and Oickle 2009; Oickle and Andreas 2011; Ricketts and Ton-That 2000).

4. *Unoptimized pairing among the electrolyte and electrode and the influence on entire EC performance:* Optimized matching of the electrolyte and electrode is cardinal in obtaining the ultracapacitors enhanced performance. Few research works were concentrated on the electrochemical capacitors rate and power performance, as compared with the extensive studies on optimizing the specific capacitance via the matching of the electrolyte's ionic size with the porous structure of electrode materials. The matching strategy is a successful approach to improve the specific capacitance of ECs because the approach does not compromise the power density which is the key merit of ultracapacitors.

   Also, various results concerning the pore size and surface-area-standard capacitance have been obtained (Largeot et al. 2008; Jiang et al. 2012; Chmiola et al. 2006; Centeno, Sereda, and Stoeckli 2011). Chmiola et al. (2006) and Largeot et al. (2008) presented that capacitance could be elevated if mean pore size of the electrode approached the ionic size for ionic liquids and organic electrolytes. Centeno, Sereda, and Stoeckli (2011) examined 28 porous carbon electrodes in TEABF$_4$/ACN electrolytes and discovered that capacitance was fairly steady between 0.7 and 15 nm.

Moreover, the design of novel electrolytes consisting of multivalent ions, combinations of ILs, combinations of ILs and organic solvents, and solid-state electrolytes, recently made the optimization of pairing of electrode substances and the electrolytes less attractive.

5. *Undesirable temperature range for ECs working for various real-life usages:* Generally, utilized ultracapacitor organic solvents like ACN and PC have a fairly narrow working temperature window because of the elevated volatility at elevated temperatures or low viscosity at low temperatures. Although adiponitrile and IL electrolytes that have higher temperature limits have been examined, however, they have associated concerns.

6. *Expensiveness of some electrolytes:* The reduction of ECs' cost to the barest minimum is essential for their commercialization. Aqueous electrolytes are not expensive, but they usually yield a low energy density because of their low ESPW values, especially in symmetrical ECs. Ultracapacitors with organic or ionic liquid electrolytes have improved working voltages and therefore improved energy and power densities, but they increase greatly the overall cost of ultracapacitors.

There is an inadequate basic understanding of the electrolyte processes for EC performance in devices with new electrolyte material, design, and optimization of interactions between electrode and the electrolyte materials. There have recently been reasonable achievements in theories and in situ experiments, but with the design of novel electrolytes like ionic liquids and solid-state electrolytes and novel electrode substances, sound knowledge of charge and discharge mechanisms for the novel systems are still a problem. For example, there are some controversies concerning the influence of temperature (Fedorov and Kornyshev 2014) and introduced organic solvents (Feng et al. 2011; Shim, Jung, and Kim 2011) on the EDL form and capacitance at the interface of the ionic liquids and electrode. Density functional theory examinations suggested that the capacitance of nano-porous electrodes via IL electrolytes swing with the increase in pore size (Jiang, Jin, and Wu 2011), but direct experiments are needed for the confirmation. Comprehension of the movements of ions in rigorous electrode formations when charging and discharging, as well as a basic understanding like modelling equations of pseudocapacitors and hybrid ultracapacitors for a given electrolyte are still poorly appreciated or understood (Ozoliņš, Zhou, and Asta 2013).

7. *Appropriate and standardized approaches of characterization to determine performance of EC electrolytes:* Because of several novel electrolytes developed, there is a need to get a reliable approach appropriate to distinguish their performance and the related ECs. It is currently challenging to match experimental results in various literature because performance indices were normally obtained under varying situations. Also, it is difficult to recognize suitable electrolytes from various literature, and ultracapacitors gravimetric capacitance, energy, and power densities were usually determined on the basis that the device mass is the mass of only the active electrode substances. The majority of works in literature ignore electrolyte mass, which constitutes a substantial portion of the entire EC's mass and cost. Adulterants and remnant water significantly reduce the ESPW of organic electrolytes, thereby causing the evolution of gas. Achieving next-generation ultracapacitors having enhanced energy and power densities through cell voltage enlargement is still a real challenge.

11.10 Summary

The influences of the properties of electrolytes, such as the electrochemical stability potential window, ionic conductivity, viscosity, and thermal stability on electrochemical capacitors' performance–like its capacitance, energy and power densities, ESR, cycling stability, energy retention, and degradation were fully discussed. The likely interactivity among electrolytes and ECs inactive constituents like current collectors, binders, and separators, were studied. Upon reasonable success in the area of research, several difficulties like low energy density, un-optimized pairing of novel electrolytes and electrode substances, lack of excellent ways to determine electrolyte performance in ECs, capital intensiveness of many electrolytes, and a lack of basic knowledge of charge storage phenomena for novel electrolytes in practical and rigorous electrodes are in existence.

The following research directions are proposed to overcome the above stated problems:

1. *Enhancement of electrochemical stability potential window to enhance energy density of ultracapacitors:* ESPWs of organic electrolytes depend upon the ions in the conducting salt and organic solvent, such as single or a mixture of solvents. The electrochemical stability potential windows can therefore be enhanced by using novel organic solvents and/or novel conducting salts by optimizing and tuning the mainly used organic electrolytes. The application of IL electrolytes in ECs has the capacity to enhance the device voltage and thus the energy and power densities, although increased ESPWs could lead to degradation of other properties like ionic conductivity, ESR, and viscosity. The development of a blueprint of ions or utilizing combinations of electrolytes and ionic liquids having high electrochemical stability potential windows and suitable ionic conductivities and viscosity is achievable, although it is challenging for electrolytes to satisfy every demand at the same time. Therefore, some compromise which greatly depends on the electrodes and electrolyte materials, as well as the requirements of the ECs application is appropriate in solving the real challenge.

2. *Use of a pseudocapacitive contribution to enhance the charge capacity:* Under this condition, electrochemical capacitors using aqueous electrolytes can attain a reasonable high energy density particularly with hybrid or asymmetric configurations like lithium ion capacitors. Hybrid ultracapacitors with aqueous electrolytes are likely to give a high energy density because of their favourable merits like low cost, high safety, and simple to fabricate without rigorous and stringent drying/purification.

3. *Increasing the electrolyte's purity:* Because adulterants in electrolytes affect the ESPW and lead to fast self-discharge, it is more effective to produce an appropriate purification system to reduce the concentration of adulterants in electrolytes. The purification of organic solvents could be understood from the knowledge of Li-ion batteries, since the area of Li-ion batteries is clearly understood. It is proposed that great attention be given to ionic liquid electrolytes with hydrophobic anions, which will significantly reduce the quantity of water in electrolytes and make the electrolytes stable in the long run.

4. *Reduction of ESR values to enhance the ultracapacitors power density:* The design of electrolytes that have high ionic conductivities and low viscosity will certainly help to reduce the ESR values. In organic electrolytes, very low ESR values are obtained by using novel conducting salts like $SBPBF_4$, thus producing a combination of solvents that reduce viscosity and optimize the solutions. Concerning the IL electrolytes, the blueprint or choice of suitable IL ions could help to increase the electrolyte's ionic conductivity and reduce the viscosity.

5. *Optimizing the pairing of electrolyte with electrode substances to enhance entire performance:* Since various outcomes of how pore size and surface-area-normalized capacitance are related were presented (Largeot et al. 2008; Chmiola et al. 2006; Centeno, Sereda, and Stoeckli 2011), more basic experimental results are needed to absolutely comprehend how they are related, and that will give a guideline to optimizing pairing electrode pore size with the ionic size. Moreover, enhancing the ultracapacitors' capacitance and specific energy while maintaining the power density should lead to more examinations dedicated to the production of novel electrodes having a unique form and capacity to give high capacitance via a high interfacial area without compromising perfect movement of ions.

6. *Increasing the ultracapacitors operating temperature range:* Different approaches can be used to increase the working temperature range of electrochemical capacitors depending on the type of electrolytes employed. Ethylene glycol could be utilized as a salt to reduce the temperature range in aqueous electrolytes (Roberts, Namor, and Slade 2013), whereas design or formulation of novel organic solvent mixtures has the potential to widen the operating temperature window more than the single solvent electrolytes. Concerning ionic liquid electrolytes, using an eutectic mixture of ionic liquids and the production of novel electrode substances can successfully

increase the temperature window of ECs using ionic liquid electrolytes, particularly the lower temperature boundary (Lin et al. 2011).

7. *More basic knowledge via outcomes of theories and experiments:* Concerning blueprint and optimization of novel electrolytes, choice of electrolytes to enhance the electrochemical stability potential windows, ionic conductivity, thermal stability, reduced viscosity, and clearer basic comprehension via experiments and theories are inevitable. It is quite essential to get the basic knowledge of processes of ionic movements, solvation/desolvation, and the mechanisms of charge storage in practical electrode structures during charging and discharging.

   Particular attention should be given to the newly produced electrolytes with many valent ions, ionic liquid mixtures, ionic liquid-organic solvent mixtures, and solid-state. Basic studies on the interaction between electrolytes and pseudocapacitive or hybrid electrodes in the newly developed pseudocapacitive electrode–electrolyte systems and hybrid ECs are important. The basic understanding will give a guideline for choosing or producing novel electrolytes and enhance the production of electrode substances to match a given kind of electrolyte. The understanding of electrolyte degradation and failure modes, which is achievable via theoretical and experimental modelling methods is needed in order to suppress electrolyte degradation.

8. *Development of standard approaches to evaluate electrolyte performance:* The mass and volume of the electrolytes used should be considered while characterizing the ultracapacitor performance. A standard description and approach for evaluating the mechanical properties and performance of flexible ultracapacitors using solid-state electrolytes under applied stress is required. Also, greater output and quick selection approaches to evaluate electrolyte performance need to be prioritized in order to hasten the design, formulation, and production of novel electrolytes.

In order to overcome these challenges, functional group reduction or generation of passivation film on electrode surfaces via tuning are suggested as successful in suppressing the concerns like reduction of ESPWs through decomposition of electrolytes and gas evolution on prolonged elevated voltage charging and discharging. Furthermore, hybrid configurations are suggested to get the highest applied voltage so as to improve ultracapacitor energy density. Research and development by industries and academia showed that lithium-ion capacitors could function with an applied voltage of 3.7–4.1 V while maintaining high cycle-life and perfect power density, which are regarded as real electrochemical capacitors. While it is true that many examinations have shown that their methods can increase the ultracapacitors applied voltage to improve specific energy, many other factors like functioning temperature window and ageing of EC's binders, current collectors, and separators might cause reduced capacitance, increased ESR, and elevated internal pressure, which degrade devices in the long run.

ACKNOWLEDGEMENTS

The financial assistance of the African Centre of Excellence in Future Energies and Electrochemical Systems (ACE-FUELS), Federal University of Technology, Owerri, Nigeria, National Research Foundation (NRF), and DST/NRF Centre of Excellence in Strong Materials (COE-SM), University of the Witwatersrand, Private Bag 3, Johannesburg 2050, South Africa, towards this research is hereby acknowledged. Opinions expressed and conclusions arrived at, are those of the authors and are not necessarily to be attributed to the African Centre of Excellence in Future Energies and Electrochemical Systems (ACE-FUELS), Federal University of Technology, Owerri, Nigeria, and NRF, DST/NRF Centre of Excellence in Strong Materials.

REFERENCES

Agrawal, R. C., and G. P. Pandey. 2008. "Solid Polymer Electrolytes: Materials Designing and All-Solid-State Battery Applications: An Overview." *Journal of Physics D: Applied Physics* 41 (22): 223001. doi:10.1088/0022-3727/41/22/223001.

Aida, Taira, Ichiro Murayama, Koji Yamada, and Masayuki Morita. 2007. "Analyses of Capacity Loss and Improvement of Cycle Performance for a High-Voltage Hybrid Electrochemical Capacitor." *Journal of the Electrochemical Society* 154 (8): A798–804. doi:10.1149/1.2746562.

Amatucci, Glenn G., Fadwa Badway, Aurelien Du Pasquier, and Tao Zheng. 2001. "An Asymmetric Hybrid Nonaqueous Energy Storage Cell." *Journal of the Electrochemical Society* 148 (8): A930–39. doi:10.1149/1.1383553.

Andreas, Heather A., Kate Lussier, and Alicia M. Oickle. 2009. "Effect of Fe-Contamination on Rate of Self-Discharge in Carbon-Based Aqueous Electrochemical Capacitors." *Journal of Power Sources* 187 (1): 275–83. doi:10.1016/j.jpowsour.2008.10.096.

Armand, Michel, Frank Endres, Douglas R. MacFarlane, Hiroyuki Ohno, and Bruno Scrosati. 2009. "Ionic-Liquid Materials for the Electrochemical Challenges of the Future." *Nature Materials* 8 (8): 621–29. doi:10.1038/nmat2448.

Augustyn, Veronica, Patrice Simon, and Bruce Dunn. 2014. "Pseudocapacitive Oxide Materials for High-Rate Electrochemical Energy Storage." *Energy & Environmental Science* 7 (5): 1597–614. doi:10.1039/C3EE44164D.

Baldelli, Steven. 2008. "Surface Structure at the Ionic Liquid–Electrified Metal Interface." *Accounts of Chemical Research* 41 (3): 421–31. doi:10.1021/ar700185h.

Balducci, Andrea. 2016. "Electrolytes for High Voltage Electrochemical Double Layer Capacitors: A Perspective Article." *Journal of Power Sources* 326 (September): 534–40. doi:10.1016/j.jpowsour.2016.05.029.

Balducci, Andrea, R. Dugas, P. L. Taberna, P. Simon, D. Plée, M. Mastragostino, and S. Passerini. 2007. "High Temperature Carbon–Carbon Supercapacitor Using Ionic Liquid as Electrolyte." *Journal of Power Sources* 165 (2): 922–27. doi:10.1016/j.jpowsour.2006.12.048.

Balducci, Andrea, Wesley A. Henderson, Marina Mastragostino, Stefano Passerini, Patrice Simon, and Francesca Soavi. 2005. "Cycling Stability of a Hybrid Activated Carbon//Poly(3-Methylthiophene) Supercapacitor with N-Butyl-N-Methylpyrrolidinium Bis(Trifluoromethanesulfonyl)Imide Ionic Liquid as Electrolyte." *Electrochimica Acta* 50 (11): 2233–37. doi:10.1016/j.electacta.2004.10.006.

Ban, Shuai, Jiujun Zhang, Lei Zhang, Ken Tsay, Datong Song, and Xinfu Zou. 2013. "Charging and Discharging Electrochemical Supercapacitors in the Presence of Both Parallel Leakage Process and Electrochemical Decomposition of Solvent." *Electrochimica Acta* 90 (February): 542–49. doi:10.1016/j.electacta.2012.12.056.

Banerjee, Abhik, Kush Kumar Upadhyay, Dhanya Puthusseri, Vanchiappan Aravindan, Srinivasan Madhavi, and Satishchandra Ogale. 2014. "MOF-Derived Crumpled-Sheet-Assembled Perforated Carbon Cuboids as Highly Effective Cathode Active Materials for Ultra-High Energy Density Li-Ion Hybrid Electrochemical Capacitors (Li-HECs)." *Nanoscale* 6 (8): 4387–94. doi:10.1039/C4NR00025K.

Barranco, V., M. A. Lillo-Rodenas, A. Linares-Solano, A. Oya, F. Pico, J. Ibañez, F. Agullo-Rueda, J. M. Amarilla, and J. M. Rojo. 2010. "Amorphous Carbon Nanofibers and Their Activated Carbon Nanofibers as Supercapacitor Electrodes." *The Journal of Physical Chemistry C* 114 (22): 10302–7. doi:10.1021/jp1021278.

Béguin, François, Volker Presser, Andrea Balducci, and Elzbieta Frackowiak. 2014. "Carbons and Electrolytes for Advanced Supercapacitors." *Advanced Materials* 26 (14): 2219–51. doi:10.1002/adma.201304137.

Bichat, M. P., E. Raymundo-Piñero, and F. Béguin. 2010. "High Voltage Supercapacitor Built with Seaweed Carbons in Neutral Aqueous Electrolyte." *Carbon* 48 (15): 4351–61. doi:10.1016/j.carbon.2010.07.049.

Bittner, A. M., M. Zhu, Y. Yang, H. F. Waibel, M. Konuma, U. Starke, and C. J. Weber. 2012. "Ageing of Electrochemical Double Layer Capacitors." *Journal of Power Sources* 203 (April): 262–73. doi:10.1016/j.jpowsour.2011.10.083.

Blanc, Frédéric, Michal Leskes, and Clare P. Grey. 2013. "In Situ Solid-State NMR Spectroscopy of Electrochemical Cells: Batteries, Supercapacitors, and Fuel Cells." *Accounts of Chemical Research* 46 (9): 1952–63. doi:10.1021/ar400022u.

Boltersdorf, Jonathan, Samuel A. Delp, Jin Yan, Ben Cao, Jim P. Zheng, T. Richard Jow, and Jeffrey A. Read. 2018. "Electrochemical Performance of Lithium-Ion Capacitors Evaluated under High Temperature and High Voltage Stress Using Redox Stable Electrolytes and Additives." *Journal of Power Sources* 373 (January): 20–30. doi:10.1016/j.jpowsour.2017.10.084.

Bonhomme, F., J. C. Lassègues, and L. Servant. 2001. "Raman Spectroelectrochemistry of a Carbon Supercapacitor." *Journal of the Electrochemical Society* 148 (11): E450–58. doi:10.1149/1.1409546.

Boukhalfa, Sofiane, Daniel Gordon, Lilin He, Yuri B. Melnichenko, Naoki Nitta, Alexandre Magasinski, and Gleb Yushin. 2014. "In Situ Small Angle Neutron Scattering Revealing Ion Sorption in Microporous Carbon Electrical Double Layer Capacitors." *ACS Nano* 8 (3): 2495–503. doi:10.1021/nn406077n.

Brandon, Erik J., William C. West, Marshall C. Smart, Larry D. Whitcanack, and Gary A. Plett. 2007. "Extending the Low Temperature Operational Limit of Double-Layer Capacitors." *Journal of Power Sources* 170 (1): 225–32. doi:10.1016/j.jpowsour.2007.04.001.

Brandt, A., and A. Balducci. 2012. "The Influence of Pore Structure and Surface Groups on the Performance of High Voltage Electrochemical Double Layer Capacitors Containing Adiponitrile-Based Electrolyte." *Journal of the Electrochemical Society* 159 (12): A2053–59. doi:10.1149/2.074212jes.

Brandt, A., S. Pohlmann, A. Varzi, A. Balducci, and S. Passerini. 2013. "Ionic Liquids in Supercapacitors." *MRS Bulletin* 38 (7): 554–59. doi:10.1557/mrs.2013.151.

Budarin, Vitaly L., James H. Clark, Sergey V. Mikhalovsky, Alina A. Gorlova, Nataly A. Boldyreva, and Vitaly K. Yatsimirsky. 2000. "The Hydrophobisation of Activated Carbon Surfaces by Organic Functional Groups." *Adsorption Science & Technology* 18 (1): 55–64. doi:10.1260/0263617001493279.

Burke, Andrew. 2000. "Ultracapacitors: Why, How, and Where Is the Technology." *Journal of Power Sources* 91 (1): 37–50. doi:10.1016/S0378-7753(00)00485-7.

Burke, Andrew, and Marshall Miller. 2011. "The Power Capability of Ultracapacitors and Lithium Batteries for Electric and Hybrid Vehicle Applications." *Journal of Power Sources* 196 (1): 514–22. doi:10.1016/j.jpowsour.2010.06.092.

Burt, Ryan, Greg Birkett, and X. S. Zhao. 2014. "A Review of Molecular Modelling of Electric Double Layer Capacitors." *Physical Chemistry Chemical Physics* 16 (14): 6519–38. doi:10.1039/C3CP55186E.

Cai, Yong, Bote Zhao, Jie Wang, and Zongping Shao. 2014. "Non-Aqueous Hybrid Supercapacitors Fabricated with Mesoporous TiO_2 Microspheres and Activated Carbon Electrodes with Superior Performance." *Journal of Power Sources* 253 (May): 80–89. doi:10.1016/j.jpowsour.2013.11.097.

Cao, W. J., and J. P. Zheng. 2012. "Li-Ion Capacitors with Carbon Cathode and Hard Carbon/Stabilized Lithium Metal Powder Anode Electrodes." *Journal of Power Sources* 213 (September): 180–85. doi:10.1016/j.jpowsour.2012.04.033.

Cao, W. J., and J. P. Zheng. 2013. "The Effect of Cathode and Anode Potentials on the Cycling Performance of Li-Ion Capacitors." *Journal of the Electrochemical Society* 160 (9): A1572–76. doi:10.1149/2.114309jes.

Centeno, Teresa A., Olha Sereda, and Fritz Stoeckli. 2011. "Capacitance in Carbon Pores of 0.7 to 15 Nm: A Regular Pattern." *Physical Chemistry Chemical Physics* 13 (27): 12403–6. doi:10.1039/C1CP20748B.

Chandrasekaran, Ramasamy, Meiten Koh, Akiyoshi Yamauchi, and Masashi Ishikawa. 2010. "Electrochemical Cell Studies Based on Non-Aqueous Magnesium Electrolyte for Electric Double Layer Capacitor Applications." *Journal of Power Sources* 195 (2): 662–66. doi:10.1016/j.jpowsour.2009.07.043.

Chen, Libin, Hua Bai, Zhifeng Huang, and Lei Li. 2014. "Mechanism Investigation and Suppression of Self-Discharge in Active Electrolyte Enhanced Supercapacitors." *Energy of Environmental Science* 7 (5): 1750–59. doi:10.1039/C4EE00002A.

Chen, Libin, Yanru Chen, Jifeng Wu, Jianwei Wang, Hua Bai, and Lei Li. 2014. "Electrochemical Supercapacitor with Polymeric Active Electrolyte." *Journal of Materials Chemistry A* 2 (27): 10526–31. doi:10.1039/C4TA01319K.

Chen, Tao, Huisheng Peng, Michael Durstock, and Liming Dai. 2014. "High-Performance Transparent and Stretchable All-Solid Supercapacitors Based on Highly Aligned Carbon Nanotube Sheets." *Scientific Reports* 4 (January). doi:10.1038/srep03612.

Chen, Yao, Xiong Zhang, Dacheng Zhang, Peng Yu, and Yanwei Ma. 2011. "High Performance Supercapacitors Based on Reduced Graphene Oxide in Aqueous and Ionic Liquid Electrolytes." *Carbon* 49 (2): 573–80. doi:10.1016/j.carbon.2010.09.060.

Chen, Yi-Cheng, and Lu-Yin Lin. 2019. "Investigating the Redox Behavior of Activated Carbon Supercapacitors with Hydroquinone and P-Phenylenediamine Dual Redox Additives in the Electrolyte." *Journal of Colloid and Interface Science* 537 (March): 295–305. doi:10.1016/j.jcis.2018.11.026.

Chi, Yu-Wen, Chi-Chang Hu, Hsiao-Hsuan Shen, and Kun-Ping Huang. 2016. "New Approach for High-Voltage Electrical Double-Layer Capacitors Using Vertical Graphene Nanowalls with and without Nitrogen Doping." *Nano Letters* 16 (9): 5719–27. doi:10.1021/acs.nanolett.6b02401.

Chiba, Kazumi, Tsukasa Ueda, Yoji Yamaguchi, Yusuke Oki, Fumiya Saiki, and Katsuiko Naoi. 2011. "Electrolyte Systems for High Withstand Voltage and Durability II. Alkylated Cyclic Carbonates for Electric Double-Layer Capacitors." *Journal of the Electrochemical Society* 158 (12): A1320–27. doi:10.1149/2.038112jes.

Chiba, Kazumi, Tsukasa Ueda, Yoji Yamaguchi, Yusuke Oki, Fumitaka Shimodate, and Katsuhiko Naoi. 2011. "Electrolyte Systems for High Withstand Voltage and Durability I. Linear Sulfones for Electric Double-Layer Capacitors." *Journal of the Electrochemical Society* 158 (8): A872–82. doi:10.1149/1.3593001.

Chmiola, John, Celine Largeot, Pierre-Louis Taberna, Patrice Simon, and Yury Gogotsi. 2008. "Desolvation of Ions in Subnanometer Pores and Its Effect on Capacitance and Double-Layer Theory." *Angewandte Chemie International Edition* 47 (18): 3392–95. doi:10.1002/anie.200704894.

Chmiola, John, G. Yushin, Y. Gogotsi, C. Portet, P. Simon, and P. L. Taberna. 2006. "Anomalous Increase in Carbon Capacitance at Pore Sizes Less Than 1 Nanometer." *Science* 313 (5794): 1760–63. doi:10.1126/science.1132195.

Cho, Min-Young, Mok-Hwa Kim, Hyun-Kyung Kim, Kwang-Bum Kim, Jung Rag Yoon, and Kwang Chul Roh. 2014. "Electrochemical Performance of Hybrid Supercapacitor Fabricated Using Multi-Structured Activated Carbon." *Electrochemistry Communications* 47 (October): 5–8. doi:10.1016/j.elecom.2014.07.012.

Choi, Hong Soo, and Chong Rae Park. 2014. "Theoretical Guidelines to Designing High Performance Energy Storage Device Based on Hybridization of Lithium-Ion Battery and Supercapacitor." *Journal of Power Sources* 259 (August): 1–14. doi:10.1016/j.jpowsour.2014.02.001.

Choudhury, N. A., S. Sampath, and A. K. Shukla. 2008. "Hydrogel-Polymer Electrolytes for Electrochemical Capacitors: An Overview." *Energy & Environmental Science* 2 (1): 55–67. doi:10.1039/B811217G.

Chu, Andrew, and Paul Braatz. 2002. "Comparison of Commercial Supercapacitors and High-Power Lithium-Ion Batteries for Power-Assist Applications in Hybrid Electric Vehicles: I. Initial Characterization." *Journal of Power Sources* 112 (1): 236–46. doi:10.1016/S0378-7753(02)00364-6.

Conway, B. E. 1999. *Electrochemical Supercapacitors: Scientific Fundamentals and Technological Applications.* New York; London: Kluwer Academic/Plenum.

Conway, B. E., and W. G. Pell. 2003. "Double-Layer and Pseudocapacitance Types of Electrochemical Capacitors and Their Applications to the Development of Hybrid Devices." *Journal of Solid State Electrochemistry* 7 (9): 637–44. doi:10.1007/s10008-003-0395-7.

Dagousset, Laure, Grégory Pognon, Giao T. M. Nguyen, Frédéric Vidal, Sébastien Jus, and Pierre-Henri Aubert. 2017. "Electrochemical Characterisations and Ageing of Ionic Liquid/γ-Butyrolactone Mixtures as Electrolytes for Supercapacitor Applications over a Wide Temperature Range." *Journal of Power Sources* 359 (August): 242–49. doi:10.1016/j.jpowsour.2017.05.068.

Demarconnay, L., E. Raymundo-Piñero, and F. Béguin. 2010. "A Symmetric Carbon/Carbon Supercapacitor Operating at 1.6 V by Using a Neutral Aqueous Solution." *Electrochemistry Communications* 12 (10): 1275–78. doi:10.1016/j.elecom.2010.06.036.

Deschamps, Michaël, Edouard Gilbert, Philippe Azais, Encarnación Raymundo-Piñero, Mohammed Ramzi Ammar, Patrick Simon, Dominique Massiot, and François Béguin. 2013. "Exploring Electrolyte Organization in Supercapacitor Electrodes with Solid-State NMR." *Nature Materials* 12 (4): 351–58. doi:10.1038/nmat3567.

Ding, M. S., K. Xu, J. P. Zheng, and T. R. Jow. 2004. "γ-Butyrolactone-Acetonitrile Solution of Triethylmethylammonium Tetrafluoroborate as an Electrolyte for Double-Layer Capacitors." *Journal of Power Sources* 138 (1): 340–50. doi:10.1016/j.jpowsour.2004.06.039.

Duay, Jonathon, Eleanor Gillette, Ran Liu, and Sang Bok Lee. 2012. "Highly Flexible Pseudocapacitor Based on Freestanding Heterogeneous MnO_2/Conductive Polymer Nanowire Arrays." *Physical Chemistry Chemical Physics* 14 (10): 3329–37. doi:10.1039/C2CP00019A.

Fan, Le-Qing, Ji Zhong, Ji-Huai Wu, Jian-Ming Lin, and Yun-Fang Huang. 2014. "Improving the Energy Density of Quasi-Solid-State Electric Double-Layer Capacitors by Introducing Redox Additives into Gel Polymer Electrolytes." *Journal of Materials Chemistry A* 2 (24): 9011–14. doi:10.1039/C4TA01408A.

Fang, B., Y. Z. Wei, K. Maruyama, and M. Kumagai. 2005. "High Capacity Supercapacitors Based on Modified Activated Carbon Aerogel." *Journal of Applied Electrochemistry* 35 (3): 229–33. doi:10.1007/s10800-004-3462-6.

Fang, Baizeng, and Leo Binder. 2006. "A Novel Carbon Electrode Material for Highly Improved EDLC Performance." *The Journal of Physical Chemistry B* 110 (15): 7877–82. doi:10.1021/jp060110d.

Fedorov, Maxim V., and Alexei A. Kornyshev. 2014. "Ionic Liquids at Electrified Interfaces." *Chemical Reviews* 114 (5): 2978–3036. doi:10.1021/cr400374x.

Feng, Guang, Jingsong Huang, Bobby G. Sumpter, Vincent Meunier, and Rui Qiao. 2011. "A 'Counter-Charge Layer in Generalized Solvents' Framework for Electrical Double Layers in Neat and Hybrid Ionic Liquid Electrolytes." *Physical Chemistry Chemical Physics* 13 (32): 14723. doi:10.1039/c1cp21428d.

Feng, Guang, Song Li, Volker Presser, and Peter T. Cummings. 2013. "Molecular Insights into Carbon Supercapacitors Based on Room-Temperature Ionic Liquids." *The Journal of Physical Chemistry Letters* 4 (19): 3367–76. doi:10.1021/jz4014163.

Fic, Krzysztof, Elzbieta Frackowiak, and François Béguin. 2012. "Unusual Energy Enhancement in Carbon-Based Electrochemical Capacitors." *Journal of Materials Chemistry* 22 (46): 24213–23. doi:10.1039/C2JM35711A.

Fic, Krzysztof, Grzegorz Lota, Mikolaj Meller, and Elzbieta Frackowiak. 2012. "Novel Insight into Neutral Medium as Electrolyte for High-Voltage Supercapacitors." *Energy & Environmental Science* 5 (2): 5842–50. doi:10.1039/C1EE02262H.

Forse, Alexander C., John M. Griffin, Hao Wang, Nicole M. Trease, Volker Presser, Yury Gogotsi, Patrice Simon, and Clare P. Grey. 2013. "Nuclear Magnetic Resonance Study of Ion Adsorption on Microporous Carbide-Derived Carbon." *Physical Chemistry Chemical Physics* 15 (20): 7722–30. doi:10.1039/C3CP51210J.

Francisco, Brian E., Christina M. Jones, Se-Hee Lee, and Conrad R. Stoldt. 2012. "Nanostructured All-Solid-State Supercapacitor Based on Li_2S-P_2S_5 Glass-Ceramic Electrolyte." *Applied Physics Letters* 100 (10): 103902. doi:10.1063/1.3693521.

Freemantle, Michael. 1998. "Designer Solvents." *Chemical & Engineering News Archive* 76 (13): 32–37. doi:10.1021/cen-v076n013.p032.

Galiński, Maciej, Andrzej Lewandowski, and Izabela Stępniak. 2006. "Ionic Liquids as Electrolytes." *Electrochimica Acta* 51 (26): 5567–80. doi:10.1016/j.electacta.2006.03.016.

Gao, Han, and Keryn Lian. 2014. "Proton-Conducting Polymer Electrolytes and Their Applications in Solid Supercapacitors: A Review." *RSC Advances* 4 (62): 33091–113. doi:10.1039/C4RA05151C.

Gao, Han, Alvin Virya, and Keryn Lian. 2014. "Monovalent Silicotungstate Salts as Electrolytes for Electrochemical Supercapacitors." *Electrochimica Acta* 138 (August): 240–46. doi:10.1016/j.electacta.2014.06.127.

Hahn, M., A. Würsig, R. Gallay, P. Novák, and R. Kötz. 2005. "Gas Evolution in Activated Carbon/Propylene Carbonate Based Double-Layer Capacitors." *Electrochemistry Communications* 7 (9): 925–30. doi:10.1016/j.elecom.2005.06.015.

Hanlon, Damien, Claudia Backes, Thomas M. Higgins, Marguerite Hughes, Arlene O'Neill, Paul King, Niall McEvoy, et al. 2014. "Production of Molybdenum Trioxide Nanosheets by Liquid Exfoliation and Their Application in High-Performance Supercapacitors." *Chemistry of Materials* 26 (4): 1751–63. doi:10.1021/cm500271u.

Hardwick, Laurence J., Matthias Hahn, Patrick Ruch, Michael Holzapfel, Werner Scheifele, Hilmi Buqa, Frank Krumeich, Petr Novák, and Rüdiger Kötz. 2006. "An in Situ Raman Study of the Intercalation of Supercapacitor-Type Electrolyte into Microcrystalline Graphite." *Electrochimica Acta* 52 (2): 675–80. doi:10.1016/j.electacta.2006.05.053.

He, Tieshi, Xiangling Meng, Junping Nie, Yujin Tong, and Kedi Cai. 2016. "Thermally Reduced Graphene Oxide Electrochemically Activated by Bis-Spiro Quaternary Alkyl Ammonium for Capacitors." *ACS Applied Materials & Interfaces* 8 (22): 13865–70. doi:10.1021/acsami.6b00885.

He, Yan-Bing, Baohua Li, Ming Liu, Chen Zhang, Wei Lv, Cheng Yang, Jia Li, et al. 2012. "Gassing in Li(4) Ti(5)O(12)-Based Batteries and Its Remedy." *Scientific Reports* 2: 913. doi:10.1038/srep00913.

Hess, Lars H., and Andrea Balducci. 2018. "1,2-Butylene Carbonate as Solvent for EDLCs." *Electrochimica Acta* 281 (August): 437–44. doi:10.1016/j.electacta.2018.05.168.

Heß, Lars Henning, Ladyna Wittscher, and Andrea Balducci. 2019. "The Impact of Carbonate Solvents on the Self-Discharge, Thermal Stability and Performance Retention of High Voltage Electrochemical Double Layer Capacitors." *Physical Chemistry Chemical Physics*, April. doi:10.1039/C9CP00483A.

Hsu, Chun-Tsung, Chi-Chang Hu, Tzu-Ho Wu, Jia-Cing Chen, and Muniyandi Rajkumar. 2014. "How the Electrochemical Reversibility of a Battery-Type Material Affects the Charge Balance and Performances of Asymmetric Supercapacitors." *Electrochimica Acta* 146 (November): 759–68. doi:10.1016/j.electacta.2014.09.041.

Huang, Cheng-Wei, Ching-An Wu, Sheng-Shu Hou, Ping-Lin Kuo, Chien-Te Hsieh, and Hsisheng Teng. 2012. "Gel Electrolyte Derived from Poly(Ethylene Glycol) Blending Poly(Acrylonitrile) Applicable to Roll-to-Roll Assembly of Electric Double Layer Capacitors." *Advanced Functional Materials* 22 (22): 4677–85. doi:10.1002/adfm.201201342.

Huang, Jingsong, Bobby G. Sumpter, and Vincent Meunier. 2008. "Theoretical Model for Nanoporous Carbon Supercapacitors." *Angewandte Chemie International Edition* 47 (3): 520–24. doi:10.1002/anie.200703864.

Huang, Po-Ling, Xu-Feng Luo, You-Yu Peng, Nen-Wen Pu, Ming-Der Ger, Cheng-Hsien Yang, Tzi-Yi Wu, and Jeng-Kuei Chang. 2015. "Ionic Liquid Electrolytes with Various Constituent Ions for Graphene-Based Supercapacitors." *Electrochimica Acta* 161 (April): 371–77. doi:10.1016/j.electacta.2015.02.115.

Huang, Xiaodan, Bing Sun, Shuangqiang Chen, and Guoxiu Wang. 2014. "Self-Assembling Synthesis of Free-Standing Nanoporous Graphene–Transition-Metal Oxide Flexible Electrodes for High-Performance Lithium-Ion Batteries and Supercapacitors." *Chemistry – An Asian Journal* 9 (1): 206–11. doi:10.1002/asia.201301121.

Huddleston, Jonathan G., Ann E. Visser, W. Matthew Reichert, Heather D. Willauer, Grant A. Broker, and Robin D. Rogers. 2001. "Characterization and Comparison of Hydrophilic and Hydrophobic Room Temperature Ionic Liquids Incorporating the Imidazolium Cation." *Green Chemistry* 3 (4): 156–64. doi:10.1039/B103275P.

Hulicova-Jurcakova, Denisa, Alexander M. Puziy, Olga I. Poddubnaya, Fabian Suárez-García, Juan M. D. Tascón, and Gao Qing Lu. 2009. "Highly Stable Performance of Supercapacitors from Phosphorus-Enriched Carbons." *Journal of the American Chemical Society* 131 (14): 5026–27. doi:10.1021/ja809265m.

Iamprasertkun, Pawin, Atiweena Krittayavathananon, and Montree Sawangphruk. 2016. "N-Doped Reduced Graphene Oxide Aerogel Coated on Carboxyl-Modified Carbon Fiber Paper for High-Performance Ionic-Liquid Supercapacitors." *Carbon* 102 (June): 455–61. doi:10.1016/j.carbon.2015.12.092.

Ike, Innocent S. 2017. "Mathematical Modelling, Design, and Optimization of Electrochemical Capacitor Cells." Thesis. http://wiredspace.wits.ac.za/handle/10539/24089.

Ike, Innocent S., Iakovos Sigalas, and Sunny Iyuke. 2018. "The Effects of Self-Discharge on the Performance of Asymmetric/Hybrid Electrochemical Capacitors with Redox-Active Electrolytes: Insights from Modeling and Simulation." *Journal of Electronic Materials* 47 (1): 470–92. doi:10.1007/s11664-017-5796-y.

Ike, Innocent S., Iakovos Sigalas, and Sunny E. Iyuke. 2016. "The Influences of Operating Conditions and Design Configurations on the Performance of Symmetric Electrochemical Capacitors." *Physical Chemistry Chemical Physics* 18 (41): 28626–47. doi:10.1039/C6CP06214H.

Ike, Innocent S., Iakovos Sigalas, and Sunny E. Iyuke. 2017a. "The Effects of Self-Discharge on the Performance of Symmetric Electric Double-Layer Capacitors and Active Electrolyte-Enhanced Supercapacitors: Insights from Modeling and Simulation." *Journal of Electronic Materials* 46 (2): 1163–89. doi:10.1007/s11664-016-5053-9.

Ike, Innocent S., Iakovos Sigalas, and Sunny E. Iyuke. 2017b. "Optimization of Design Parameters and Operating Conditions of Electrochemical Capacitors for High Energy and Power Performance." *Journal of Electronic Materials* 46 (3): 1692–713. doi:10.1007/s11664-016-5213-y.

Ike, Innocent S., Iakovos Sigalas, and Sunny E. Iyuke. 2017c. "Modelling and Optimization of Electrodes Utilization in Symmetric Electrochemical Capacitors for High Energy and Power." *Journal of Energy Storage* 12 (August): 261–75. doi:10.1016/j.est.2017.05.006.

Ionica-Bousquet, C. M., W. J. Casteel, R. M. Pearlstein, G. GirishKumar, G. P. Pez, P. Gómez-Romero, M. R. Palacín, and D. Muñoz-Rojas. 2010. "Polyfluorinated Boron Cluster – $[B_{12}F_{11}H]_{2-}$ – Based Electrolytes for Supercapacitors: Overcharge Protection." *Electrochemistry Communications* 12 (5): 636–39. doi:10.1016/j.elecom.2010.02.018.

Ishimoto, Shuichi, Yuichiro Asakawa, Masanori Shinya, and Katsuhiko Naoi. 2009. "Degradation Responses of Activated-Carbon-Based EDLCs for Higher Voltage Operation and Their Factors." *Journal of the Electrochemical Society* 156 (7): A563–71. doi:10.1149/1.3126423.

Iwama, Etsuro, Tsukasa Ueda, Yoko Ishihara, Kenji Ohshima, Wako Naoi, McMahon Thomas Homer Reid, and Katsuhiko Naoi. 2019. "High-Voltage Operation of $Li_4Ti_5O_{12}$/AC Hybrid Supercapacitor Cell in Carbonate and Sulfone Electrolytes: Gas Generation and Its Characterization." *Electrochimica Acta* 301 (April): 312–18. doi:10.1016/j.electacta.2019.01.088.

Jänes, Alar, J. Eskusson, T. Thomberg, and E. Lust. 2014. "Supercapacitors Based on Propylene Carbonate with Small Addition of Different Sulfur Containing Organic Solvents." *Journal of the Electrochemical Society* 161 (9): A1284–90. doi:10.1149/2.0691409jes.

Jänes, Alar, T. Thomberg, J. Eskusson, and E. Lust. 2013. "Fluoroethylene Carbonate as Co-Solvent for Propylene Carbonate Based Electrical Double Layer Capacitors." *Journal of the Electrochemical Society* 160 (8): A1025–30. doi:10.1149/2.016308jes.

Jänes, Alar, and Enn Lust. 2006. "Use of Organic Esters as Co-Solvents for Electrical Double Layer Capacitors with Low Temperature Performance." *Journal of Electroanalytical Chemistry* 588 (2): 285–95. doi:10.1016/j.jelechem.2006.01.003.

Jiang, De-en, Zhehui Jin, Douglas Henderson, and Jianzhong Wu. 2012. "Solvent Effect on the Pore-Size Dependence of an Organic Electrolyte Supercapacitor." *The Journal of Physical Chemistry Letters* 3 (13): 1727–31. doi:10.1021/jz3004624.

Jiang, De-en, Zhehui Jin, and Jianzhong Wu. 2011. "Oscillation of Capacitance inside Nanopores." *Nano Letters* 11 (12): 5373–77. doi:10.1021/nl202952d.

Jiménez-Cordero, Diana, Francisco Heras, Miguel A. Gilarranz, and Encarnación Raymundo-Piñero. 2014. "Grape Seed Carbons for Studying the Influence of Texture on Supercapacitor Behaviour in Aqueous Electrolytes." *Carbon* 71 (May): 127–38. doi:10.1016/j.carbon.2014.01.021.

Karthikeyan, Kaliyappan, Samuthirapandian Amaresh, Sol Nip Lee, Vanchiappan Aravindan, and Yun Sung Lee. 2014. "Fluorine-Doped Fe_2O_3 as High Energy Density Electroactive Material for Hybrid Supercapacitor Applications." *Chemistry – An Asian Journal* 9 (3): 852–57. doi:10.1002/asia.201301289.

Komaba, Shinichi, Tomoya Tsuchikawa, Masataka Tomita, Naoaki Yabuuchi, and Atsushi Ogata. 2013. "Efficient Electrolyte Additives of Phosphate, Carbonate, and Borate to Improve Redox Capacitor Performance of Manganese Oxide Electrodes." *Journal of the Electrochemical Society* 160 (11): A1952–61. doi:10.1149/2.019311jes.

Kou, Liang, Tieqi Huang, Bingna Zheng, Yi Han, Xiaoli Zhao, Karthikeyan Gopalsamy, Haiyan Sun, and Chao Gao. 2014. "Coaxial Wet-Spun Yarn Supercapacitors for High-Energy Density and Safe Wearable Electronics." *Nature Communications* 5 (May): 3754. doi:10.1038/ncomms4754.

Krause, A., and A. Balducci. 2011. "High Voltage Electrochemical Double Layer Capacitor Containing Mixtures of Ionic Liquids and Organic Carbonate as Electrolytes." *Electrochemistry Communications* 13 (8): 814–17. doi:10.1016/j.elecom.2011.05.010.

Kruusma, Jaanus, Arvo Tõnisoo, Rainer Pärna, Ergo Nõmmiste, and Enn Lust. 2014. "In Situ XPS Studies of Electrochemically Positively Polarized Molybdenum Carbide Derived Carbon Double Layer Capacitor Electrode." *Journal of the Electrochemical Society* 161 (9): A1266–77. doi:10.1149/2.0641409jes.

Kühnel, Ruben-Simon, and Andrea Balducci. 2014. "Comparison of the Anodic Behavior of Aluminum Current Collectors in Imide-Based Ionic Liquids and Consequences on the Stability of High Voltage Supercapacitors." *Journal of Power Sources* 249 (March): 163–71. doi:10.1016/j.jpowsour.2013.10.072.

Kumar, Yogesh, Sangeeta Rawal, Bhawana Joshi, and S. A. Hashmi. 2019. "Background, Fundamental Understanding and Progress in Electrochemical Capacitors." *Journal of Solid State Electrochemistry* 23 (3): 667–92. doi:10.1007/s10008-018-4160-3.

Kunze, Miriam, Sangsik Jeong, Elie Paillard, Monika Schönhoff, Martin Winter, and Stefano Passerini. 2011. "New Insights to Self-Aggregation in Ionic Liquid Electrolytes for High-Energy Electrochemical Devices." *Advanced Energy Materials* 1 (2): 274–81. doi:10.1002/aenm.201000052.

Kunze, Miriam, Elie Paillard, Sangsik Jeong, Giovanni B. Appetecchi, Monika Schönhoff, Martin Winter, and Stefano Passerini. 2011. "Inhibition of Self-Aggregation in Ionic Liquid Electrolytes for High-Energy Electrochemical Devices." *The Journal of Physical Chemistry C* 115 (39): 19431–36. doi:10.1021/jp2055969.

Kuo, Shin-Liang, and Nae-Lih Wu. 2006. "Investigation of Pseudocapacitive Charge-Storage Reaction of $MnO_2 \cdot NH_2O$ Supercapacitors in Aqueous Electrolytes." *Journal of the Electrochemical Society* 153 (7): A1317–24. doi:10.1149/1.2197667.

Kurzweil, P., and M. Chwistek. 2008. "Electrochemical Stability of Organic Electrolytes in Supercapacitors: Spectroscopy and Gas Analysis of Decomposition Products." *Journal of Power Sources* 176 (2): 555–67. doi:10.1016/j.jpowsour.2007.08.070.

Laheäär, Ann, Alar Jänes, and Enn Lust. 2011. "Electrochemical Properties of Carbide-Derived Carbon Electrodes in Non-Aqueous Electrolytes Based on Different Li-Salts." *Electrochimica Acta* 56 (25): 9048–55. doi:10.1016/j.electacta.2011.05.126.

Laheäär, Ann, Heisi Kurig, Alar Jänes, and Enn Lust. 2009. "LiPF6 Based Ethylene Carbonate–Dimethyl Carbonate Electrolyte for High Power Density Electrical Double Layer Capacitor." *Electrochimica Acta* 54 (19): 4587–94. doi:10.1016/j.electacta.2009.03.059.

Largeot, Celine, Cristelle Portet, John Chmiola, Pierre-Louis Taberna, Yury Gogotsi, and Patrice Simon. 2008. "Relation between the Ion Size and Pore Size for an Electric Double-Layer Capacitor." *Journal of the American Chemical Society* 130 (9): 2730–31. doi:10.1021/ja7106178.

Łatoszyńska, Anna A., Grażyna Zofia Żukowska, Iwona A. Rutkowska, Pierre-Louis Taberna, Patrice Simon, Pawel J. Kulesza, and Władysław Wieczorek. 2015. "Non-Aqueous Gel Polymer Electrolyte with Phosphoric Acid Ester and Its Application for Quasi Solid-State Supercapacitors." *Journal of Power Sources* 274 (January): 1147–54. doi:10.1016/j.jpowsour.2014.10.094.

Lee, Juhan, Pattarachai Srimuk, Simon Fleischmann, Xiao Su, T. Alan Hatton, and Volker Presser. 2019. "Redox-Electrolytes for Non-Flow Electrochemical Energy Storage: A Critical Review and Best Practice." *Progress in Materials Science* 101 (April): 46–89. doi:10.1016/j.pmatsci.2018.10.005.

Lee, Seung Woo, Betar M. Gallant, Hye Ryung Byon, Paula T. Hammond, and Yang Shao-Horn. 2011. "Nanostructured Carbon-Based Electrodes: Bridging the Gap between Thin-Film Lithium-Ion Batteries and Electrochemical Capacitors." *Energy & Environmental Science* 4 (6): 1972–85. doi:10.1039/C0EE00642D.

Levi, Mikhael D., Grigory Salitra, Naomi Levy, Doron Aurbach, and Joachim Maier. 2009. "Application of a Quartz-Crystal Microbalance to Measure Ionic Fluxes in Microporous Carbons for Energy Storage." *Nature Materials* 8 (11): 872–75. doi:10.1038/nmat2559.

Lewandowski, Andrzej, and Maciej Galinski. 2007. "Practical and Theoretical Limits for Electrochemical Double-Layer Capacitors." *Journal of Power Sources* 173 (2): 822–28. doi:10.1016/j.jpowsour.2007.05.062.

Lewandowski, Andrzej, Angelika Olejniczak, Maciej Galinski, and Izabela Stepniak. 2010. "Performance of Carbon–Carbon Supercapacitors Based on Organic, Aqueous and Ionic Liquid Electrolytes." *Journal of Power Sources* 195 (17): 5814–19. doi:10.1016/j.jpowsour.2010.03.082.

Leyva-García, Sarai, Dolores Lozano-Castelló, Emilia Morallón, Thomas Vogl, Christoph Schütter, Stefano Passerini, Andrea Balducci, and Diego Cazorla-Amorós. 2016. "Electrochemical Performance of a Superporous Activated Carbon in Ionic Liquid-Based Electrolytes." *Journal of Power Sources* 336 (December): 419–26. doi:10.1016/j.jpowsour.2016.11.010.

Li, Qi, Xiaoxi Zuo, Jiansheng Liu, Xin Xiao, Dong Shu, and Junmin Nan. 2011. "The Preparation and Properties of a Novel Electrolyte of Electrochemical Double Layer Capacitors Based on $LiPF_6$ and Acetamide." *Electrochimica Acta* 58 (December): 330–35. doi:10.1016/j.electacta.2011.09.059.

Li, Shin-Ming, Shin-Yi Yang, Yu-Sheng Wang, Hsiu-Ping Tsai, Hsi-Wen Tien, Sheng-Tsung Hsiao, Wei-Hao Liao, Chien-Liang Chang, Chen-Chi M. Ma, and Chi-Chang Hu. 2015. "N-Doped Structures and Surface Functional Groups of Reduced Graphene Oxide and Their Effect on the Electrochemical Performance of Supercapacitor with Organic Electrolyte." *Journal of Power Sources* 278 (March): 218–29. doi:10.1016/j.jpowsour.2014.12.025.

Li, Siheng, Li Qi, Lehui Lu, and Hongyu Wang. 2012. "Facile Preparation and Performance of Mesoporous Manganese Oxide for Supercapacitors Utilizing Neutral Aqueous Electrolytes." *RSC Advances* 2 (8): 3298–308. doi:10.1039/C2RA00991A.

Li, Ziyu, Kaichang Kou, Jing Xue, Chen Pan, and Guanglei Wu. 2019. "Study of Triazine-Based-Polyimides Composites Working as Gel Polymer Electrolytes in ITO-Glass Based Capacitor Devices." *Journal of Materials Science: Materials in Electronics* 30 (4): 3426–31. doi:10.1007/s10854-018-00617-x.

Lian, C., K. Liu, K. L. Van Aken, Y. Gogotsi, D. J. Wesolowski, H. L. Liu, D. E. Jiang, and J. Z. Wu. 2016. "Enhancing the Capacitive Performance of Electric Double-Layer Capacitors with Ionic Liquid Mixtures." *ACS Energy Letters* 1 (1): 21–26. doi:10.1021/acsenergylett.6b00010.

Lian, Cheng, Haiping Su, Honglai Liu, and Jianzhong Wu. 2018. "Electrochemical Behavior of Nanoporous Supercapacitors with Oligomeric Ionic Liquids." *The Journal of Physical Chemistry C* 122 (26): 14402–7. doi:10.1021/acs.jpcc.8b04464.

Lian, Keryn, and Chang Ming Li. 2009. "Asymmetrical Electrochemical Capacitors Using Heteropoly Acid Electrolytes." *Electrochemical and Solid-State Letters* 12 (1): A10–12. doi:10.1149/1.3007424.

Liaw, Horng-Jang, Chan-Cheng Chen, Yi-Chien Chen, Jenq-Renn Chen, Shih-Kai Huang, and Sheng-Nan Liu. 2012. "Relationship between Flash Point of Ionic Liquids and Their Thermal Decomposition." *Green Chemistry* 14 (7): 2001–8. doi:10.1039/C2GC35449G.

Lim, Eunho, Haegyeom Kim, Changshin Jo, Jinyoung Chun, Kyojin Ku, Seongseop Kim, Hyung Ik Lee, et al. 2014. "Advanced Hybrid Supercapacitor Based on a Mesoporous Niobium Pentoxide/Carbon as High-Performance Anode." *ACS Nano* 8 (9): 8968–78. doi:10.1021/nn501972w.

Lin, Rongying, Pierre-Louis Taberna, Sébastien Fantini, Volker Presser, Carlos R. Pérez, François Malbosc, Nalin L. Rupesinghe, Kenneth B. K. Teo, Yury Gogotsi, and Patrice Simon. 2011. "Capacitive Energy Storage from −50°C to 100°C Using an Ionic Liquid Electrolyte." *The Journal of Physical Chemistry Letters* 2 (19): 2396–2401. doi:10.1021/jz201065t.

Lin, Zifeng, Pierre-Louis Taberna, and Patrice Simon. 2016. "Graphene-Based Supercapacitors Using Eutectic Ionic Liquid Mixture Electrolyte." *Electrochimica Acta* 206 (July): 446–51. doi:10.1016/j.electacta.2015.12.097.

Liu, Chenguang, Zhenning Yu, David Neff, Aruna Zhamu, and Bor Z. Jang. 2010. "Graphene-Based Supercapacitor with an Ultrahigh Energy Density." *Nano Letters* 10 (12): 4863–68. doi:10.1021/nl102661q.

Liu, Ching-Fang, Yu-Chien Liu, Tien-Yu Yi, and Chi-Chang Hu. 2019. "Carbon Materials for High-Voltage Supercapacitors." *Carbon* 145 (April): 529–48. doi:10.1016/j.carbon.2018.12.009.

Liu, Ping, Mark Verbrugge, and Souren Soukiazian. 2006. "Influence of Temperature and Electrolyte on the Performance of Activated-Carbon Supercapacitors." *Journal of Power Sources* 156 (2): 712–18. doi:10.1016/j.jpowsour.2005.05.055.

Liu, Shuangyu, Jian Xie, Haibo Li, Ye Wang, Hui Ying Yang, Tiejun Zhu, Shichao Zhang, Gaoshao Cao, and Xinbing Zhao. 2014. "Nitrogen-Doped Reduced Graphene Oxide for High-Performance Flexible All-Solid-State Micro-Supercapacitors." *Journal of Materials Chemistry A* 2 (42): 18125–31. doi:10.1039/C4TA03192J.

Liu, Xiaorong, and Peter G. Pickup. 2008. "Performance and Low Temperature Behaviour of Hydrous Ruthenium Oxide Supercapacitors with Improved Power Densities." *Energy & Environmental Science* 1 (4): 494–500. doi:10.1039/B809939A.

Lota, Grzegorz, and Elzbieta Frackowiak. 2009. "Striking Capacitance of Carbon/Iodide Interface." *Electrochemistry Communications* 11 (1): 87–90. doi:10.1016/j.elecom.2008.10.026.

Lota, Grzegorz, and Grzegorz Milczarek. 2011. "The Effect of Lignosulfonates as Electrolyte Additives on the Electrochemical Performance of Supercapacitors." *Electrochemistry Communications* 13 (5): 470–73. doi:10.1016/j.elecom.2011.02.023.

Lu, Max, F. Beguin, and E. Frackowiak. 2019. "Supercapacitors: Materials, Systems, and Applications." *Wiley.Com.* Accessed April 3. https://www.wiley.com/en-us/Supercapacitors%3A+Materials%2C+Systems%2C+and+Applications-p-9783527328833.

Ma, Guofu, Enke Feng, Kanjun Sun, Hui Peng, Jiajia Li, and Ziqiang Lei. 2014. "A Novel and High-Effective Redox-Mediated Gel Polymer Electrolyte for Supercapacitor." *Electrochimica Acta* 135 (July): 461–66. doi:10.1016/j.electacta.2014.05.045.

Ma, Guofu, Jiajia Li, Kanjun Sun, Hui Peng, Jingjing Mu, and Ziqiang Lei. 2014. "High Performance Solid-State Supercapacitor with PVA–KOH–$K_3[Fe(CN)_6]$ Gel Polymer as Electrolyte and Separator." *Journal of Power Sources* 256 (June): 281–87. doi:10.1016/j.jpowsour.2014.01.062.

Mai, Li-Qiang, Aamir Minhas-Khan, Xiaocong Tian, Kalele Mulonda Hercule, Yun-Long Zhao, Xu Lin, and Xu Xu. 2013. "Synergistic Interaction between Redox-Active Electrolyte and Binder-Free Functionalized Carbon for Ultrahigh Supercapacitor Performance." *Nature Communications* 4 (December): 2923. doi:10.1038/ncomms3923.

Mansour, Azzam N., Juanjuan Zhou, and Xiangyang Zhou. 2014. "X-Ray Absorption Spectroscopic Study of Sodium Iodide and Iodine Mediators in a Solid-State Supercapacitor." *Journal of Power Sources* 245 (January): 270–76. doi:10.1016/j.jpowsour.2013.06.129.

McDonough, John K., Andrey I. Frolov, Volker Presser, Junjie Niu, Christopher H. Miller, Teresa Ubieto, Maxim V. Fedorov, and Yury Gogotsi. 2012. "Influence of the Structure of Carbon Onions on Their Electrochemical Performance in Supercapacitor Electrodes." *Carbon* 50 (9): 3298–309. doi:10.1016/j.carbon.2011.12.022.

McEwen, Alan B., Stephen F. McDevitt, and Victor R. Koch. 1997. "Nonaqueous Electrolytes for Electrochemical Capacitors: Imidazolium Cations and Inorganic Fluorides with Organic Carbonates." *Journal of the Electrochemical Society* 144 (4): L84–86. doi:10.1149/1.1837561.

Meng, Chuizhou, Jimin Maeng, Simon W. M. John, and Pedro P. Irazoqui. 2014. "Ultrasmall Integrated 3D Micro-Supercapacitors Solve Energy Storage for Miniature Devices." *Advanced Energy Materials* 4 (7): 1301269. doi:10.1002/aenm.201301269.

Menzel, Jakub, Elzbieta Frackowiak, and Krzysztof Fic. 2019. "Electrochemical Capacitor with Water-Based Electrolyte Operating at Wide Temperature Range." *Journal of Power Sources* 414 (February): 183–91. doi:10.1016/j.jpowsour.2018.12.080.

Merlet, Céline, Benjamin Rotenberg, Paul A. Madden, Pierre-Louis Taberna, Patrice Simon, Yury Gogotsi, and Mathieu Salanne. 2012. "On the Molecular Origin of Supercapacitance in Nanoporous Carbon Electrodes." *Nature Materials* 11 (4): 306–10. doi:10.1038/nmat3260.

Michan, Alison L., Bharathy S. Parimalam, Michal Leskes, Rachel N. Kerber, Taeho Yoon, Clare P. Grey, and Brett L. Lucht. 2016. "Fluoroethylene Carbonate and Vinylene Carbonate Reduction: Understanding Lithium-Ion Battery Electrolyte Additives and Solid Electrolyte Interphase Formation." *Chemistry of Materials* 28 (22): 8149–59. doi:10.1021/acs.chemmater.6b02282.

Morita, M., M. Goto, and Y. Matsuda. 1992. "Ethylene Carbonate-Based Organic Electrolytes for Electric Double Layer Capacitors." *Journal of Applied Electrochemistry* 22 (10): 901–8. doi:10.1007/BF01024137.

Mysyk, R., E. Raymundo-Piñero, and F. Béguin. 2009. "Saturation of Subnanometer Pores in an Electric Double-Layer Capacitor." *Electrochemistry Communications* 11 (3): 554–56. doi:10.1016/j.elecom.2008.12.035.

Naoi, Katsuhiko. 2010. "'Nanohybrid Capacitor': The Next Generation Electrochemical Capacitors." *Fuel Cells* 10 (5): 825–33. doi:10.1002/fuce.201000041.

Naoi, Katsuhiko, Syuichi Ishimoto, Jun-ichi Miyamoto, and Wako Naoi. 2012. "Second Generation 'Nanohybrid Supercapacitor': Evolution of Capacitive Energy Storage Devices." *Energy & Environmental Science* 5 (11): 9363–73. doi:10.1039/C2EE21675B.

Naoi, Katsuhiko, and Patrice Simon. 2008. "New Materials and New Configurations for Advanced Electrochemical Capacitors." *Journal of the Electrochemical Society (JES)* 17 (April): 34–37.

Niu, Zhiqiang, Haibo Dong, Bowen Zhu, Jinzhu Li, Huey Hoon Hng, Weiya Zhou, Xiaodong Chen, and Sishen Xie. 2013. "Highly Stretchable, Integrated Supercapacitors Based on Single-Walled Carbon Nanotube Films with Continuous Reticulate Architecture." *Advanced Materials* 25 (7): 1058–64. doi:10.1002/adma.201204003.

Noori, Abolhassan, Maher F. El-Kady, Mohammad S. Rahmanifar, Richard B. Kaner, and Mir F. Mousavi. 2019. "Towards Establishing Standard Performance Metrics for Batteries, Supercapacitors and Beyond." *Chemical Society Reviews* 48 (5): 1272–341. doi:10.1039/C8CS00581H.

Oickle, Alicia M., and Heather A. Andreas. 2011. "Examination of Water Electrolysis and Oxygen Reduction as Self-Discharge Mechanisms for Carbon-Based, Aqueous Electrolyte Electrochemical Capacitors." *The Journal of Physical Chemistry C* 115 (10): 4283–88. doi:10.1021/jp1067439.

Orita, A., K. Kamijima, and M. Yoshida. 2010. "Allyl-Functionalized Ionic Liquids as Electrolytes for Electric Double-Layer Capacitors." *Journal of Power Sources* 195 (21): 7471–79. doi:10.1016/j.jpowsour.2010.05.066.

Ozoliņš, Vidvuds, Fei Zhou, and Mark Asta. 2013. "Ruthenia-Based Electrochemical Supercapacitors: Insights from First-Principles Calculations." *Accounts of Chemical Research* 46 (5): 1084–93. doi:10.1021/ar3002987.

Patil, U. M., R. R. Salunkhe, K. V. Gurav, and C. D. Lokhande. 2008. "Chemically Deposited Nanocrystalline NiO Thin Films for Supercapacitor Application." *Applied Surface Science* 255 (5, Part 2): 2603–7. doi:10.1016/j.apsusc.2008.07.192.

Pell, Wendy G., and Brian E. Conway. 2001a. "Voltammetry at a de Levie Brush Electrode as a Model for Electrochemical Supercapacitor Behaviour." *Journal of Electroanalytical Chemistry* 500 (1–2): 121–33. doi:10.1016/S0022-0728(00)00423-X.

Pell, Wendy G., and Brian E. Conway. 2001b. "Analysis of Power Limitations at Porous Supercapacitor Electrodes under Cyclic Voltammetry Modulation and Dc Charge." *Journal of Power Sources* 96 (1): 57–67. doi:10.1016/S0378-7753(00)00682-0.

Peng, Lele, Xu Peng, Borui Liu, Changzheng Wu, Yi Xie, and Guihua Yu. 2013. "Ultrathin Two-Dimensional MnO_2/Graphene Hybrid Nanostructures for High-Performance, Flexible Planar Supercapacitors." *Nano Letters* 13 (5): 2151–57. doi:10.1021/nl400600x.

Perret, Philippe, Zohreh Khani, Thierry Brousse, Daniel Bélanger, and Daniel Guay. 2011. "Carbon/PbO_2 Asymmetric Electrochemical Capacitor Based on Methanesulfonic Acid Electrolyte." *Electrochimica Acta* 56 (24): 8122–28. doi:10.1016/j.electacta.2011.05.125.

Perricone, E., M. Chamas, L. Cointeaux, J.-C. Leprêtre, P. Judeinstein, P. Azais, F. Béguin, and F. Alloin. 2013. "Investigation of Methoxypropionitrile as Co-Solvent for Ethylene Carbonate Based Electrolyte in Supercapacitors. A Safe and Wide Temperature Range Electrolyte." *Electrochimica Acta* 93 (March): 1–7. doi:10.1016/j.electacta.2013.01.084.

Perricone, E., M. Chamas, J.-C. Leprêtre, P. Judeinstein, P. Azais, E. Raymundo-Pinero, F. Béguin, and F. Alloin. 2013. "Safe and Performant Electrolytes for Supercapacitor. Investigation of Esters/Carbonate Mixtures." *Journal of Power Sources* 239 (October): 217–24. doi:10.1016/j.jpowsour.2013.03.123.

Pilon, Laurent, Hainan Wang, and Anna d'Entremont. 2015. "Recent Advances in Continuum Modeling of Interfacial and Transport Phenomena in Electric Double Layer Capacitors." *Journal of the Electrochemical Society* 162 (5): A5158–78. doi:10.1149/2.0211505jes.

Pohlmann, S., C. Ramirez-Castro, and A. Balducci. 2015. "The Influence of Conductive Salt Ion Selection on EDLC Electrolyte Characteristics and Carbon-Electrolyte Interaction." *Journal of the Electrochemical Society* 162 (5): A5020–30. doi:10.1149/2.0041505jes.

Pohlmann, Sebastian, Ruben-Simon Kühnel, Teresa A. Centeno, and Andrea Balducci. 2014. "The Influence of Anion–Cation Combinations on the Physicochemical Properties of Advanced Electrolytes for Supercapacitors and the Capacitance of Activated Carbons." *ChemElectroChem* 1 (8): 1301–11. doi:10.1002/celc.201402091.

Qu, Wen-Hui, Fei Han, An-Hui Lu, Chao Xing, Mo Qiao, and Wen-Cui Li. 2014. "Combination of a SnO2–C Hybrid Anode and a Tubular Mesoporous Carbon Cathode in a High Energy Density Non-Aqueous Lithium Ion Capacitor: Preparation and Characterisation." *Journal of Materials Chemistry A* 2 (18): 6549–57. doi:10.1039/C4TA00670D.

Rajkumar, Muniyandi, Chun-Tsung Hsu, Tzu-Ho Wu, Ming-Guan Chen, and Chi-Chang Hu. 2015. "Advanced Materials for Aqueous Supercapacitors in the Asymmetric Design." *Progress in Natural Science: Materials International* 25 (6): 527–44. doi:10.1016/j.pnsc.2015.11.012.

Ramasamy, C., J. S. Palma del vel, and M. Anderson. 2014. "An Activated Carbon Supercapacitor Analysis by Using a Gel Electrolyte of Sodium Salt-Polyethylene Oxide in an Organic Mixture Solvent." Text. https://www.ingentaconnect.com/content/ssam/14328488/2014/00000018/00000008/art00017;jsession id=1xd2gegfrbr6p.x-ic-live-03.

Ren, Jing, Wenyu Bai, Guozhen Guan, Ye Zhang, and Huisheng Peng. 2013. "Flexible and Weaveable Capacitor Wire Based on a Carbon Nanocomposite Fiber." *Advanced Materials* 25 (41): 5965–70. doi:10.1002/adma.201302498.

Ricketts, B. W., and C. Ton-That. 2000. "Self-Discharge of Carbon-Based Supercapacitors with Organic Electrolytes." *Journal of Power Sources* 89 (1): 64–69. doi:10.1016/S0378-7753(00)00387-6.

Roberts, Alexander J., Angela F. Danil de Namor, and Robert C. T. Slade. 2013. "Low Temperature Water Based Electrolytes for MnO2/Carbon Supercapacitors." *Physical Chemistry Chemical Physics* 15 (10): 3518–26. doi:10.1039/C3CP50359C.

Rogers, Robin D., and Gregory A. Voth. 2007. "Ionic Liquids." *Accounts of Chemical Research* 40 (11): 1077–78. doi:10.1021/ar700221n.

Roldán, Silvia, Zoraida González, Clara Blanco, Marcos Granda, Rosa Menéndez, and Ricardo Santamaría. 2011. "Redox-Active Electrolyte for Carbon Nanotube-Based Electric Double Layer Capacitors." *Electrochimica Acta* 56 (9): 3401–5. doi:10.1016/j.electacta.2010.10.017.

Roldán, Silvia, Marcos Granda, Rosa Menéndez, Ricardo Santamaría, and Clara Blanco. 2011. "Mechanisms of Energy Storage in Carbon-Based Supercapacitors Modified with a Quinoid Redox-Active Electrolyte." *The Journal of Physical Chemistry C* 115 (35): 17606–11. doi:10.1021/jp205100v.

Roldán, Silvia, Marcos Granda, Rosa Menéndez, Ricardo Santamaría, and Clara Blanco. 2012. "Supercapacitor Modified with Methylene Blue as Redox Active Electrolyte." *Electrochimica Acta* 83 (November): 241–46. doi:10.1016/j.electacta.2012.08.026.

Romann, T., E. Anderson, P. Pikma, H. Tamme, P. Möller, and E. Lust. 2017. "Reactions at Grapheneltetracyanoborate Ionic Liquid Interface – New Safety Mechanisms for Supercapacitors and Batteries." *Electrochemistry Communications* 74 (January): 38–41. doi:10.1016/j.elecom.2016.11.016.

Romann, T., O. Oll, P. Pikma, H. Tamme, and E. Lust. 2014. "Surface Chemistry of Carbon Electrodes in 1-Ethyl-3-Methylimidazolium Tetrafluoroborate Ionic Liquid – an In Situ Infrared Study." *Electrochimica Acta* 125 (April): 183–90. doi:10.1016/j.electacta.2014.01.077.

Ruiz, V., T. Huynh, S. R. Sivakkumar, and A. G. Pandolfo. 2012. "Ionic Liquid–Solvent Mixtures as Supercapacitor Electrolytes for Extreme Temperature Operation." *RSC Advances* 2 (13): 5591–98. doi:10.1039/C2RA20177A.

Ruther, Rose E., Che-Nan Sun, Adam Holliday, Shiwang Cheng, Frank M. Delnick, Thomas A. Zawodzinski, and Jagjit Nanda. 2017. "Stable Electrolyte for High Voltage Electrochemical Double-Layer Capacitors." *Journal of the Electrochemical Society* 164 (2): A277–83. doi:10.1149/2.0951702jes.

Salinas-Torres, David, Soshi Shiraishi, Emilia Morallón, and Diego Cazorla-Amorós. 2015. "Improvement of Carbon Materials Performance by Nitrogen Functional Groups in Electrochemical Capacitors in Organic Electrolyte at Severe Conditions." *Carbon* 82 (February): 205–13. doi:10.1016/j.carbon.2014.10.064.

Samui, A. B., and P. Sivaraman. 2010. "Solid Polymer Electrolytes for Supercapacitors." *Polymer Electrolytes*, January, 431–70. doi:10.1533/9781845699772.2.431.

Schroeder, M., P. Isken, M. Winter, S. Passerini, A. Lex-Balducci, and A. Balducci. 2013. "An Investigation on the Use of a Methacrylate-Based Gel Polymer Electrolyte in High Power Devices." *Journal of the Electrochemical Society* 160 (10): A1753–58. doi:10.1149/2.067310jes.

Senthilkumar, S. T., R. Kalai Selvan, J. S. Melo, and C. Sanjeeviraja. 2013. "High Performance Solid-State Electric Double Layer Capacitor from Redox Mediated Gel Polymer Electrolyte and Renewable Tamarind Fruit Shell Derived Porous Carbon." *ACS Applied Materials & Interfaces* 5 (21): 10541–50. doi:10.1021/am402162b.

Senthilkumar, S. T., R. Kalai Selvan, N. Ponpandian, and J. S. Melo. 2012. "Redox Additive Aqueous Polymer Gel Electrolyte for an Electric Double Layer Capacitor." *RSC Advances* 2 (24): 8937–40. doi:10.1039/C2RA21387G.

Shellikeri, A., S. Yturriaga, J. S. Zheng, W. Cao, M. Hagen, J. A. Read, T. R. Jow, and J. P. Zheng. 2018. "Hybrid Lithium-Ion Capacitor with LiFePO$_4$/AC Composite Cathode – Long Term Cycle Life Study, Rate Effect and Charge Sharing Analysis." *Journal of Power Sources* 392 (July): 285–95. doi:10.1016/j.jpowsour.2018.05.002.

Shen, Hsiao-Hsuan, and Chi-Chang Hu. 2014. "Capacitance Enhancement of Activated Carbon Modified in the Propylene Carbonate Electrolyte." *Journal of the Electrochemical Society* 161 (12): A1828–35. doi:10.1149/2.0681412jes.

Shim, Youngseon, YounJoon Jung, and Hyung J. Kim. 2011. "Graphene-Based Supercapacitors: A Computer Simulation Study." *The Journal of Physical Chemistry C* 115 (47): 23574–83. doi:10.1021/jp203458b.

Shulga, Y. M., S. A. Baskakov, V. A. Smirnov, N. Y. Shulga, K. G. Belay, and G. L. Gutsev. 2014. "Graphene Oxide Films as Separators of Polyaniline-Based Supercapacitors." *Journal of Power Sources* 245 (January): 33–36. doi:10.1016/j.jpowsour.2013.06.094.

Si, Wenping, Chenglin Yan, Yao Chen, Steffen Oswald, Luyang Han, and Oliver G. Schmidt. 2013. "On Chip, All Solid-State and Flexible Micro-Supercapacitors with High Performance Based on MnOx/Au Multilayers." *Energy & Environmental Science* 6 (11): 3218–23. doi:10.1039/C3EE41286E.

Simon, Patrice, and Yury Gogotsi. 2013. "Capacitive Energy Storage in Nanostructured Carbon–Electrolyte Systems." *Accounts of Chemical Research* 46 (5): 1094–103. doi:10.1021/ar200306b.

Simon, Patrice, and Yury Gogotsi. 2008. "Materials for Electrochemical Capacitors." *Nature Materials* 7 (11): 845–54. doi:10.1038/nmat2297.

Su, Linghao, Liangyu Gong, Haitao Lü, and Qiang Xü. 2014. "Enhanced Low-Temperature Capacitance of MnO2 Nanorods in a Redox-Active Electrolyte." *Journal of Power Sources* 248 (February): 212–17. doi:10.1016/j.jpowsour.2013.09.047.

Suárez-Guevara, J., V. Ruiz, and P. Gomez-Romero. 2013. "Hybrid Energy Storage: High Voltage Aqueous Supercapacitors Based on Activated Carbon–Phosphotungstate Hybrid Materials." *Journal of Materials Chemistry A* 2 (4): 1014–21. doi:10.1039/C3TA14455K.

Sudhakar, Y. N., M. Selvakumar, and D. Krishna Bhat. 2013. "LiClO$_4$-Doped Plasticized Chitosan and Poly(Ethylene Glycol) Blend as Biodegradable Polymer Electrolyte for Supercapacitors." *Ionics* 19 (2): 277–85. doi:10.1007/s11581-012-0745-5.

Sun, Gengzhi, Jia An, Chee Kai Chua, Hongchang Pang, Jie Zhang, and Peng Chen. 2015. "Layer-by-Layer Printing of Laminated Graphene-Based Interdigitated Microelectrodes for Flexible Planar Micro-Supercapacitors." *Electrochemistry Communications* 51 (February): 33–36. doi:10.1016/j.elecom.2014.11.023.

Sun, Gengzhi, Juqing Liu, Xiao Zhang, Xuewan Wang, Hai Li, Yang Yu, Wei Huang, Hua Zhang, and Peng Chen. 2014. "Fabrication of Ultralong Hybrid Microfibers from Nanosheets of Reduced Graphene Oxide and Transition-Metal Dichalcogenides and Their Application as Supercapacitors." *Angewandte Chemie (International Ed. in English)* 53 (46): 12576–80. doi:10.1002/anie.201405325.

Sun, Guohua, Kaixi Li, Lijing Xie, Jianlong Wang, and Yanqiu Li. 2012. "Preparation of Mesoporous Carbon Spheres with a Bimodal Pore Size Distribution and Its Application for Electrochemical Double Layer Capacitors Based on Ionic Liquid as the Electrolyte." *Microporous and Mesoporous Materials* 151 (March): 282–86. doi:10.1016/j.micromeso.2011.10.023.

Tanahashi, I. 2005. "Capacitance Enhancement of Activated Carbon Fiber Cloth Electrodes in Electrochemical Capacitors with a Mixed Aqueous Solution of H$_2$SO$_4$ and AgNO$_3$." *Electrochemical and Solid-State Letters* 8 (12): A627–29. doi:10.1149/1.2087187.

Tanaka, Akimi, Taku Iiyama, Tomonori Ohba, Sumio Ozeki, Koki Urita, Toshihiko Fujimori, Hirofumi Kanoh, and Katsumi Kaneko. 2010. "Effect of a Quaternary Ammonium Salt on Propylene Carbonate Structure in Slit-Shape Carbon Nanopores." *Journal of the American Chemical Society* 132 (7): 2112–13. doi:10.1021/ja9087874.

Tian, Meng, Jiawen Wu, Ruihan Li, Youlin Chen, and Donghui Long. 2019. "Fabricating a High-Energy-Density Supercapacitor with Asymmetric Aqueous Redox Additive Electrolytes and Free-Standing Activated-Carbon-Felt Electrodes." *Chemical Engineering Journal* 363 (May): 183–91. doi:10.1016/j.cej.2019.01.070.

Tian, Qifeng, and Keryn Lian. 2010. "In Situ Characterization of Heteropolyacid Based Electrochemical Capacitors." *Electrochemical and Solid-State Letters* 13 (1): A4–6. doi:10.1149/1.3247071.

Tian, Shengfeng, Li Qi, Masaki Yoshio, and Hongyu Wang. 2014. "Tetramethylammonium Difluoro(Oxalato) Borate Dissolved in Ethylene/Propylene Carbonates as Electrolytes for Electrochemical Capacitors." *Journal of Power Sources* 256 (June): 404–9. doi:10.1016/j.jpowsour.2014.01.101.

Tõnurist, K., T. Thomberg, A. Jänes, I. Kink, and E. Lust. 2012. "Specific Performance of Electrical Double Layer Capacitors Based on Different Separator Materials in Room Temperature Ionic Liquid." *Electrochemistry Communications* 22 (August): 77–80. doi:10.1016/j.elecom.2012.05.029.

Tooming, T., T. Thomberg, L. Siinor, K. Tõnurist, A. Jänes, and E. Lust. 2014. "A Type High Capacitance Supercapacitor Based on Mixed Room Temperature Ionic Liquids Containing Specifically Adsorbed Iodide Anions." *Journal of the Electrochemical Society* 161 (3): A222–27. doi:10.1149/2.014403jes.

Torchała, Kamila, Krzysztof Kierzek, and Jacek Machnikowski. 2012. "Capacitance Behavior of KOH Activated Mesocarbon Microbeads in Different Aqueous Electrolytes." *Electrochimica Acta* 86 (December): 260–67. doi:10.1016/j.electacta.2012.07.062.

Tsay, Keh-Chyun, Lei Zhang, and Jiujun Zhang. 2012. "Effects of Electrode Layer Composition/Thickness and Electrolyte Concentration on Both Specific Capacitance and Energy Density of Supercapacitor." *Electrochimica Acta* 60 (January): 428–36. doi:10.1016/j.electacta.2011.11.087.

Ue, Makoto. 2007. "Chemical Capacitors and Quaternary Ammonium Salts." *Electrochemistry* 75 (8): 565–72. doi:10.5796/electrochemistry.75.565.

Ue, Makoto, Kazuhiko Ida, and Shoichiro Mori. 1994. "Electrochemical Properties of Organic Liquid Electrolytes Based on Quaternary Onium Salts for Electrical Double-Layer Capacitors." *Journal of the Electrochemical Society* 141 (11): 2989–96. doi:10.1149/1.2059270.

Ulihin, A. S., Yu. G. Mateyshina, and N. F. Uvarov. 2013. "All-Solid-State Asymmetric Supercapacitors with Solid Composite Electrolytes." *Solid State Ionics* 251 (November): 62–65. doi:10.1016/j.ssi.2013.03.014.

Väli, R., A. Laheäär, A. Jänes, and E. Lust. 2014. "Characteristics of Non-Aqueous Quaternary Solvent Mixture and Na-Salts Based Supercapacitor Electrolytes in a Wide Temperature Range." *Electrochimica Acta* 121 (March): 294–300. doi:10.1016/j.electacta.2013.12.149.

Vaquero, Susana, Raul Díaz, Marc Anderson, Jesus Palma, and Rebeca Marcilla. 2012. "Insights into the Influence of Pore Size Distribution and Surface Functionalities in the Behaviour of Carbon Supercapacitors." *Electrochimica Acta* 86 (December): 241–47. doi:10.1016/j.electacta.2012.08.006.

Varzi, Alberto, Andrea Balducci, and Stefano Passerini. 2014. "Natural Cellulose: A Green Alternative Binder for High Voltage Electrochemical Double Layer Capacitors Containing Ionic Liquid-Based Electrolytes." *Journal of the Electrochemical Society* 161 (3): A368–75. doi:10.1149/2.063403jes.

Wang, Chunlei, Ying Zhou, Li Sun, Peng Wan, Xu Zhang, and Jieshan Qiu. 2013. "Sustainable Synthesis of Phosphorus- and Nitrogen-Co-Doped Porous Carbons with Tunable Surface Properties for Supercapacitors." *Journal of Power Sources* 239 (October): 81–88. doi:10.1016/j.jpowsour.2013.03.126.

Wang, Guoping, Lei Zhang, and Jiujun Zhang. 2012. "A Review of Electrode Materials for Electrochemical Supercapacitors." *Chemical Society Reviews* 41 (2): 797–828. doi:10.1039/c1cs15060j.

Wang, Hao, Alexander C. Forse, John M. Griffin, Nicole M. Trease, Lorie Trognko, Pierre-Louis Taberna, Patrice Simon, and Clare P. Grey. 2013. "In Situ NMR Spectroscopy of Supercapacitors: Insight into the Charge Storage Mechanism." *Journal of the American Chemical Society* 135 (50): 18968–80. doi:10.1021/ja410287s.

Wang, Hao, Thomas K.-J. Köster, Nicole M. Trease, Julie Ségalini, Pierre-Louis Taberna, Patrice Simon, Yury Gogotsi, and Clare P. Grey. 2011. "Real-Time NMR Studies of Electrochemical Double-Layer Capacitors." *Journal of the American Chemical Society* 133 (48): 19270–73. doi:10.1021/ja2072115.

Wang, Qian, Jun Yan, Yanbo Wang, Tong Wei, Milin Zhang, Xiaoyan Jing, and Zhuangjun Fan. 2014. "Three-Dimensional Flower-like and Hierarchical Porous Carbon Materials as High-Rate Performance Electrodes for Supercapacitors." *Carbon* Complete (67): 119–27. doi:10.1016/j.carbon.2013.09.070.

Wang, Xianfu, Bin Liu, Qiufan Wang, Weifeng Song, Xiaojuan Hou, Di Chen, Yi-bing Cheng, and Guozhen Shen. 2013. "Three-Dimensional Hierarchical GeSe$_2$ Nanostructures for High Performance Flexible All-Solid-State Supercapacitors." *Advanced Materials* 25 (10): 1479–86. doi:10.1002/adma.201204063.

Wang, Yuan, Zheng Chang, Meng Qian, Zhichao Zhang, Jie Lin, and Fuqiang Huang. 2019. "Enhanced Specific Capacitance by a New Dual Redox-Active Electrolyte in Activated Carbon-Based Supercapacitors." *Carbon* 143 (March): 300–308. doi:10.1016/j.carbon.2018.11.033.

Wasiński, Krzysztof, Mariusz Walkowiak, and Grzegorz Lota. 2014. "Humic Acids as Pseudocapacitive Electrolyte Additive for Electrochemical Double Layer Capacitors." *Journal of Power Sources* 255 (June): 230–34. doi:10.1016/j.jpowsour.2013.12.140.

Wen, Z. B., Q. T. Qu, Q. Gao, X. W. Zheng, Z. H. Hu, Y. P. Wu, Y. F. Liu, and X. J. Wang. 2009. "An Activated Carbon with High Capacitance from Carbonization of a Resorcinol–Formaldehyde Resin." *Electrochemistry Communications* 11 (3): 715–18. doi:10.1016/j.elecom.2009.01.015.

Wu, Hao, Xianyou Wang, Lanlan Jiang, Chun Wu, Qinglan Zhao, Xue Liu, Ben'an Hu, and Lanhua Yi. 2013. "The Effects of Electrolyte on the Supercapacitive Performance of Activated Calcium Carbide-Derived Carbon." *Journal of Power Sources* 226 (March): 202–9. doi:10.1016/j.jpowsour.2012.11.014.

Wu, Jihuai, Haijun Yu, Leqing Fan, Genggeng Luo, Jianming Lin, and Miaoliang Huang. 2012. "A Simple and High-Effective Electrolyte Mediated with p-Phenylenediamine for Supercapacitor." *Journal of Materials Chemistry* 22 (36): 19025–30. doi:10.1039/C2JM33856D.

Xu, Bin, Feng Wu, Renjie Chen, Gaoping Cao, Shi Chen, Zhiming Zhou, and Yusheng Yang. 2008. "Highly Mesoporous and High Surface Area Carbon: A High Capacitance Electrode Material for EDLCs with Various Electrolytes." *Electrochemistry Communications* 10 (5): 795–97. doi:10.1016/j.elecom.2008.02.033.

Xu, Bin, Feng Wu, Yuefeng Su, Gaoping Cao, Shi Chen, Zhiming Zhou, and Yusheng Yang. 2008. "Competitive Effect of KOH Activation on the Electrochemical Performances of Carbon Nanotubes for EDLC: Balance between Porosity and Conductivity." *Electrochimica Acta* 53 (26): 7730–35. doi:10.1016/j.electacta.2008.05.033.

Xu, Chao, Fredrik Lindgren, Bertrand Philippe, Mihaela Gorgoi, Fredrik Björefors, Kristina Edström, and Torbjörn Gustafsson. 2015. "Improved Performance of the Silicon Anode for Li-Ion Batteries: Understanding the Surface Modification Mechanism of Fluoroethylene Carbonate as an Effective Electrolyte Additive." *Chemistry of Materials* 27 (7): 2591–99. doi:10.1021/acs.chemmater.5b00339.

Xu, Jun, Chunting Chris Mi, Binggang Cao, and Junyi Cao. 2013. "A New Method to Estimate the State of Charge of Lithium-Ion Batteries Based on the Battery Impedance Model." *Journal of Power Sources* 233 (July): 277–84. doi:10.1016/j.jpowsour.2013.01.094.

Xu, Kang, Michael S. Ding, and T. Richard Jow. 2001b. "Quaternary Onium Salts as Nonaqueous Electrolytes for Electrochemical Capacitors." *Journal of the Electrochemical Society* 148 (3): A267–74. doi:10.1149/1.1350665.

Xu, Kang, Michael S. Ding, and T. Richard Jow. 2001a. "A Better Quantification of Electrochemical Stability Limits for Electrolytes in Double Layer Capacitors." *Electrochimica Acta* 46 (12): 1823–27. doi:10.1016/S0013-4686(01)00358-9.

Yamazaki, Shigeaki, Tatsuya Ito, Masaki Yamagata, and Masashi Ishikawa. 2012. "Non-Aqueous Electrochemical Capacitor Utilizing Electrolytic Redox Reactions of Bromide Species in Ionic Liquid." *Electrochimica Acta* 86 (December): 294–97. doi:10.1016/j.electacta.2012.01.031.

Yan, Jun, Qian Wang, Tong Wei, and Zhuangjun Fan. 2014. "Recent Advances in Design and Fabrication of Electrochemical Supercapacitors with High Energy Densities." *Advanced Energy Materials* 4 (4). doi:10.1002/aenm.201300816.

Yang, Cheol-Min, Yong-Jung Kim, Morinobu Endo, Hirofumi Kanoh, Masako Yudasaka, Sumio Iijima, and Katsumi Kaneko. 2007. "Nanowindow-Regulated Specific Capacitance of Supercapacitor Electrodes of Single-Wall Carbon Nanohorns." *Journal of the American Chemical Society* 129 (1): 20–21. doi:10.1021/ja065501k.

Yang, Lu, Brian H. Fishbine, Albert Migliori, and Lawrence R. Pratt. 2009. "Molecular Simulation of Electric Double-Layer Capacitors Based on Carbon Nanotube Forests." *Journal of the American Chemical Society* 131 (34): 12373–76. doi:10.1021/ja9044554.

Yin, Yijing, Juanjuan Zhou, Azzam N. Mansour, and Xiangyang Zhou. 2011. "Effect of NaI/I2 Mediators on Properties of PEO/LiAlO2 Based All-Solid-State Supercapacitors." *Journal of Power Sources* 196 (14): 5997–6002. doi:10.1016/j.jpowsour.2011.02.079.

Yu, Haijun, Leqing Fan, Jihuai Wu, Youzhen Lin, Miaoliang Huang, Jianming Lin, and Zhang Lan. 2012. "Redox-Active Alkaline Electrolyte for Carbon-Based Supercapacitor with Pseudocapacitive Performance and Excellent Cyclability." *RSC Advances* 2 (17): 6736–40. doi:10.1039/C2RA20503C.

Yu, Haijun, Qunwei Tang, Jihuai Wu, Youzhen Lin, Leqing Fan, Miaoliang Huang, Jianming Lin, Yan Li, and Fuda Yu. 2012. "Using Eggshell Membrane as a Separator in Supercapacitor." *Journal of Power Sources* 206 (May): 463–68. doi:10.1016/j.jpowsour.2012.01.116.

Yu, Xuewen, Dianbo Ruan, Changcheng Wu, Jing Wang, and Zhiqiang Shi. 2014. "Spiro-(1,1′)-Bipyrrolidinium Tetrafluoroborate Salt as High Voltage Electrolyte for Electric Double Layer Capacitors." *Journal of Power Sources* 265 (November): 309–16. doi:10.1016/j.jpowsour.2014.04.144.

Yu, Xuewen, Jing Wang, Chenglin Wang, and Zhiqiang Shi. 2015. "A Novel Electrolyte Used in High Working Voltage Application for Electrical Double-Layer Capacitor Using Spiro-(1,1′)-Bipyrrolidinium Tetrafluoroborate in Mixtures Solvents." *Electrochimica Acta* 182 (November): 1166–74. doi:10.1016/j.electacta.2015.09.013.

Yuan, Longyan, Bin Yao, Bin Hu, Kaifu Huo, Wen Chen, and Jun Zhou. 2013. "Polypyrrole-Coated Paper for Flexible Solid-State Energy Storage." *Energy & Environmental Science* 6 (2): 470–76. doi:10.1039/C2EE23977A.

Zhai, Yunpu, Yuqian Dou, Dongyuan Zhao, Pasquale F. Fulvio, Richard T. Mayes, and Sheng Dai. 2011. "Carbon Materials for Chemical Capacitive Energy Storage." *Advanced Materials* 23 (42): 4828–50. doi:10.1002/adma.201100984.

Zhang, Fan, Tengfei Zhang, Xi Yang, Long Zhang, Kai Leng, Yi Huang, and Yongsheng Chen. 2013. "A High-Performance Supercapacitor-Battery Hybrid Energy Storage Device Based on Graphene-Enhanced Electrode Materials with Ultrahigh Energy Density." *Energy & Environmental Science* 6 (5): 1623–32. doi:10.1039/C3EE40509E.

Zhang, Hao, Gaoping Cao, Yusheng Yang, and Zhennan Gu. 2008. "Capacitive Performance of an Ultralong Aligned Carbon Nanotube Electrode in an Ionic Liquid at 60°C." *Carbon* 46 (1): 30–34. doi:10.1016/j.carbon.2007.10.023.

Zhang, Lei, Ken Tsay, Christina Bock, and Jiujun Zhang. 2016. "Ionic Liquids as Electrolytes for Non-Aqueous Solutions Electrochemical Supercapacitors in a Temperature Range of 20°C–80°C." *Journal of Power Sources* 324 (August): 615–24. doi:10.1016/j.jpowsour.2016.05.008.

Zhang, Li Li, and X. S. Zhao. 2009. "Carbon-Based Materials as Supercapacitor Electrodes." *Chemical Society Reviews* 38 (9): 2520–31. doi:10.1039/B813846J.

Zhang, Qing, Kathryn Scrafford, Mingtao Li, Zeyuan Cao, Zhenhai Xia, Pulickel M. Ajayan, and Bingqing Wei. 2014. "Anomalous Capacitive Behaviors of Graphene Oxide Based Solid-State Supercapacitors." *Nano Letters* 14 (4): 1938–43. doi:10.1021/nl4047784.

Zhang, Xiaoyan, Xianyou Wang, Lanlan Jiang, Hao Wu, Chun Wu, and Jingcang Su. 2012. "Effect of Aqueous Electrolytes on the Electrochemical Behaviors of Supercapacitors Based on Hierarchically Porous Carbons." *Journal of Power Sources* 216 (October): 290–96. doi:10.1016/j.jpowsour.2012.05.090.

Zhang, Xuan, Dandan Zhao, Yongqing Zhao, Pengyi Tang, Yinglin Shen, Cailing Xu, Hulin Li, and Yu Xiao. 2013. "High Performance Asymmetric Supercapacitor Based on MnO2 Electrode in Ionic Liquid Electrolyte." *Journal of Materials Chemistry A* 1 (11): 3706–12. doi:10.1039/C3TA00981E.

Zheng, Jim P. 2003. "The Limitations of Energy Density of Battery/Double-Layer Capacitor Asymmetric Cells." *Journal of The Electrochemical Society* 150 (4): A484–92. doi:10.1149/1.1559067.

Zheng, Jim P. 2005. "Theoretical Energy Density for Electrochemical Capacitors with Intercalation Electrodes." *Journal of The Electrochemical Society* 152 (9): A1864–69. doi:10.1149/1.1997152.

Zheng, Jim P., J. Huang, and T. R. Jow. 1997. "The Limitations of Energy Density for Electrochemical Capacitors." *Journal of the Electrochemical Society* 144 (6): 2026–31. doi:10.1149/1.1837738.

Zheng, Jim P., and T. R. Jow. 1997. "The Effect of Salt Concentration in Electrolytes on the Maximum Energy Storage for Double Layer Capacitors." *Journal of the Electrochemical Society* 144 (7): 2417–20. doi:10.1149/1.1837829.

Zhong, Cheng, Yida Deng, Wenbin Hu, Jinli Qiao, Lei Zhang, and Jiujun Zhang. 2015. "A Review of Electrolyte Materials and Compositions for Electrochemical Supercapacitors." *Chemical Society Reviews* 44 (21): 7484–7539. doi:10.1039/C5CS00303B.

Zhou, Juanjuan, Yijing Yin, Azzam N. Mansour, and Xiangyang Zhou. 2011. "Experimental Studies of Mediator-Enhanced Polymer Electrolyte Supercapacitors." *Electrochemical and Solid-State Letters* 14 (3): A25–28. doi:10.1149/1.3526094.

Zhou, Zhi-Bin, Hajime Matsumoto, and Kuniaki Tatsumi. 2004. "Low-Melting, Low-Viscous, Hydrophobic Ionic Liquids: 1-Alkyl(Alkyl Ether)-3-Methylimidazolium Perfluoroalkyltrifluoroborate." *Chemistry – A European Journal* 10 (24): 6581–91. doi:10.1002/chem.200400533.

Zhu, Yanwu, Shanthi Murali, Meryl D. Stoller, K. J. Ganesh, Weiwei Cai, Paulo J. Ferreira, Adam Pirkle, et al. 2011. "Carbon-Based Supercapacitors Produced by Activation of Graphene." *Science (New York, N.Y.)* 332 (6037): 1537–41. doi:10.1126/science.1200770.

Index

Note: Page numbers in italic and bold refer to figures and tables, respectively.

A

acetonitrile (ACN), 217, 221
activated carbon (AC), 137, 224, 229
alloying-type anode materials, 48–50, 94–97
all phenyl complex (APC) electrolytes, 83, *84*, 97
anthraquinone, 92, *93*
aqueous electrolytes, 111, 219
    with redox-active species, 224
artificial interphase, 85–86, *87*
asymmetric EC, 226
asymmetric supercapacitor, 136
asymmetric ultracapacitors, 156, 162
atomistic models, 159–161

B

battery type supercapacitor, 136
binary metal oxides, 141
Bi/RGO composites, 97, *97*
bismuth oxyfluoride (BiOF), 96, *96*

C

capacitance, 109, 216
carbide-derived carbon, 137
carbohydrates, 44
carbon
    aerogel, 137
    composites, 143
    conducting polymer composites, 143
    metal oxides composites, 143
carbonaceous materials, 24, 34
carbon-coated $LiMnPO_4$ (S–LMP/C), 9–10, *10*
carbon-encapsulated Sn@N-doped carbon nanotube
    composite, 49–50, *49*
carbon nanomaterials (CNs), 137, 143
carbon nanotubes (CNTs), 132, 137–138, 160, 184
cationic (H^+) movement, 205
c-DFT (classical DFT) method, 159
Ce^{3+} cations, 182, 184
ceria, 181
    hollow spheres, 187, *187*
    nanocubes, 186, *187*
    nanoplates, 185, *185*
    nanorods, *183*, 184
    nanowire, 184, *184*
cerium oxide (CeO_2), 181
    crystal structure, *182*, 182–183
    electrochemical process, *188*
    in fuel cells, 191–192, *192*
    on graphene, 189
    morphology and synthesis, 183–187
    nanoflakes, *186*
    in photocatalytic water splitting, 192–193, *193*
    in supercapacitors, 188–191, *189*, *190*
    XRD, *182*, 182–183
charge-discharge performance, SnO_2/GO composite,
    25–26, *26*
charge storage mechanism, 151, *152*
chemical vapour deposition (CVD) graphene growth,
    105–106
clarified systematic models, 161–163
classical DFT (c-DFT) method, 159
CNTs, *see* carbon nanotubes (CNTs)
Co_2SnO_4, 29
Co_3O_4, *33*, 33–34, 46–47, *48*, 140
    nanowires, 3D graphene foam, 119–120, *121*
cobalt oxide (CoO), 32, 46–47, *48*
cobalt oxyhydroxide (CoOOH), 120–121, *122*
cobalt sulphides, 142
commercial scale applications, supercapacitors, 143
composite supercapacitors, 136
compressive force, 205–206
compressive stress, 204
conductive polymer, 143
continuum equations, 158–159
conventional separator, *see* separator
copper oxide (CuO), 32–33, 45–46, *47*
CuS, 90–92
CVD graphene growth, *see* chemical vapour deposition
    (CVD) graphene growth

D

density functional theory (DFT), 156, 160, 168
dielectric materials, 201, *201*
dimethoxyethane (DME), 83, *84*
dipole oscillation, 201
direct alcohol fuel cells (DAFCs), 191–192, *192*
disordered CNTs (D-CNTs), 113–115, *115*
dissipation transmission line models, 157–158
double-wall carbon nanotube, *138*

E

EDLCs, *see* electric double-layer capacitors (EDLCs);
    electrochemical double-layer capacitor (EDLC);
    electrostatic double-layer capacitors (EDLCs)
electrical behaviour models, 157–158
electric double-layer capacitors (EDLCs), 151, 219–220, 226
electric double layers (EDLs), 215

electrochemical capacitors (ECs), 109, 111, 215, 217; *see also* ultracapacitors
  asymmetric/hybrid, 152, *153*, 226
  cell voltage, 228–229
  design and fabrication, 163
  electrode-electrolyte interface, *166*
  electrolyte effects, *218*
  energy density, 153
  high capacitance, 153
  modelling and simulation, 154
  next-generation, 155–156, *225*, 225–226
  performance, 230
  physical modelling, 167
  self-discharges and charge redistributions, 163–167, *165*
  solid-state, 223
electrochemical double-layer capacitor (EDLC), 101–102, *102*, 188
electrochemical equations, 158–159
electrodes, 202
  active materials, 204–205
  capacitance, 110
  cathodes and anodes, 62, *62*
  electron transfer, *206*
  layered oxides, 62–72
  phosphates polyanions, 72–74
  ultracapacitors, 217
electrolytes, 202, 205, 215
  aqueous, 219
  conducting salts for, 222
  on EC performance, *218*
  EC/ultracapacitors, 217, 227–229, 230–231
  electrical conductivity, 217
  expensiveness, 231
  IL-based, 222–223
  Mg batteries, 83–85, *84*
  neutral, 219–220
  in next-generation EC, *225*, 225–226
  organic, 220–221
  redox-active, 224
  solid-/quasi-solid-state, 216, 223
  strong acid, 219
  strong alkaline, 219
  types, 216–217
electrolyte stable potential window (ESPW), 216, 230, 232
electronics, 144
electrostatic double-layer capacitors (EDLCs), 134–135, *135*
  activated carbon, 137
  electrodes for, 136
  hybrid supercapacitor, 136
  pseudocapacitors, 135
electro-thermal equation, 163
energy harvesters, 198
energy storage, 132
  and conversion devices, *134*
  mechanism, 205–206
ESPW (electrolyte stable potential window), 216, 230, 232
ethylenediamine solvents, 184

F

faradaic mechanisms, 134
Fe_3O_4, 34
Fe_3O_4-rGO nanocomposite powder, 112
field emission scanning electron microscopy (FESEM) images, VGNS, 105, *106*
fractional-order models, 161
fuel cells, 191–192, *192*

G

galvanostatic charge-discharge (GCD) profile, 109
gel polymer electrolyte (GPE), 223–224
Gouy-Chapman equation, 158
graphene, 42–43, *43*, 189
  -based electrodes, 111–124
  carbon allotropes, 102, *103*, **103**
  electronic properties, 102
  Raman spectroscopy, 106–107, *107*
  shapes, 137, *138*
  synthesis methods, *104*, 104–106
  TEM and SEM, 107–108, *108*
  XPS, 108, *109*
graphene aerogel (GA), 21
graphene nanosheets (GNSs), 119, *120*
graphene oxide (GO), 104–105, *105*, 112
graphene oxide (GO)-coated polystyrene (PS@GO) microspheres, 42

H

halide-free electrolytes, 83
hard carbon, 43–44
Helmholtz equation, 156, 158
hexadecyltrimethylammonium bromide (CTAB), 185
hexamethyldisilazide magnesium chloride (HMDSMgCl), 88
high-resolution transmission electron microscopy (HRTEM), 186, *186*
hybrid electric vehicles (HEVs), 145
hybrid electrochemical capacitor (HEC), 229
hybrid supercapacitor, 136
hydrogen economy, 192
hydrothermal synthesis method, 184

I

ideal electrolyte, 216
integrated system, 203
ion diffusion, 164
ionic liquids (ILs) electrolytes, 111, 216, 222–223, 227–228, 232
iron 2 and 3 oxides (Fe_2O_3/Fe_3O_4), 141
iron oxide (FeO), 32, 45

J

Jahn-Teller effect, 9

L

$La_{0.6}Sr_{0.4}Co_{0.2}Fe_{0.8}O_{3-\delta}$ (LSCF) cathode, 191, *192*
layered lithium metal oxide (LiMO), 60
Li_2FeSiO_4, 11, *12*
Li_2MnSiO_4, 11–12, *13*
$Li_4Ti_5O_{12}$ (LTO), 21–24, *23*
LIB, *see* lithium-ion battery (LIB)
$LiCo_{0.33}Mn_{0.33}Ni_{0.33}O_2$, 6
$LiCoO_2$ (LCO), 3–5, *4*, *5*
$LiFePO_4$ (LFP), 8–9, *9*
light-emitting diode (LED), 144
LightScribe CD/DVD optical drive, 105
$LiMn_{1.5}Ni_{0.5}O_4$, 7–8
$LiMn_2O_4$, 7
$LiMnO_2$, *5*, 5–6
$LiMnPO_4$, 9–10, *10*
$LiNiO_2$ (LNO), 6
$LiNiPO_4$, 10
liquid electrolytes, 216
Li-rich and manganese-rich materials, 6–7
lithium-ion battery (LIB), 19–20, 59, 217, 229
    energy density, 1, *2*
    layered cathode materials, 3–7
    metal oxide anodes, 20–27
    olivine cathodes, 8–10
    silicate cathode, 11–13
    spinel cathode, 7–8
    ternary tin oxides, 27–29
    TM_3O_4, 33–35
    TMO, 30–33
    working principles, 2–3
low-power equipment power buffer, 144

M

magnesium (Mg) batteries
    alloying-type anode materials, 94–97
    challenges, 81
    Chevrel-phase structure, 81–82
    CuS, 90–92
    electrolytes, 83–85, *84*
    vs. lithium (Li) batteries, 81, **82**
    Mg^{2+} ions movement, Mo_6S_8 structure, 82, *82*
    organic cathodes, 92–94
    requirements, 98
    sulphur, 87–90
    V_2O_5, 85–87, *86*, *87*
magnesium polysulphides (Mg-PSs), 88
magnesium–sulphur (Mg/S) battery
    Mg_2 electrolytes, 88–89
    obstacles, *88*
    OMBB electrolyte, 88, *89*
    S-CNT cathode, 88, *89*
    theoretical capacity, 87
    XPS, sulphur (2p) cathodes, 89, *90*
manganese (IV) oxide, 140
manganese oxide (MnO), 31–32
maricite $NaFePO_4$, 72, *73*
mass-balancing, 110

matching strategy approach, 230
mechanical stress, 199
medical devices, 144
membrane electron assembly (MEA), 191, *192*
metallics, impurity of, 164
metal organic frameworks, 190
metal phosphides (MPs), 53
$Mg(AlCl_2(butyl)(ethyl))_2$/THF electrolyte solution, 82–83
micro train computer (MITRAC) energy saver system, 145
Mn_2SnO_4@GS nanocomposites, 28, *29*
Mn_2SnO_4/Sn/C 3D composite cubes, *28*, 28–29
Mn_3O_4, 34–35, *35*
$MnFe_2O_4$@C (MFO@C) materials, 45, *46*
molecular dynamics (MDs) simulation, 156
    MC, 159–161
monoclinic $LiMnO_2$, *5*, 5–6
Monte Carlo (MC)
    computations, 221
    MD simluations, 159–161
multifunctional devices, 197
multiple metal layered oxides, 70–72
multi-walled carbon nanotube (MWCNT), 34, 112, 117, *118*, 138, *138*

N

Na_2TP@GE, 51, *51*
$NaCrO_2$
    capacity, 66, **70**
    and $LiCrO_2$, crystal structure, 66, 68
    $Na_{0.5}CrO_2$, 68
    $NaNiO_2$, 68, *70*
    O3-$NaVO_2$ and P2-$Na_{0.7}VO_2$, 70
    P2-$Na_{2/3}Ni_{1/3}Mn_{2/3}O_2$ cathode, *71*
    phase evolutions, $Na_{1-x}CrO_2$, *69*
$NaFeO_2$, 66
    crystal structures, *68*
    phases of, **68**
Na-ion battery
    capacity of, 61–62
    energy, 60
Na-ion storage mechanism, 44
nanoceria
    1-dimensional, 183–184
    3-dimensional, 186–187
    2-dimensional, 185
    ultra-fine, *186*
nanogenerators, 199
    piezoelectric, 199
    pyroelectric, 200–201
    triboelectric, 199–200
nanopores, 161
nanorods formation, 184
NASICON structured phosphates, 72–73, *73*
Na_xCoO_2, 64
    compositions/phase types/specific capacity, **65**
    crystal structure, *64*
Na_xMnO_2, 64–66
    compositions/phase types/specific capacity, **67**
    crystal structures, *65*

Na_xMO_2 layered oxides, 62–63
  crystal structures, 63, *63*
N-doped-graphitized hard carbon (N-GHC), 44, *44*
neutral electrolytes, 219–220
next-generation EC, 155–156
  electrolytes in, *225*, 225–226
nickel foam graphene (Ni-FG), 120–121, *122*
nickel oxide (NiO), 30–31, *31*, 141
nickel-rich $LiNi_xCo_yMn_zO_2$ cathode materials, 7
nickel sulphides, 142
NiO nanowire foam (NWF), 30–31, *31*
nitrogen-doped graphene materials, 43
non-aqueous electrolyte, redox-active species, 219
non-integrated system, 202–203

O

olivine cathodes, 8–10
olivine $NaFePO_4$, 72
1D carbon nanotubes (1D CNTs), 112–118
1-dimensional nanoceria, 183–184
1,4-polyanthraquinone (14PAQ), 92, *93*
organic cathodes, 92–94
organic electrolytes, 216–217, 220–221
organic materials-based anodes, 51
organic solvents, 221–222
orthorhombic $LiMnO_2$, *5*, 5–6
oxidation reactions, 206

P

PANI/SWCNT/cloth electrode
  CV and GCD curves, 115–116, *117*
  Nyquist and Ragone plots, 116, *117*
  preparation, 115, *116*
PANI/VA-CNTs electrodes, *see* polyaniline/
    verticalaligned carbon nanotubes
    (PANI/VA-CNTs) electrodes
PAQS cathode, 92–93, *94*
phosphates polyanions, 72–74
phosphorus (P), *52*, 52–53
phosphotungstic acid, 224
photocatalytic water splitting, 192–193, *193*
piezoelectric energy harvester/harvesting, 197–199
  traditional circuit, 202, *202*
piezoelectric nanogenerator, 199
  materials, 199, **200**
piezoelectric self-charging cell, **208**; *see also* self-charging
    power cell (SCPC)
  challenges, 208–209
  components, 202
  mechanism, 203–204
  non-integrated system, 202–203
  working principle, 203
Poisson-Boltzmann equation, 158
polarization charge density, 199
poly (2,5-dimercapto-1,3,4-thiadiazole (PDMcT)), 43
polyaniline/verticalaligned carbon nanotubes
    (PANI/VA-CNTs) electrodes
  and D-CNTs, 113–115, *115*
  SEM images of, *114*
  structure of, 112–113

polyvinylidene difluoride/fluoride (PVDF) film, 202–203
  piezoelectricity and charge distribution, 204–205, *205*
  strain-free, 204
  unpolarized, 205
power buffer, 144
Prussian blue analogues, 74
pseudocapacitors, 101–102, *102*, 135, 151, 188
  binary metal oxides, 141
  charge storage mechanism, *152*
  cobalt oxide (Co_3O_4), 140
  development, 162
  electrode, 152
  iron oxides, 141
  manganese (IV) oxide, 140
  nickel oxide/hydroxide, 141
  ruthenium oxide (RuO_2), 139–140
  self-discharge, 164
  transition metal oxides and hydroxides, 138–139
PVDF film, *see* polyvinylidene difluoride/fluoride
    (PVDF) film
pyroelectric nanogenerator, 200–201, *201*
pyrophosphate, 73

Q

quantum equations, ultracapacitors, 159
quasi-solid-state electrolytes, 223

R

Raman spectroscopy, graphene, 106–107, *107*
RC (resistor capacitor) systems, 157
redox-active electrolytes, 224
redox pair electrodes, 162
redox-type electrolytes, 217
reduced graphene oxide (RGO), 97, *97*
reduction reactions, 206
rock-salt-LTO ($Li_7Ti_5O_{12}$), 22–23
room temperature ionic-liquids (RTILs), 111, 160, 167
ruthenium oxide (RuO_2), 139–140

S

SAED (selected area electron diffraction), 95, *95*, 182
samarium-doped cerium oxide (SDC), 191, *192*
scanning electron microscopy (SEM)
  CoOOH/Ni-FG, 120–121, *122*
  GNS and SnO_2/GNS, 119, *120*
  graphene, 108, *108*
  Mn_2SnO_4/Sn/carbon cubes, *28*
  NiO NWF, *31*
  PANI/VA-CNTs electrodes, *114*
SCPC, *see* self-charging power cell (SCPC)
SEI (solid-electrolyte interface), 21, 24, 33–34, 84
selected area electron diffraction (SAED), 95, *95*, 182
self-charging power cell (SCPC), 202–203
  advantages, 208
  applications, 208
  performance, 206–207, *207*
  processes, 206
self-charging process, 206, *206*

self-charging supercapacitor power cells (SCSPCs), *203*, 207
  piezoelectric, **207**
self-discharge process, 164
self-powered systems, 197
SEM, *see* scanning electron microscopy (SEM)
separator, 154, 202–203, 225
SIB, *see* sodium-ion battery (SIB)
silicate cathode, 11–13
silicotungstic acid, 224
single metal layered oxides, 64–70
single-walled carbon nanotube (SWCNT), 138, *138*, 165–166
smart devices, 145
SnO_2 anode materials, 24, *27*
  lithium ions (Li^+), 24–25
  SnO_2/GNS electrode, 119, *120*
  SnO_2/GO composite, 25–26, *26*
  SnO_2@RGO-1 composite, 26, *27*
  SnS/SnO_2 heterostructures, 25, *25*
SnSb nanocrystals, 50, *50*
sodiation/lithiation mechanism process, 47
sodium fluorophosphates, 73–74
sodium-ion battery (SIB), 59–60
  alloy-based materials, 48–50
  carbon-based materials, 42–44
  cathodes materials, 74
  cost effect, 42
  electrode materials, 62–74
  and LIBs, 41–42
  lithium, potassium and magnesium ions, **60**
  operation of, *61*
  organic materials-based anodes, 51
  phosphorus (P), *52*, 52–53
  storage, 42
  TMOs, 45–48
sodium metal oxide (NaMO), 60
sodium terephthalate $Na_2C_8H_4O_4$ (Na_2TP), 51, *51*
solid electrolyte, redox-active species, 219
solid-electrolyte interface (SEI), 21, 24, 33–34, 84
solid oxides fuel cells (SOFCs), 191, *192*
solid polymer electrolyte (SPE), 84–85, 223
solid/quasi-solid-state electrolytes, 216, 223
solvothermal method, 43, 96, 183, *183*
SPE (solid polymer electrolyte), 84–85, 223
spinel cathode, 7–8
Stern surface, 158–159
streetlights, 144
strong acid electrolytes, 219
strong alkaline electrolytes, 219
supercapacitors, *151*, 188–191, *189*, *190*, 199; *see also*
      graphene
  applications, 143–145
  asymmetric, 136
  battery type, 136
  composite, 136
  development, 145
  EDLCs, 134–135, *135*
  electrode materials, 136–138
  energy storage system, 0D to 3D nanostructures, 132, *132*
  hybrid, 136
  ideal system, 205
  multifunctional, 198

  performance, strategies, 132, *133*
  performance evaluation, **135**
  principle and performance, 109–111
  pseudocapacitors, 101–102, *102*, 135, 138–141
  recommendations, 146
  types, 133–134
  0D/1D/2D/3D graphene-based nanocomposites, **123–124**
surface tuning approach, 229
surfactant-free approach, 184–185
SWCNT, *see* single-walled carbon nanotube (SWCNT)
SWNT (single-walled carbon nanotube), 138, *138*, 165–166
symmetric capacitors, 162

T

$TEABF_4$ electrolyte, 222
TEM, *see* transmission electron microscopy (TEM)
temperature programmed desorption (TPD) model, 193, *193*
ternary tin oxides, 27–29
tetrahydrofuran (THF), 82–83, 97
3D graphene foams and aerogels, 119–122
3-dimensional nanoceria, 186–187
TiO_2 anode, 20–21, *22*
TM_3O_4, 33–35
TPD (temperature programmed desorption) model, 193, *193*
transition metal oxides (TMOs), 138–139
  CoO, 32, 46–47, *48*
  CuO, 32–33, 45–46, *47*
  FeO, 32, 45
  MnO, 31–32
  NiO, 30–31, *31*
  structure of, 30, *30*
transition metal phosphides, 141–142
transition metal sulphides (TMSs), 142
transmission electron microscopy (TEM), 182
  bilayer graphene, 107–108, *108*
  CNT and MnO_2-CNT, *118*
  CuS-I nanoparticles, *91*
  NP-Bi_6Sn_4, *95*
  rGO/Co_3O_4 composites, *113*
  SnO_2/GNS, 119, *120*
triboelectric nanogenerator, 199–200
triphylite $NaFePO_4$, 72, *73*
tunable nanoporous carbon, 137
2D graphene sheets/films, 118–119, *120*
2-dimensional nanoceria, 185
2,6-polyanthraquinone (26PAQ), 92, *93*

U

ultracapacitors, 151, 153, 215–216
  asymmetric, 156, 162
  challenges, 230–231
  charge-storage features, 217
  charging and discharging process, 167
  difficulties in, 167
  electrical behaviour, 157
  electrolytes, *227*, 227–229
  energy and power densities, 218
  energy efficiency, 162
  ESR values, 230

ultracapacitors (*Continued*)
    features, 154
    hybrid, 232
    modelling and simulation, 156–163
    organic electrolytes in, 220
    quantum equations, 159
    self-discharge, 163–165
    symmetric and asymmetric, 227
    thermal modelling, 163
    3D equation, 159, 162
ultra-fine nanoceria, *186*
ultra-thin ceria nanoribbons, 185

V

V_2O_5 cathode, 85–87, *86*, *87*, 98
vertical graphene nanosheets (VGNSs), 105, *106*
voltage stabilizer, 144

W

water splitting process, 192–193
world electrical energy production, *198*

X

X-ray diffraction (XRD), *182*, 182–183
    Bi/RGO, 97, *97*
    Mg/V_2O_5, 86, *87*
    NP-Bi_6Sn_4, *95*
X-ray photoelectron spectroscopy (XPS), 89, *90*,
        108, *109*

Z

0D graphene dots/powders, 112
$ZnSn_2O_4$, 29

Printed in the United States
by Baker & Taylor Publisher Services